Construction Law

From beginner to practitioner

Jim Mason

Routledge
Taylor & Francis Group

LONDON AND NEW YORK

First published 2016
by Routledge
2 Park Square, Milton Park, Abingdon, Oxon OX14 4RN

and by Routledge
711 Third Avenue, New York, NY 10017

Routledge is an imprint of the Taylor & Francis Group, an informa business

British Library Cataloguing-in-Publication Data
A catalogue record for this book is available from the British Library

Library of Congress Cataloging in Publication Data
A catalog record for this book has been requested

ISBN: 978-1-138-93331-6 (hbk)
ISBN: 978-1-138-93332-3 (pbk)
ISBN: 978-1-315-67863-4 (ebk)

Typeset in Bembo
by Fakenham Prepress Solutions, Fakenham, Norfolk NR21 8NN
Printed and bound in Great Britain by
Ashford Colour Press Ltd, Gosport, Hampshire

For Becky

Thanks

Contents

PART 3
Construction contract law 175

Figures

Tables

Foreword

The purpose of this book is to guide the student from the first tentative steps in a law subject through to the development of a detailed understanding of this fascinating field of study. The aim is to give the student knowledge of construction law and confidence to rely on their own ability to research and resolve any issues arising in this area. As such, an appreciation of the matters discussed forms a starting point for a wider study into all aspects of construction law.

The layout of the book takes the student through subject areas broken down into sections. The sections build on the concepts introduced in the preceding chapters to allow students to expand their knowledge at their own pace. The sections range from basic legal concepts in Part 1 through to a more detailed analysis of the background to construction law in Part 2. The notion that the construction industry is a specialist user of legal services is explored in Part 3. Part 4 takes the student through the dispute resolution mechanisms available to the construction industry. The distinctive and innovative features of the construction industry and the legal ramifications are considered in Part 5.

The book is intended to support the study of construction law as a component of both undergraduate and postgraduate degree courses in the built environment. Equally, the work supports a study of construction law as a discrete discipline. In part, the style adopted follows the textbook approach of disseminating information in a methodical manner. Elsewhere, the writing is intended to be more academic in terms of making connections between different parts of the law and the specific character of the construction sector through exploration of developing themes. In some areas, consistency of approach has given way to the desire to put across my approach on construction law for the reader's benefit. As a result, this is more than just a law book. Aspects of construction and project management are drawn on when necessary to give the reader a holistic view of the applications of law in this field. This book also provides an opportunity to reflect on recent initiatives within the built environment sector. I have set out, therefore, to add my contribution, seeking to encourage the compelling case made by the agents of change who would see the construction industry continue to take measures to improve its practices.

The main stimulus for writing this book is a desire to plug a gap I perceive to be evident between the 'black letter law' approach taken by many legal writers and the expectation of professionals working and studying in this area. I have sought, throughout my career as a solicitor and then as an academic, to make the law accessible to everyone. Any failures on my part to follow the norms of legal writing are acknowledged as a conscious decision to present the law in a more accessible way than is the case in a good number of legal textbooks. The expectation, which I recall vividly from my time as a law student, that the reader will invest the time needed to locate, print and read the law cases referred to belongs to a different era. I hope that the reader will be able to follow and make the necessary connections and reflections on what you already know, thereby unlocking a deeper appreciation of construction law.

Jim Mason

Department of Architecture and the Built Environment, University of the West of England, Bristol, UK

September 2015

Preface

This book is intended to be a stand-alone reference point for those studying construction law. I have included further reading sections at the end of chapters which identify textbooks containing more information for the student to follow up on should they wish. Where multiple chapters refer to the same books, I have identified the sections to which particular attention should be paid.

This book refers to the 'employer', by which reference is intended to the client or buyer of construction services. The 'contractor' refers to the building company supplying the construction services. References to 'he' and 'him' should be taken to imply equally to 'she' and 'her'.

This book refers to the 'AEC industry', by which reference is made to the Architecture, Engineering and Construction industry. The integration of these previously separate fields of practice is one of the major changes in the sector in the last 20 years.

The numbering system used by the book is intended to assist navigation around the various chapters, and there are cross-references made in the text.

Acknowledgements

Thanks to the construction law research community of BEL-NET (Built Environment Lawyer Network) and W113 CIB. The friendships and mutual support have been extremely valuable in realising what it is possible to achieve.

Thanks to Kevin Burnside and other colleagues from the University of the West of England for their support and contribution to this work.

Thanks to Paul Revell for his help with the figures.

Thanks to Andrew Mitchell for the cover design.

Thanks to my co-authors and dissertation students for their assistance and valued input.

Chapter 2, 'The Law of Contract', is reproduced with kind permission of Wiley-Blackwell. It was first published in *Law and the Built Environment*, © 2011, Douglas Wood, Paul Chynoweth, Julie Adshead and James Mason.

Part 1

The background law

1   Legal systems

1.1 Introduction to construction law

Most people understand the role that architects, engineers and surveyors perform in the construction industry. Less is known about the role of construction lawyers and how they contribute to the building process. Construction law can be regarded as the underpinning beneath all other fields of practice, enabling other roles to be performed safe in the knowledge that actions are regulated.

At the project inception stage, construction lawyers play an essential role in ensuring that all parties comply with government regulation and that the parties' rights and responsibilities are written into contracts reflecting what they hope to achieve. If any party infringes their rights then the contract must allow that party the right to pursue a remedy through the courts or such other dispute resolution technique as may apply.

The drafting of the construction contract is one of the key tasks performed by the construction lawyer and is essential to the success of any construction project. The contract outlines the roles and responsibilities of the parties, allocates project risks and sets our procedures for the avoidance and resolution of any disputes arising. The contract attempts to anticipate everything that might happen during the project and directs the parties as to how they should deal with such developments.

Construction projects rarely end up being constructed exactly in accordance with the original drawings and specifications. During construction, changes almost inevitably occur. For example, the employer may change his mind about an aspect of the design and ask the contractor to perform extra or different work. Alternatively, the contractor might start excavations and discover what lies beneath the ground is different to what was envisaged. For example, the soil might be contaminated and require remediation experts to be called upon, requiring extra time and money.

These sorts of issues can result in the parties making claims against each other during the course of a project. For example, the contractor may believe he is entitled to extra time and money and will seek the assistance of a construction lawyer to prepare a claim. Alternatively, the owner may believe he is entitled to withhold money from the contractor because the project is behind schedule. The making of claims during the course of a construction project can strain the relationships between parties. Construction projects have earned an unenviable reputation for being highly adversarial and prone to many conflicts and disputes. This is due in part to the uncertain nature of construction and the involvement of multiple parties in the process, each with their own agenda and needs.

All too often, disputes that arise during the project are not resolved until after the completion of the project. Construction lawyers help the parties to resolve these disputes using a variety of methods such as alternative dispute resolution, litigation and arbitration.

This portrait of lawyers only being on hand to assist in the unwelcome incidence of dispute belies the true importance of the law. It is necessary to first return to the notion of why law is needed. Would it not be possible to operate without construction law and lawyers? The first requirement before a study of construction law can be undertaken is to appreciate the function that law serves, how law is made and the different sources of law.

1.2 Introduction to law

Law is an inescapable feature of daily life. Most democratic countries promote the freedom of their citizens to do as they please. However, the populace is nevertheless bound to obey the rule of law in their dealings with one another. The unpalatable alternative is a lawless and anarchical state. 'Man is born free, and everywhere he is in chains.'[1] This quote from Jean-Jacques Rousseau is typical of some of the thinking that gave rise to the French Revolution of 1789. The 'chains' that Rousseau was referring to would doubtless include the restrictive nature of the laws passed to keep the unfranchised in their place. The Revolution had its origins in the displeasure of the majority of citizens with these existing arrangements, including the legal system. The Revolution completely overturned the legal system and rewrote the law book. Many of our modern laws can be traced back to the sentiments enshrined in such writings as the *Declaration of the Rights of Man and of the Citizen*.[2] The point to note here is that lawmaking remained indispensable even at one of the most revolutionary junctures in recent history. In this, we can observe the universal recognition that laws are needed to provide a set of absolute rules to govern life and our interaction with one another.

The law is sometimes described as being 'black and white', by which it is meant that the law is clear and that there is no room for any argument around its meaning. This is true of some laws but not all. Law can deliberately provide room for argument on and around the provisions it introduces. Judges are given discretion as to how the law is interpreted. This is evident in criminal law in suggested ranges for prison sentences to be imposed (e.g. from two to five years) and the identification of a band within which fines can be imposed. Neither is the stated law always accessible for the layman to discern its meaning, and it often needs explaining by legal advisers or judges. One of the key notions underpinning law is that legal argument is a valid end in itself even if it does not provide definitive answers. The law stated in this book is a summation of what the law is. Conversely, law texts can themselves be sources of law inasmuch as they represent the embodiment and codification of the law.

Legal argument and success in a contested case therefore depends on the skill of the advocate but also on the facts of any given situation, and opposing interpretations frequently occur. Which view and version of events should prevail? The answer depends on which legal argument the Tribunal are more persuaded by on the day. The importance of legal argument based on the presentation of a strong case cannot be overstated.

The legal system considered in this chapter is primarily the English legal system made up in part of the common law. This represents one country's attempt to develop, interpret and protect legal rights. The English approach has been exported to many other countries around the globe, mainly through the legacy of colonisation. Many readers will also be familiar with other approaches to law, most notably the civil law approach founded on Romano-German principles and Islamic approaches that incorporate religious teaching in the legal system. These different approaches are discussed at points throughout the book to compare and contrast with the general common law approach covered.

1.3 Civil and common law distinguished

National law systems, sometimes known as jurisdictions, can be described as being 'common' or 'civil'. The key difference is shown in Figure 1.1 and can be described as follows:

- *Common law* places great emphasis on the importance of case law. Cases can create principles of law that apply across the legal system. In this sense, common law can be seen as being a 'bottom-up' system. Cases, sometimes starting in a lowly ranked court, can, usually through a system of appeal, become the dominant source of law on any given issue in the jurisdiction. Common law jurisdictions use elements of civil law to complement the common law approach. Common law countries can therefore be described as taking a dual approach to their lawmaking.
- *Civil law* involves law being written down in a legal proclamation. The types of proclamations vary greatly and are known by such terms as statute, code, constitution, declaration, regulation and directive. Civil law can be seen as being a 'top-down' approach to lawmaking. Law is created for the politicians by draftsmen who capture the purpose of the required law in a document which everyone must obey. Changes to civil law can be influenced by case law; for example, amendments to the US Constitution. Amendments to the Constitution are sometimes required where cases have been brought in court requiring a decision on an aspect not previously

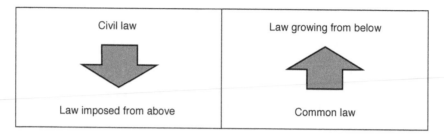

Civil law	Law growing from below
Law imposed from above	Common law

Figure 1.1 Civil and common law systems

covered by the written constitution. The constitution is therefore amended (changed) so that it takes account of the new issue raised. However, this reaction to a case brought in the court does not amount to the same power that judges have in a common law country. Put simply, civil law judges do not normally have the same ability to make new law in their judgements as is the case in common law jurisdictions.

1.4 Why law is needed

Law can be described as society's rule book – our attempt to regulate ourselves. This was summed up by Thomas Hobbes in his seventeenth-century work[3]: 'Law is the formal glue that holds fundamentally disorganised societies together'.

Consider a queue at the check-in desk at the local airport. People queue because they know that, in the normal course of events, their turn in the queue will come and this will allow them to catch the aeroplane in good time. If there was no queue and no airline staff to process the tickets, this might result in a free-for-all. Only the pushiest and loudest people will have an opportunity to catch the aeroplane. The queuing system is fairer and more just – a key theme in any debate about the need for law.

The airport example shows one aspect of the purpose of law. A more complete list of the benefits of a legal system is considered in the next sections.[4]

1.4.1 Providing guidance on how to behave and interact with others

If there were no laws (especially written laws) then people would be without guidance on what is acceptable behaviour in society and what is not. The relationship between laws and customs is a grey area and the two concepts interrelate to a large degree. Travellers visiting foreign countries should be mindful that customs in those countries may dictate that behaviours considered acceptable in their own countries are unacceptable to their hosts (such as allowing the head to be uncovered or entering a temple whilst wearing shoes). If the custom/law was written down then a visitor would have a better chance of knowing about it and therefore being able to avoid inadvertent offence.

Breaking customs is unlikely to be regarded as seriously as breaking the law. People are largely forgiven for breaking customs and ignorance, in this case, is a defence. Ignorance of the law is no defence. Customs, unlike laws, do not usually involve a system of enforcement. In some legal traditions, this distinction between laws and customs is not made, meaning that extra care is needed to observe customs with the force of law. Another term of use in this area is 'legal norm', which involves a mandatory rule of social behaviour established by the state.

1.4.2 Providing protection of certain interests

It is high on the list of importance for most people to want to own property and use property free from molestation. This is not the only view; political ideology has sometimes sought to impose a different approach, most strikingly the anarchist view that 'all property is theft'.[5] Property in this sense does not simply mean real estate (land and buildings) but can also extend to a person's belongings and intangible property. The role of law here is to protect all property and recognise forms of ownership. The following list demonstrates the range of property interests with the relevant protecting/enabling law shown in brackets:

- physical interests (tort law);
- dignitary interests (tort law);
- property interests
  - real property (property law)
  - chattels or belongings (tort law)
  - intangible property interests (intellectual property law); and
- financial interests (contract law).

Tort, property and contract law are considered separately in later chapters in Part 1. The desirability of the protection afforded by these laws is felt very keenly in the world of construction as can be demonstrated in the example given below.

Landowner A needs to establish that he owns the land on which he intends to build (*property interest*). The builder B wants to know that he may occupy the land for the duration of the project and that the contract he signs with the landowner will be enforceable in the event of non-payment (*financial interest*). Landowner A wishes to rely on the designs that have been prepared with due professionalism by architect C (*property interest*). Architect C wishes to ensure that ownership over his designs stays within his organisation and will not be copied either by the client or a third party (*intangible property interest*). The builder B wants to ensure that his good name is not libelled in newspaper articles complaining about his standard of work (*dignitary interest*).

The persons involved at each stage of the example given above are able to consult the law to apprise themselves of their rights in the situation. They may each, either by themselves or through a lawyer, take steps to enforce their rights in the event that they perceive that there has been or will be an infringement against those rights.

1.4.3   *Expressing disapproval (criminal law)*

The analogy of the check-in queue at the airport has already been used. Most people will feel annoyed if queue-jumping is allowed and may want to see the offenders sent to the back of the queue. This sentiment involves aspects of punishment for improper actions. Looked at another way, interests need to be protected and rule breakers to be admonished. This is one of the functions of criminal law.

There are several different approaches to criminal law, ranging from a strict authoritarian *zero tolerance* approach to more reconciliatory views. Rehabilitation of offenders focuses on the desire to reintegrate people who have erred into society once they have reflected on and realised the negative impacts of their actions. The lawmaking process is more typically concerned with the punitive element. The criminal law involves imposing tariffs for fines, bans and prison sentences on offenders. The underlying theme here is the expression of disapproval for transgressing legal norms or acceptable behaviour at whatever societal level the offence is encountered. This is often referred to as white-collar crime (professional crime such as fraud) or blue-collar crime (offences related to injury to people or property).

There are many examples of the application of the criminal law to situations in and around construction law. Breaches of health and safety procedures can lead to criminal proceedings and the imposition of fines. A great deal of effort has been made both domestically and internationally to curb corruption in construction. In the UK, this included an investigation by the Office of Fair Trading culminating in prosecutions against 103 contracting firms found guilty of cover pricing. Internationally, the work of Transparency International and the World Bank has been concerned with reducing the incidence of bribery. The resource wasted through these nefarious activities runs into billions of pounds every year. These themes are considered in more detail in Chapter 22.

It is not always straightforward to ascertain the difference between criminal and civil law. 'Civil' law in this context means both common and civil law systems. This book is primarily concerned with civil law – the law that offers private protection and a means whereby the aggrieved party can be compensated for any loss suffered. The successful outcome of a civil case is usually an award of damages (money) to the winning party.

Civil law has its own court system and the parties are referred to as the claimant (the person bringing the claim) and the defendant (the person against whom the claim is brought). It is standard practice to refer to one party versus or -v- the other party to show that a court case is involved. The outcome of the case is of primary interest to the parties involved. However, if the case is reported, it is of interest to others given its status as precedent.

The purpose of the criminal law is to protect the public as a whole rather than any particular individual. The successful outcome of a criminal action is that the offender is convicted of a crime and punished. This punishment might be financial or involve taking away the offender's liberty by means of a custodial sentence. There is a slight crossover with civil law in the sense that financial compensation is sometimes available to the victim or their family through a government-sponsored scheme.

Criminal law has its own court system, kept separate from the civil law system. In keeping with the notion that it is society bringing the action against the rule breaker, the parties are known, in England and Wales, as *Regina* (Latin for Queen) v the defendant (rule breaker). In other jurisdictions, the public are referred to collectively by such terms as the People v (USA) and *La République* v (France).

This reference to the Queen in criminal action shows the historical importance of the Crown in the development of the English legal system. Table 1.1 helps summarise the differences between criminal and civil law.

1.4.4 Providing a means of resolving disputes

In any society, people occasionally fall out. This occurs in many different scenarios including, for example, squabbling siblings disputing their inheritance in the absence of a will, neighbouring landowners arguing over a boundary or a dispute between an employer and a disgruntled employee who was recently dismissed. In the construction context, the parties disagreeing might include a developer and builder with divergent views on a final account payment or a supplier and subcontractor disputing the quality of items supplied. Differences are commonly dealt with by negotiation. However, when parties cannot resolve their differences through consensus, they need recourse to the courts for a decision on who is right and who is wrong. Courts and other tribunals that are available frequently act as a deterrent in that people are rightly nervous about the exorbitant costs involved in a fully contested hearing. The types of dispute resolution procedures used in the AEC industry are covered in Part 4.

Table 1.1 The differences between criminal law and civil law

Criminal law	Civil law
For public protection	For private protection
Offender punished	Aggrieved party compensated
Criminal courts	Civil courts
Regina v Martin	Smith v Jones

1.4.5 Supplanting morality with law

Morality largely stems from religious teaching. Historically, religious laws were, and in some cases remain, very important in the development of legal systems. In the Christian tradition, biblical pronouncements such as the Ten Commandments are expressed as laws being handed down from a higher authority. However, in most secular societies, the law and religious teachings have become distinct to the extent that one has supplanted the other. For example, taking another biblical reference, 'an eye for an eye, and a tooth for a tooth'[6] might be a religious teaching but it is not part of our legal system. If you are wronged then you must rely on the legal system to mete out justice on your behalf rather than taking matters into your own hands as the religious teaching might be interpreted.

This separation of religious and secular law is not universal. Sometimes the religious and/or political will of a country precludes the formation of a separate legal system. An example of this is Sharia law where the law is regarded as an expression or extension of religious principles. This manifests itself in ways such as a prohibition in Middle Eastern countries on adding interest to debts, which is seen as being against the scriptures. Contracting firms operating in Arabic countries need to be aware of such local variances when undertaking projects in these areas. The links between religious and modern laws are still present in Western societies, albeit more diluted and indirect.

These sections have involved a wide-ranging discussion about the role of law and may seem a long way removed from the day-to-day operation of construction contracts. It is only by appreciating the context and purpose of law that we can understand the specifics as they apply to the chosen field of study.

1.5 Legal systems

The previous section looked at why law is needed and the interests law protects. The next issue to consider is how law is delivered. The protection afforded by the law is only as good as the legal system's ability to satisfy the demands placed upon it. Consideration is now given to how the legal system meets these requirements and its performance.

1.5.1 Introduction

Confidence in the ability of the legal system to uphold the law is essential. There are three component requirements to deliver this confidence in the performance of the legal system. The system must be:

- consistent in its approach;
- consistent in its results;
- accountable for its actions.

Consistency involves reassuring the users of a legal system that the same law will be applied to the same type of cases. The outcome of similar cases must be consistent. Put another way, judges must use the same general criteria when arriving at their decisions. If new rules are introduced, there should be a clear explanation of the reasoning and evolution behind their creation. Third, the legal system should be accountable in the sense that it can be held responsible for its own actions. If something goes wrong with the legal system, there needs to be a means for redress. This redress is provided through the system of appeal.

1.5.2 Historical development

The English legal system (this refers strictly to England and Wales – Scotland and Northern Ireland have their own legal systems) has developed in such a way as to meet the requirements of consistency and accountability. Table 1.2 lists some historical events playing an important role in the development of the modern-day legal system.

The consistent and accountable approach desired by legal users was on its way to being in place by the end of the seventeenth century. Access to justice

Table 1.2   Key historical dates in the development of the English legal system

Pre-Norman Conquest 1066	Before the Normans arrived, English law was a mixture of local customs. The law dispensed by local lords would differ from one region to the next. The monarch had nominal overall authority but rarely had the time or inclination to be concerned with legal matters other than raising taxes to fund their military campaigns.
Rule of Henry II 1154–89	Henry II was the first Norman king to take real interest in organising the legal system. Norman rule had settled down in England and Wales to the extent that a decent attempt was now made to establish a legal system including a clearly defined hierarchy of the courts and lawmaking bodies.
Magna Carta 1215	Not all Norman kings took as great an interest in lawmaking as Henry II. King John was a notoriously bad king and the barons/landowners became frustrated with (amongst other shortcomings) his lack of attention to maintaining the law. The barons therefore forced him to sign a charter whereby he pledged to bring a properly constituted legal system into being and to enshrine certain rights. This document was known as the Magna Carta, or 'big charter', and represents the end of absolute monarchy in England and Wales inasmuch as the power of the sovereign was now circumscribed in the area of lawmaking.
Bill of Rights 1689	Fast-forwarding several centuries, this date is picked out because it represented further inroads into the monarch's absolute power. Absolute power had first been compromised by the signing of the Magna Carta. From 1689, a constitutional monarchy was set up which basically ensured that the legal system was now fully accountable for its actions. Parliament now reigned supreme and the monarchy became largely subservient.

and the right to appeal a bad decision were established as basic human rights. These rights continue to evolve and are set out in such statutes as the Human Rights Act 1998.

1.5.3 The separation of powers

Any discussion around the historical development of law demonstrates the political worth of the lawmaking function and its powerful nature. The ability to make law can have an intoxicating effect on those wielding it, and this has led to efforts to contain its use. This sentiment is summed up in the oft-quoted phrase of Lord Acton from 1887: 'Power tends to corrupt, and absolute power corrupts absolutely. Great men are almost always bad men'.[7] Democracies (and other legal systems) have developed ways of ensuring that power is separated into different lawmaking institutions, or estates. In this way, we can attempt to keep the 'bad men' under control.

The attempt to restrict the excess of power has been a central theme running through the historical legal development known as the separation of the powers. The central notion is to separate the functions of law to ensure that its proper exercise can be checked by other institutions.

In the context of the English legal system, separation is achieved by keeping the judiciary (judges) separate from and not under the control of the executive (monarch and later Parliament). The executive is itself kept separate from the legislature, often referred to as the upper chamber. This separation ensures each legal estate performs its own role and, as far as the legal system is concerned, gives the judiciary a free hand to dispense justice regardless of whether this compromises or undermines the actions of the executive and legislature. Each lawmaking institution is therefore balancing the power of the others. This is known as the system of checks and balances, which ensures power is not abused.

This model of the separation of powers is one of the key elements in democracy as reinterpreted in many other countries. The European Union has its own version of the separation of powers. The executive, known as the European Commission, is made up of the Council of Ministers, which

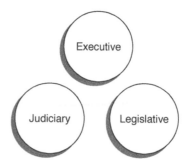

Figure 1.2   The separation of powers

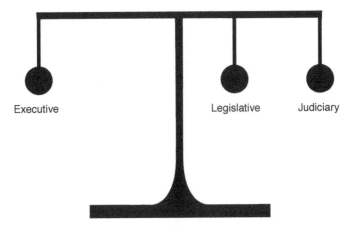

Figure 1.3 The system of checks and balances

involves the ministers of the member states coming together to agree policy. The legislative role is mirrored in part by the European Parliament. This institution merely debates the laws proposed by the Commission; it does not have the same power as national parliaments, such as the UK Parliament, to veto proposed laws. The European Court of Justice fulfils the role of the policeman.

The constitution of the judiciary, the executive and the legislature is covered in the next section where their role in creating law is described.

1.6 Sources of law

The three most important sources of law are:

- case law (made by judges);
- legislation (made by Parliament);
- European law (operating through a mixture of case law and legislative instruments).

The distinction and interplay between common law and civil law has already been discussed (see Section 1.3). Essentially law involves directing individuals and organisations on how to act. Direction on what to do is given either by statute (legislation) or by a judge (case law). On some occasions, both sources of law are involved as well as the need to comply with European law.

1.6.1 Case law

Case law operates through a system known as precedent. The function of the courts is to decide disputes. Precedent is the name given to the system where a decision of a judge has a binding effect on a different judge in similar circumstances. In short, the judge has to take into account the previous authority. This is not the same as being required to follow the decision to the letter.

In presenting a legal case, the advocate will usually seek to rely on a list of cases in support of the decision they wish the judge to make. It is a truism to say that no two cases are exactly alike, and it is, therefore, by building a legal argument based on a number of related cases that the advocate hopes to win. Some cases are distinguishable from the present case either on their facts or the law applied. It is open to either advocate to seek to distinguish a potentially relevant case submitted by the other party on the grounds that it is not relevant to the case being discussed.

In later chapters, this work quotes cases. The cases are presented as precedent and essentially as the argument or illustration to support the point of law being made. It is possible that other cases could be cited to present an alternative view of the law. However, most areas of law are settled and the authorities which apply are well rehearsed. Consensus exists in most areas as to what the law is. Some areas of law are less clear, and divergent, yet equally valid, views are possible. Examples of this include the law surrounding concurrent delay analysis and the meaning of good faith provisions (see Sections 14.5 and 22.9). However, the common law is always evolving and new precedents and points of law can be created at any time. This state of perpetual evolution is necessary for the legal system to continue to meet society's needs, and it provides interest and intrigue for lawyers and scholars alike.

For example, I see my neighbour walking through my garden and ask him to use a different route. He refuses. I could then say: 'It was decided in the case of *Smith* v *Jones* 1903 in the Court of Appeal that neighbours should not walk through each other's gardens if there is an alternative route'. If we then failed to resolve the issue and commenced court proceedings (civil law of trespass), I could rely on this case as a precedent and seek to direct the judge to follow this case if the facts are the same or similar. This earlier similar case is binding on the judge who should pay due attention in reaching a decision in my case. The decision is binding only to the extent that it applies directly to my case. The other side may establish that the case I have relied upon needs to be distinguished from the current case and will seek to address the judge on cases where a finding was made consistent with their argument and interpretation of events. The cases brought by the other side will therefore favour their desired outcome. They may argue that the alternative route is not practical for my neighbour to use as it involves a two-mile detour. In a contested hearing, both parties will arm themselves with a number of cases to seek to persuade the judge as to the veracity of their case.

To function properly, precedent needs two things:

- *Proper law reporting* – What exactly did the earlier judge say? This has only been reliably available from the nineteenth century onwards. However, some laws still in use today stretch back to cases made several hundred years earlier. For example, an important authority on the law of consideration dates from Pinnel's case[8] of 1602 when the first Queen Elizabeth was on the throne of England. The use of these cases requires that reliable records had been taken at the time. This would usually involve a case or court of some importance being involved and the careful noting down of a transcript of the judgement itself. The status and level of the decision made should be easily identifiable in the law report. Later in this work when cases are referred to, there is an accompanying reporting reference. The most common reports are taken from All England Reports (ALL ER) and Weekly Law Reports (WLR).[9]

- *A hierarchy of courts* – Which judges' decisions are binding on which other judges? This question is now considered.

1.6.2 Court systems

Precedent is the system whereby an earlier judgement in one court can be binding on another. For this system to work effectively, there needs to be a hierarchy of the courts to give clarity on which court's decisions are binding on which other court(s).

The court in which an action is commenced will depend on either the severity of the crime (criminal law) or the amount of money/importance or the specialist nature of the issue involved (civil courts). The lower courts are the starting point for minor cases. Access to the higher courts will depend on your ability to appeal against a decision in a lower court related to the amount/importance of the matter concerned. Appeal means the ability to take the matter out of the hands of the original court and ask for a higher court to rehear the case. The right to appeal is a fundamental right underpinning the English legal system. An appeal to a higher court is restricted to instances where one party contends that the judge has made an error in law rather than in the facts of the case.

The lower courts

Smaller cases start in the lower courts, which are called the magistrates' court for criminal matters and the county court for civil matters. Magistrates' courts and county courts are found in most towns in England and Wales. In the civil sector, higher-worth cases and specialist courts (e.g. employment tribunals) start in the high court. Serious offences in criminal matters are tried in the Crown court although *interlocutory* matters (such as confirming details and setting a timetable for trial) take place in the magistrates' court. Crown courts and high courts are only found in cities in England and Wales.

The higher courts

The higher courts are known as the appeal courts – it is only if the losing party in the trial at first instance wishes to challenge the decision that leave may be given to appeal. The appeal is heard in the higher court by judges who are more senior.

Cases involving construction law issues are commenced in a branch of the high court known as the Technology and Construction Court. This branch of the high court was purposefully created in recognition of the need for specialist knowledge in the field of construction law. It also serves as a reminder of the contentious nature of construction and the frequent need to have recourse to the courts. Other branches include family court and insolvency court. Case reports from the Technology and Construction Court sometimes carry the suffix TCC.

The Court of Appeal and the Supreme Court are both located in the Royal Courts of Justice in London.

The criminal court hierarchy

The hierarchy of criminal courts is shown in Table 1.3.

Table 1.3   The criminal court hierarchy

Court	Personnel involved	Precedent status
1 Magistrates' court	Either one stipendiary or three magistrates hear the cases. These usually involve minor traffic offences and breaches of the peace.	Bound by decisions of all the higher courts
2 Crown court	One judge sits in the Crown court. There is a right to jury trial for some offences where the accused pleads 'not guilty'.	Bound by the decisions of the Court of Appeal and the Supreme Court
3 Court of Appeal (Criminal)	Three Law Lords chosen from the Supreme Court of Lords sit and rehear the legal arguments from Crown court decisions if leave to appeal was granted.	Bound by its own decisions and those of the Supreme Court
4 Supreme Court	Five Law Lords sit and rehear the legal arguments from the Court of Appeal.	Not bound by its own decisions

The civil court hierarchy

The Supreme Court is not the ultimate court of appeal in the country. The European Court of Justice can override the national courts on certain matters. There is a separate European Court of Human Rights, created solely to rule on abuses of human rights where the right to take a matter to Europe exists. There have been many headline cases involving this procedure over the years involving such issues as gender swapping and the rights of unborn children. Recourse to the European Court of Human Rights is often thought necessary where the national law does not recognise, or has not considered, laws in these areas.

Table 1.4   The civil court hierarchy

Court	Personnel involved	Precedent status
1 County court	Cases are heard in the county court by district judges who preside over full lists including mortgage arrears and divorce cases.	Bound by the decisions of all the higher courts
2 High court	Larger-worth cases start in the high court where there are specialist judges to hear their own lists of cases, including Family law judges and Technology and Construction law judges. The high court can hear appeals from the county court.	Bound by the decisions of the Court of Appeal and the Supreme Court
3 Court of Appeal	Three Law Lords chosen from the Supreme Court sit and rehear the legal arguments from high court decisions if leave to appeal was granted.	Bound by its own decisions and those of the Supreme Court
4 Supreme Court	Five Law Lords sit and rehear the legal arguments from the Court of Appeal.	Not bound by its own decisions

An example of the importance of European Court of Justice decisions on construction matters concerns the implementation of the public procurement regime whereby national governments are punished for practices which are seen as being against the single and open market for competing for construction contracts (see Section 7.9). The impact of European law on the construction industry also manifests itself in improvements to health and safety, consumer regulation and environmental protection. Most contractors are familiar with the Building Regulations, which have been one of the main means of ensuring compliance with European standards.

1.6.3 Legislation

One of the roles of government is to introduce new law. The separation of powers dictates that the executive (government) cannot do this on its own and requires the assistance of the legislature (Parliament). In England and Wales, Parliament consists of two houses – the House of Commons and the House of Lords. A government is made up of whichever political party (or parties) can form a majority in the House of Commons following a general election. The majority is required in order to be able to have the opportunity to vote in new law. Amongst the lords in the House of Lords are the Law Lords who make up the highest court in the land, also known as the Supreme Court. At the opening of Parliament, the government sets out its proposed new laws in the Queen's speech. The government then promotes its new laws (known as bills) and seeks to have the bills passed into law within the four-year or five-year term of Parliament. Passing the law involves debating and drafting the law before submitting the Bill to the Upper Chamber (in this case the House of Lords) for its consideration. It is common for interest groups affected by the new law to be consulted in subcommittees.

The involvement of a lower and upper house in making new law is replicated throughout the world. The upper house is often referred to as a Senate and the lower as a House of Representatives or National Assembly. The idea of a senate comes from ancient Rome and is so called as an assembly of the senior and thus wiser members of society.

It can be observed that Acts of Parliament have a very important role to play in the functioning of a legal system. When the government wants to create law, it does so through Acts of Parliament. This is the means by which the government 'steers' the country. For example, to improve on current high rates of childhood obesity, the government may wish to ban sugary foods for children. This might entail an 'Anti-Obesity (Minors) Bill'. The bill will become an Act if it makes it through the various rounds of drafting and consultation and the issue attracts sufficient support from Members of Parliament and the House of Lords.

To continue the example introduced earlier, my neighbour brings a copy of a new Act of Parliament to court. The Act is entitled 'The Right to Recycle Act 2015' and section one of the Act says that a neighbour may cross his neighbour's gardens if it is reasonably necessary for the purpose of putting out his recycling on a weekday. He says that the Act supports his case and contends it is much more recent and therefore more relevant than the old case I was relying on. The judge agrees with my neighbour.

This example seems to suggest that case law and legislation can conflict. In fact, this is rarely the case. Legislation is often used by Parliament to untangle an area of law that has been made unclear by conflicting case law. There then usually follows some cases where judges try to interpret exactly what Parliament meant in its Act of Parliament. For example, what exactly does 'reasonably necessary' mean in the context of the imaginary Right to Recycle Act 2015? It may be necessary to await a judicial pronouncement on this point so that a clearer picture emerges on what exactly was intended by Parliament and where the line should be drawn.

Case law and legislation are common features of this book and heavy reliance is placed on both in explaining and expanding on legal concepts and principles. Law students are well versed in writing down case and statue names and conducting their own legal research into the facts of the case and related pieces of law. This book attempts to represent a single source of construction law, but the law student's discipline remains a good one and students of this book are encouraged to seek out the primary sources of law when appropriate.

1.6.4 European law

The UK has been a part of the European Union since the European Communities Act 1972 was passed through Parliament. From a legal point of view, this was the date the UK handed over the ability to self-govern. European law is therefore of fundamental importance. The sources of European law are as follows:

- Treaty of Rome 1957 – one of the founding treaties of the European Community;
- regulations – laws with direct effect into national law;
- decisions – European court decisions must be obeyed and implemented by national courts; and
- directives – laws to be implemented into domestic law by member states within a given time frame.

It is beyond the scope of this book to provide more than a cursory discussion about the functioning of the European legal system. Suffice to say that very careful note has to be taken of the impact of European law on national law, which is ignored at the peril of the member states of the European Union. Examples are given later in this book of regulations with direct effect – such as the Unfair Terms in Consumer Contracts Regulations 1999 – and European Court of Justice judgements – for example, in public procurement such as *Commission of the European Communities* v *Ireland* (the Dundalk case).[10] This case involved the fairness of a tender competition and whether tenderers from across Europe had a level playing field on which to bid for the work.

1.7 Who's who in the law?

All industry sectors feature different professional roles and the law is no exception. An understanding of the roles within the law gives further insight into the workings of the legal system.

1.7.1 Solicitor

The first point of contact on legal matters will usually be a solicitor. The solicitor may advise that a barrister is also instructed, if necessary. At its simplest level, the way the two branches of the legal profession are divided is that the solicitor prepares the case for court and the barrister presents it. However, this is an oversimplification as there are solicitor advocates who appear in court. Likewise, barristers can be very useful in the preparation of a case in terms of strategy and how to approach the gathering of evidence and similar matters. The majority of legal work is not focused on the courtroom but the protection of rights as discussed above in Section 1.4.2. This work is described as non-contentious and is usually the exclusive field of practice of solicitors.

Solicitors usually group themselves in partnerships whereby the firm of solicitors will comprise partners operating in business together. The business model frequently used is to have one or more solicitors expert in several different fields to provide a range of services; for example:

- conveyancer (buying and selling houses);
- trust/probate lawyer (dealing with inheritance issues and drafting wills);
- litigator (taking cases to trial);

- commercial lawyer (drafting contracts and setting up businesses);
- commercial property lawyer (drafting leases and acquisitions of land);
- family lawyer (dealing with divorce and children issues).

Modern firms do not necessarily follow this pattern, and there are many niche and specialist firms only undertaking one form of work; for instance, specialists in road traffic accident compensation. A small firm may process many thousands of such claims each year.

Solicitors are assisted by paralegals and legal executives. The former are used for labour-intensive tasks whilst the latter are capable of undertaking any solicitor-based work.

1.7.2 Barrister

The profession of barrister is steeped in tradition and this survives today in many of the practices and procedures that have been around for centuries. The key role for the barrister is advocacy in the courtroom and advising on strategy in preparing a case for trial together with advising on the likely prospects of success. Barristers do not enter into partnerships with each other but operate from the same offices, known as barristers' chambers. Barristers or counsel, as they are also known, are self-employed, merely gathering together to share expenses such as clerks' fees and office costs.

The top barristers are known as QCs (Queen's Counsel) and have been selected by the Queen's officers as being at the top of their profession. QCs are permitted to wear special robes made of silk. Becoming a QC is therefore known as 'taking silk'. All barristers are said to be 'called to the bar' when they qualify, and part of their training involves literally eating dinners at one of the 'inns of court'. Inns of court refer to the key barristers' chambers in London, situated not far from the Royal Courts of Justice.

Members of the public are not allowed access to barristers except in limited circumstances. It is usually for the solicitor to instruct the barrister on the client's behalf. Efforts to modernise the profession have been slow to take hold. The wearing of gowns and wigs continues to give the courtroom an anachronistic feel. Moves to reform court apparel have not so far succeeded in changing the dress code.

1.7.3 The judiciary

Strictly speaking, solicitors and barristers are the two branches of the legal profession. However, the judges, the rule-makers themselves, also need to be considered. Judges are usually chosen from amongst the ranks of the barristers together with a minority of solicitor appointments. Barristers can regard appointment to the bench as career progression and are then one of the privileged few in terms of becoming a member of the judiciary. A judge is a civil servant in the same way as a politician, and their salaries are paid for by the taxpayer.

A magistrate is a type of judge dealing with minor criminal matters. Traditionally magistrates have been chosen from members of the public who have volunteered to hold the office. Being tried by one's peers – whether by another member of the public (magistrate) or by a jury (12 members of the public) – is an important principle in the English legal system. A jury trial is a legal proceeding in which a jury either makes a decision or makes findings of fact, which then direct the actions of a judge. Jury trials are used in serious criminal cases in most common law legal systems.

Many lay (untrained) magistrates have now been replaced with paid professional magistrates called stipendiaries. The professional magistrates are able to deal with cases much more quickly and effectively than lay magistrates. Most commentators see this as necessary despite the compromise it involves of the legal principle of trial by one's equals.

The office held by a judge depends on the court in which they appear. The higher the court, the more weight is given to their judgements in accordance with the precedent hierarchy discussed above.

1.8 Conclusion

This chapter has taken a broad-brush approach to a complicated and wide-ranging topic. The intention is for the reader to acquire the background knowledge and assumptions on which the later sections of this book build. The reader should now have established a proper understanding of the processes at work and their genus.

The next chapter looks at contract law and follows the legal writing discipline of making a statement and then supporting the statement by reference to statute or case law. Any statement of law made without support or evidence is open to challenge. It is by referencing the statements that authority is given to the pronouncement. This need to reference is a core skill in terms of both academic writing and any discussion or setting down of the law.

The key points to take away from this chapter include an appreciation that:

- a legal system is created by its users to suit their needs;
- a legal system must deliver the needs identified if it is to work effectively and enjoy the support of its users;
- the English legal system has evolved over the centuries to define and meet the needs of its users;
- any discussion of law involves a consideration of the source of laws – primarily case law and legislation;
- the role of Europe must not be overlooked in connection with the English legal system;
- the organisation of the legal system is closely linked with the functions of Parliament and the other organs of state, including the monarch;

- the legal profession comprises solicitors, barristers and judges;
- the hierarchy of the courts and judiciary is necessary for the system of precedent to work effectively.

1.9 Further reading

Wood, D., Chynoweth, P., Adshead, J. and Mason, J. (2011) *Law and the Built Environment*, Second Edition, London: Wiley-Blackwell, Chapter 1.

Wild, C. and Weinstein, S. (2013) *Smith and Keenan's English Law*, Seventeenth Edition, Harlow: Pearson, Part 1.

Notes

1 Rousseau cited in Bertram, C. (2004) *Routledge Philosophy Guidebook to Rousseau and the Social Contract*, London: Routledge, p. 42.
2 Approved by the National Assembly of France, August 26, 1789. Other Revolutionary laws did not stand the test of time; in one case literally – the replacement of the 12-hour working day with a 10-hour period did not catch on and was later abandoned.
3 Hobbes, T. (1651) *The Leviathan*.
4 Based on Cane, P. (1997) *The Anatomy of Tort Law*, Oxford: Hart Publishing.
5 Proudhon, J-P. (1840), *What is Property?*
6 Matthew 5:38.
7 Letter to Bishop Mandell Creighton, April 5, 1887, published in *Historical Essays and Studies*, edited by J. N. Figgis and R. V. Laurence (London: Macmillan, 1907), p. 504.
8 (1602) 5 Co. Rep 117a.
9 For a full list of courts referred to in the cases cited, see *OSCOLA: The Oxford University Standard for Citation of Legal Authorities*, Fourth Edition. Faculty of Law, University of Oxford, www.law.ox.ac.uk/oscola.
10 C- 45/87.

2 The law of contract

2.1 Introduction

The law of contract is of central importance to the study of construction law. This importance is reflected in the more detailed examination given to it in these pages than the other background legal subjects. This chapter seeks to capture the essentials of the law of contract whereas the other subjects are dealt with as 'aspects of' commercial, tort and property law. Contracts are extremely prevalent in the interrelationships between stakeholders in the construction industry. The contract execution (or signing) is the single most important stage in the process when land or buildings are transferred and when building projects are undertaken. 'Putting pen to paper' is to enter into a contract and to bring the force of contract law in regulating the agreement reached.

The 'golden age' of the law of contract was in the nineteenth century when major principles were evolved, based on free market ideologies. The embodiment of these principles is that the parties are free to contract on whatever terms they choose provided the contract is legal. In other words, lawmakers have long seen it as their role not to interfere in contracts and have been reluctant to intervene. Many of the cases referred to in this chapter date from this groundbreaking period. The case illustrations used are not limited to this period but range from the very old to the very modern.

A contract is a legally binding agreement. It is a bargain and each side, or party to the contract, must contribute something to it for it to be valid. Not every agreement is a contract, nor is it intended to be so. The legally binding element must be present before a valid contract can emerge. In other words, the parties must be able to demonstrate their intention to adhere to the agreement made. The protection afforded by entering into a contract is that if it is broken by one party, the other party must be able to take the contract-breaker to court if desired. This is the closest thing any party has to that most sought-after commodity in legal dealings – certainty.

A distinction is made between a situation where the parties exchange mutual promises, known as a bilateral contract, and where one party promises to do something in return for the other party carrying out some task, known as a unilateral contract. When the task is completed, the promise made in a unilateral contract becomes enforceable.

2.2 Formalities

A contract may be made in any form the parties wish. This is the case regardless of the sums involved or the complexity of the agreement. There are advantages in having the contract formally drafted by a solicitor, particularly if a large sum of money is involved or the matter is complicated. There is no essential requirement in English law that a contract should be in writing, and as a general rule, the parties to a contract may insert whatever provisions they wish into the agreement provided it is for a legal purpose.

Simple contracts

These are contracts made by word of mouth or in writing or a combination of both. No particular form is required. The phrase 'contract signed under hand' refers to a simple contract being formed. Mere signature by the parties is enough to evidence the formation of the agreement.

Deeds

A contract made by deed is known as a specialty contract. Until 1990, certain contracts had to be made 'under seal' and delivered up 'as a deed'. The sealing of a contract referred to a more elaborate execution of a written contract than a mere signature. Typically, this could involve an impression made by a signature ring or company motif in wax dripped onto the document. A seal is no longer necessary, but to be valid, a deed must be signed on behalf of each party to the contract.[1] The distinctive features of a contract signed as a deed rather than as a simple contract include the rule that no consideration is necessary for a deed to be valid. Another practical advantage of a contract being made by deed is that the limitation period governing such contracts is 12 years as opposed to six years in the case of a simple contract.

In the AEC industry, it is common for the building owner to insist on the longer 12-year limitation periods for the agreements entered into. Supply agreements are more likely to be signed 'under hand' with six-year limitation or less if express terms are included to this effect.

Contracts which must be made in writing or evidenced in writing

Certain types of contract cannot be enforced or are invalid unless they are in writing, although they do not have to be made in the form of a deed. This applies to consumer credit and hire agreements governed by the Consumer Credit Act 1974 (as amended by the Consumer Credit Act 2006). These Acts require that the relevant agreements are 'not properly executed' unless in the form prescribed. Failing to comply with the requirement has the effect of rendering the contract unenforceable. The requirement that a contract be in writing also applies to the sale or other disposition of an interest in land,[2] whilst contracts of guarantee need to be evidenced in writing pursuant to the Statute of Frauds 1677.

The requirement for these contracts to be in writing underlines their importance and the need for certainty in recording exactly what was agreed in the event that future scrutiny is required.

2.3 Standard form contracts

Certain transactions are governed entirely by standard terms which are predetermined. Examples are where one of the parties enjoys a monopolist position or where particular types of business are regulated by trade unions. For example, negotiating personal terms on which a bank lends money to a consumer or terms on which a road haulier delivers construction material are far from straightforward propositions. Where this arises, the customer is often not in a position to negotiate over the terms of the contract. Instead the person requiring the goods or the services contracts on a standard form prepared in advance by the dominant organisation. Such a contract departs from the concept of freedom of contract except insofar as the customer can choose whether to contract or to walk away.

The type of standard form of contract which is commonplace in construction and land transactions is much fairer. Standard clauses have been settled over the years by negotiation between the various stakeholders in the relevant industries. The purpose of these forms is to facilitate the conduct of trade and a fair allocation of risk between the parties. The types of contract invariably used in large-scale construction works include the Joint Contracts Tribunal (JCT) Standard Form of Building Contract and the New Engineering Contract (NEC). Both forms tend to govern major works along with the International Federation of Consulting Engineers (FIDIC) contract employed on international projects. Another form of contract is the Project Partnering Contract (PPC2000). These forms of contract are considered separately in Chapter 9.

2.4   The essential elements of a valid contract

A simple contract has three essential elements:

- agreement;
- consideration;
- the intention to create legal relations.

These following sections set out the law on these elements in some detail. The law is presented in accordance with the standard practice of stating the relevant cases as authority for the submissions made along with the supporting statutes where relevant. This is the first time this book uses the legal convention of citing authorities and the reader should be aware of the approach taken. The approach may seem strange at first but perseverance with the subject matter pays rewards in terms of building a legal argument.

2.4.1   Agreement

Before a formal agreement can be reached, there must be a valid offer made by the offeror and a valid acceptance of that offer by the offeree. It may be possible to ascertain with relative ease whether or not a valid agreement has come into being if a contract is entered into solely on the basis of a standard form of contract. It may be considerably more difficult to ascertain whether or not an offer and acceptance have been made where, for instance, a contract is alleged to have come into being by a combination of statements made orally together with documents in writing. The real test is whether the parties have accepted obligations to one another. If that can be established then a valid agreement may be inferred from the conduct of the parties.

In the case of *Trentham (G Percy) Ltd* v *Archital Luxfer Ltd*,[3] a building subcontract was held to have come into existence even though the parties had not reached full agreement on all terms when the subcontractor began the work. During the progress of the works, outstanding matters were resolved by further negotiations. The judge was satisfied in this case that there was sufficient evidence to conclude that there was a binding contract; the parties had clearly intended to create a legal relationship between each other and had covered enough of the basic points to be contractually bound. In other words, there was sufficient legal certainty on the contents to proceed with the contract.

Notwithstanding this case, the general rule is that arrangements which are too vague, are uncertain or are conditional will not take effect. There must be a clear and formal offer and an unequivocal acceptance of that offer. These rules surrounding the formation of an agreement can be compared to the rules of a game such as chess or the steps in a dance routine. The aim is to end with a situation where the offer and acceptance are matched as in the ancient symbol of yin and yang (see Figure 2.1). False moves or failure to read the intention of the other party may prevent an agreement being reached. The rules of offer and acceptance are set out below.

The dots in Figure 2.1 indicate that each half contains an element of the other. This is the same with contract formation where the acceptance is based on the offer that was itself made on the assumption that it would be acceptable.

Offers and invitations

An offer must be distinguished from an invitation to treat

An invitation to treat is a preliminary stage in negotiations which may or may not result in an offer being made. It is not possible to accept an invitation to treat and thereby create a valid agreement. Marked prices on articles for sale

Offer   Acceptance

Figure 2.1 Yin and yang symbol representing contract formation

will amount to invitations to treat rather than offers to sell. In the case of *Fisher v Bell*,[4] a flick knife was displayed in a shop window with a price tag attached. The seller was prosecuted under the now repealed Restriction of Offensive Weapons Act 1961, which made it an offence to offer to sell such items. The seller was acquitted on the basis that under the ordinary law of contract, the display of the article in the shop window was merely an invitation to treat.

A similar situation arose in the case of *Pharmaceutical Society of Great Britain v Boots Cash Chemists (Southern) Ltd*.[5] This case involved the display of prescription drugs on a shelf in Boots Chemist shop. The display of the goods was challenged as being in contravention of the Pharmacy and Poisons Act 1933 which required such drugs to be sold only in the presence of a qualifed pharmacist. The claimants alleged that the point of sale was when the customer helped themselves from the shelf to the drugs and that therefore the sales were illegal. Finding for the defendants, the judge found that the display of goods in this way did not constitute an offer. The offer for sale was not made when the customer selected the goods but when the goods were presented to the cashier. The court was satisfied that the cashier was supervised as required by the Act and that no contravention of the law had actually taken place.

INQUIRIES AND REPLIES TO INQUIRIES ARE NOT OFFERS

Statements of price or rates on their own are not formal offers. Most of the relevant case law is concerned with the purchase of land or buildings where inquiries are made but agreements never actually made. In *Gibson v Manchester City Council*,[6] it was decided that a statement made in a letter by the defendants that the council 'may be prepared to sell' and inviting the other party 'to make a formal application to buy' a house was an invitation to treat rather than an offer.

COMMUNICATION OF THE OFFER

It is essential that there is communication of the offer to give the offeree an opportunity to accept or reject it. Only the offeree can accept. This rule is illustrated by the case of *Boulton* v *Jones*[7] which involved an unincorporated business. The business changed hands between offer and the acceptance of the offer. The agreement failed as the offer had not been made to the offeree purporting to accept it. However, if the offer is made to the world at large as in *Carlill* v *Carbolic Smokeball Co*[8] then any person with notice of the offer may accept it. Mrs Carlill saw an advertisement in a newspaper with 'guaranteed' results and subsequently bought a smokeball which she hoped would cure her of her flu. She used the smokeball in the manner prescribed but did not see any improvement in her health. She sued the Smokeball Company and was successful. The latter had claimed that their advert was an invitation to treat and could not constitute an offer as it was made to the world at large. The key finding made by the judge was that the statement in their advert that they had deposited money in a bank account (for the purpose of issuing awards under the guarantee) showed the seriousness of their intention to be bound.

AN OFFER MAY BE REVOKED (WITHDRAWN) ANY TIME BEFORE ACCEPTANCE

Once a valid acceptance has been made, the offeror is bound by the terms of the offer. In an auction sale, each individual bid is an offer which can be accepted or rejected at any time before the auctioneer signals (usually by bringing down his hammer or gavel) his acceptance of the offer he wants. Similar rules apply to arrangements known as options to purchase. An offeror can be obliged to keep an offer open for a specific period provided the offeree has given consideration to have the offer kept open. In such circumstances, there will be a breach of contract if there is an attempt to revoke the offer within that period. Revocation of an offer only becomes effective when it is communicated to the offeree. In the case of *Byrne* v *Van Tinenhoven*[9] the defendant sought to withdraw his written offer by sending a second letter revoking the first. The first offer had already been accepted (as evidenced by a telegram from the claimant) and the second letter did not amount to a revocation. Indirect communication of revocation may be valid if from a 'reliable source' such as a mutual friend.

AN OFFER REMAINS OPEN UNTIL IT IS ACCEPTED, REVOKED, REJECTED OR LAPSES

Only proper acceptance of an offer will result in a valid agreement. No contract can be formed if there is a valid revocation of an offer or a rejection of that offer by the offeree. An offer may also lapse and become incapable of acceptance by passage of time. This occurs where there is a time limit within which acceptance may be made and that time has passed. It also occurs where no time limit is specified and there has been no acceptance within a reasonable period. The length of time is dependent upon the circumstances of the case, and the question of reasonableness may vary depending upon the nature of the contract. For example, an offer to sell perishable goods such as fruit would lapse after a relatively short time.

Acceptance and counter-offers

Once the existence of an offer has been proved, the courts must be satisfied that a valid acceptance of that offer has taken place. Where contracts arise as the result of long and complicated negotiations, it is often difficult to ascertain the exact moment when the contract was formed. In such cases the conduct of the parties and all their actions and statements will be considered to see whether or not the parties intended to contract and whether or not they were successful. Many of these rules are centred around the proposition that the acceptance must be clear and comply exactly with the terms of the offer.

THE ACCEPTANCE MUST BE IN RESPONSE TO AN OFFER AND MUST CORRESPOND PRECISELY WITH THE TERMS OF THE OFFER

In *Peter Lind Ltd* v *Mersey Docks and Harbour Board*[10] alternative tenders were submitted for the construction of a freight terminal. The offeree accepted 'your tender' without specifying which one. Consequently there was no contract between the two parties. In the case of *Pickfords Ltd* v *Celestica Ltd*,[11] the defendant sought to assert that a contract had been formed where the claimant's offer was accepted subject to a proviso that the claimant's costs for moving should not exceed £100,000. It was held that this limitation on the claimant's offer was not an acceptance and amounted instead to a counter-offer.

COUNTER-OFFERS

Rather than amounting to a valid acceptance, the insertion of new terms into a proposed agreement amounts to a counter-offer by the person who makes it. In *Hyde* v *Wrench*[12] an offer was made to sell a farm for £1,000. A counter-offer of £950 was made and refused, whereupon the buyer tried to accept the original offer of £1,000. It was decided that the original offer had been 'killed' by the counter-offer and was no longer capable of being accepted. In these circumstances, it is possible for the original offeror to accept the counter-offer and the terms of that offer become the terms of the contract. In this situation the parties have 'changed hats' – the offeror becomes the offeree and vice versa. The parties may change hats several times during the course of a negotiation before arriving at mutually acceptable terms of contract.

It might be the case that no final agreement is made as to which offer should be accepted. Both parties may have their own preferred terms of business and seek to impose their written terms on the other using the offer and counter-offer approach outlined above. This situation is known as 'the battle of the forms', a term which first came to light in the case of *Butler Machine Tool Co. Ltd* v *Ex-Cell-O Corporation (England) Ltd*.[13] It has been established that the winner of 'the battle of the forms' is the person who last submits the counter-offer which is accepted by the other party.

A counter-offer must be contrasted with an inquiry or a request for information from the offeree which will not vary the terms of the original offer. The offer can be said to survive the inquiry. In the case of *Stevenson* v *McClean*[14] the offeree enquired whether a delivery of iron could be staggered over a two-month period. The answer was in the negative but the offeree still wanted to go ahead with the contract. The offeror was obliged to accommodate this.

THE NATURE AND TIME OF ACCEPTANCE

Silence will not amount to a valid acceptance. Neither will an attempt to force the issue such as was made in the case of *Felthouse* v *Bindley*[15] by using words such as: 'If I hear no more from you on the subject I will assume acceptance'. A contract takes effect from the time that acceptance is communicated to the offeror. Where an offeror insists on a particular method of acceptance, that method must be adopted and the offeror can refuse to recognise an acceptance made in any other manner. The defendant was entitled to insist on written notice to a specified address in the case of *Manchester Diocesan Council for Education* v *Commercial and General Investments Ltd*.[16] In the case of a unilateral contract, the offeror may dispense with the need to communicate acceptance because of the terms of the offer.[17] In such circumstances, the necessary performance by the offeree will be sufficent acceptance.

It is possible for an offer to be accepted by conduct if that conduct unequivocally relates to the offer. In the case of *Brogden* v *Metropolitan Rly Co*,[18] the parties acted in accordance with a draft agreement for the supply of coal. The draft was held to be binding even though it was never formally executed. A contract can come into existence during performance even if it cannot be precisely analysed in terms of offer and acceptance.[19] A postal acceptance is effective when it is posted, even if it is late or is never actually delivered, so long as there is proof of posting. The rationale for this is that the postal service is the agent of both the parties and has authority to act on their behalf in the contract formulation. If the offeror specifies that acceptance must actually be communicated and that posting is not enough in itself, the rule does not apply. The post rule will not apply if it would give rise to 'manifest inconvenience or absurdity' as contemplated in the case of *Hollwell Securities* v *Hughes*[20] where a letter seeking to exercise an option to purchase was lost in the post. However, the situation is different where letters are delayed in the post, particularly, as in the case of *Adams* v *Lindsell*,[21] where the offer letter was wrongly addressed. In this case, the acceptance was binding notwithstanding the sale of the goods in question to a third party in the intervening period. The result was that the defendant was liable for breach of the contract formed upon the posting of the acceptance.

INSTANTANEOUS COMMUNICATION

An oral acceptance needs to be communicated to the offeror. The same rule applies to faxes and emails. These forms of communication are treated as being instantaneous and take effect, in terms of acceptance, when received. The issue of where the contract was made was important in the case of *Entores* v *Miles Far East Corporation*.[22] The offeree was based in the Netherlands and the offeror in London. The acceptance was sent back by facsimile from the Netherlands and was received in the London office. The issue of where the contract was made was important as this would dictate which country's laws would govern the operation of the contract. It was held that the contract was subject to English law – the key event being the receipt of the facsimile, not its sending.

The case of *Mondial Shipping and Chartering BV* v *Astarte Shipping Ltd*[23] considered a related topic concerning the status of transmissions made outside normal working hours. Here a telex message was sent at 23:41. The issue was whether this was 'sent' on the actual day or deemed to be sent on the next day. The court found that the effective communication had happened on the next working day and decided the issues in the case accordingly.

Emails and mobile telephone text messages are widely used as a means of communicating offers and counter-offers, acceptances and rejections. On the face of it, the approach in the *Entores* case seems to apply equally well in relation to email – the acceptance is binding from the time the email is received by the recipient's server and is available to be read. One difference is that the sender might not know of the failure of the message to arrive until sometime after the email has been sent. Further, the email may not be read by the recipient until the account is checked at some future time. However, similar shortcomings can be experienced with other means of communication and the basic rules of offer and acceptance can be said to apply to emails.

A last question to consider in this section is whether a course of dealings between two businesses can amount to a binding agreement that future work will be awarded on similar terms. This question is relevant to the construction industry where partnering and framework agreements are popular. The case which explored the question is *Baird Textiles Holdings Ltd* v *Marks and Spencer plc*.[24] The claimant had supplied garments to the defendants for a period of 30 years on a series of individual orders. In October 1999, the defendant terminated all supply arrangements without warning. The claimant sought damages based on the previous dealings and claimed entitlement to a notice period and further orders. The court found that the claimant was not so entitled. The relationship between the parties was only as good as the last order and the defendant had, by its actions, clearly not intended for there to be any commitment beyond this basic arrangement.

2.4.2   *Consideration*

Unless a contract is made by deed it must be supported by consideration; otherwise it will be unenforceable. The party who wishes to enforce a contract must show that consideration has been provided for the obligation which is sought to be enforced. The classic definition of consideration comes from the case of *Currie* v *Misa*:[25] 'some right, interest, profit or benefit accruing to one party, or some forbearance, detriment, loss or responsibility given, suffered or undertaken by the other'.

Consideration can be described as 'contractual glue' – something that binds the offer and acceptance together. In its absence, the agreement falls apart. A bare promise without consideration is not binding. A contract is essentially a bargain requiring reciprocal obligations between the parties. Sometimes consideration is known as value. Examples of consideration include the payment of money, the provision of goods, the performance of services or the transfer of land. A distinction is made between executed consideration, where the price is paid for the other party's act, and executory consideration, where the parties exchange promises to carry out obligations in the future.

Consideration must have some value but need not be adequate

This means that the consideration must have some worth, but the courts will not interfere with a bargain which has been made between the parties in the absence of fraud or other underhand dealings. In *Mountford* v *Scott*,[26] a token payment of £1 to secure an option to purchase a house for £10,000 was adequate consideration, while in *Chappell Co Ltd* v *Nestle Ltd Co*,[27] the wrappers from bars of chocolate were held to constitute good consideration when capable of being exchanged in part payment for goods. Token consideration of this nature is quite common on land transactions where the land is in effect gifted. It also appears in such agreements as collateral warranties. Forbearance to sue (agreeing not to pursue a claim in return for a promise by the other party) may be adequate consideration. In the *Carlill* v *Carbolic Smokeball Co* case, paying for and using the smokeball was deemed to be adequate consideration.

Performance of an existing contractual duty is not good consideration

It will not amount to consideration where a party performs an act which is simply a discharge of an existing obligation owed by contract. A similar rule applies where the existing duty is owed by the general law. Two examples of this rule in practice are the decisions in *Stilk* v *Myrick*[28] and *Hartley* v *Ponsonby*.[29] Both cases involve claims by seamen relating to work carried out under their contracts. In the former case, a claim relating to extra work carried out under a contract was dismissed because the crew were contractually bound to carry out any extra work by the terms of their contracts. In the latter case, any extra work involved in the voyage was an addition to the crew's contractual obligations and a successful claim could be made in respect of the additional work which had been carried out.

In *Williams* v *Roffey Bros Ltd*,[30] the defendants were contractors responsible for the refurbishment of a block of flats for the benefit of the employer. Mr Williams, a joiner, contracted to carry out joinery work on the flats for the sum of £20,000. The contract contained a provision that if the work was not completed by a specific date then damages would be payable by the defendants as liquidated damages to the employer. The claimant had contracted at a sum which was unprofitable to him and was in no hurry to complete the contract. Somewhat concerned that the contract might not be finalised on time, the defendant promised, by word of mouth, to pay the claimant an extra £10,300 in respect of the flats so long as the work was completed according to schedule. The joiner carried out the work and claimed the extra sums. The Court of Appeal decided that these amounts were payable notwithstanding the terms of the original agreement. Consideration had been given for the contractor's promise to make the additional payment. The contract had been completed; the damages payable to the employer in respect of late completion had been avoided; and it had not been necessary to employ others to finish off the joinery work.

Payment of a lesser sum under a contract than the amount which is due will not discharge the debt for a greater sum

The debtor is not discharged from the obligation to pay the balance where a debtor pays a lesser sum to his creditor than that which is due. The established law was that a creditor can go back on the agreement and sue for the balance, as happened in the case of *Foakes* v *Beer*.[31] The issue at stake was whether the interest on a judgement debt could be claimed by the claimant who had agreed to allow the defendant to pay in stages. It could – the defendant had given nothing of value for the claimant's promise not to enforce. There is an important exception to this waiver rule and it arises if, at the creditor's request, some new element is introduced into the arrangement, such as payment at a different time or place. Appropriate compliance with this request will amount to consideration for the waiver and the creditor will have no further claim. This is known as the rule in *Pinnel's Case*.[32]

However, in this situation, there must be a true accord. This was lacking in the case of *D & C Builders Ltd* v *Rees*[33] where impecunious builders felt they had no choice but to accept a lower sum from their client. It was decided by Lord Denning that there was no reason in law or equity why the claimant should not enforce the full amount of the debt.

Both these decisions appear to be at odds with a 2007 case heard in the Court of Appeal. The judgement in *Collier* v *Wright*[34] has effectively nullified the rule that part payment can never be sufficient by expanding the doctrine of promissory estoppel to come to the aid of the debtor in all such cases. In this case, the creditor voluntarily accepted part payment of a debt. The debtor relied on this acceptance and paid the amount agreed. The creditor was then estopped from going back on his promise to accept less notwithstanding the lack of consideration for such an agreement.

Promissory estoppel

Promissory estoppel describes the equitable remedy available in very limited circumstances where a party has waived rights under a contract and the other party has relied upon the promise not to enforce those rights. The promisor is estopped (prevented) from going back on the promise even though no consideration has been given for the waiver if certain conditions are met. The doctrine has its roots in the nineteenth century and in particular the case of *Hughes* v *Metropolitan Rlw Co.*[35] This case involved a request by a landlord to the tenant to carry out repairs on a rented property. The request was followed by a negotiation between the parties on a new lease, during which time the tenant delayed the repairs. The negotiations did not come to anything and the landlord cited the tenant's failure to carry out repairs as grounds for terminating the lease. The court did not agree – the landlord was estopped from pursuing the course of action because the tenant had relied on the landlord's implied waiver of the requirement to repair during the course of the negotiations.

The modern development of the doctrine dates from the High Trees case or, to give it its proper name, *Central London Property Trust Ltd* v *High Trees House Ltd.*[36] This case concerns the lease of a block of flats in London which turned out to be unprofitable because of the threat of enemy bombing in the Second World War. The landlord made a written promise to reduce the rent for the duration of the war. In 1945, when circumstances had changed, the landlord changed his mind and started to charge the full rent. It was decided that he could do so for the future, but the full rent could not be obtained for the period between making the promise and the end of the war since the tenants had relied on his previous promise. Before the doctrine can come into being, there must be an agreement between the parties, an unequivocal promise by one party to the other and, at least for the time being, an indication that those rights will be waived. The claimant must have given no consideration for the promise but acted upon that promise by relying upon it. The promisor is then estopped from altering his position under the contract if to do so would cause harm to the other party.

The modern instances where estoppel has been used are extremely limited. This is an area of law which has fallen mainly into disuse, and it is difficult to see it applying in a construction law setting. If a party is left running an estoppel argument as the main part of their case then this usually means they have little prospect of success in the dispute.

2.4.3 The intention to create legal relations

A contract is a legally binding agreement. The parties must intend that their agreement should be regulated and have the protection of the law for it become a contract. Where no such intention can be shown, there is no contract. Many agreements are never intended to be legally binding whilst the presumption will exist in others that the parties intended to be so bound.

Business agreements

In business and commercial agreements, it is presumed that the parties intend to create legal relations. This includes an agreement to build or to transfer land or buildings. This presumption can be challenged (rebutted) if there is evidence to the contrary or, as in the case of *Milner & Son v Percy Bilton Ltd*,[37] the words used were too vague to amount to a definite commitment to sell land. There must be strong evidence from the agreement that the parties do not intend to create legal relations. The parties are not bound where the agreement is expressed to be 'binding in honour only' or something similar. This area is akin to the gentlemen's agreement, the problem being that if one or other of the parties no longer wishes to behave in whatever fashion a gentleman should, they can consider themeselves no longer bound. A common occurrence in conveyancing transactions and correspondence is the use of the phrase 'subject to contract'. The phrase is often seen on letters sent to the parties in a conveyance by estate agents. The purpose of the letter is often to record the agreement made on price and to signal that both parties must now complete the other formalities of agreement required to bring about the transaction with the assistance of their legal advisors. The use of this phrase is sufficient to prevent any agreement being reached even if the words in the letter itself create the opposite impression.

Social and domestic agreements

In agreements of this type, it is presumed that the parties do not intend legal relations to arise. This principle may also be rebutted. An agreement which appears on the face of it to be simply a domestic agreement may be enforceable between the parties. This happened in the case of *Simpkins* v *Pays*,[38] which involved an agreement to share the winnings from a lottery syndicate. The agreement was binding notwithstanding its domestic setting.

2.5 Terms of a contract

The terms of a contract are its contents. Representations are precontractual statements and should not be confused with contractual terms. The two can be distinguished – a representation is a statement of fact which acts as an inducement to enter into a contract. A false representation may give a potential claimant a right to claim on the basis of misrepresentation. If the misrepresentation is actionable then the remedies are governed by the Misrepresentation Act 1967 and not by the rules relating to remedies for breach of contract. The intentions of the parties and what they understood the representation to mean are relevant in deciding whether a misrepresentation has taken place. If these inquiries are inconclusive, the test used by the judge will be to consider what an intelligent bystander would infer from the words and behaviour of the parties at the time the representation was made.

Contractual terms, on the other hand, may be 'express' or 'implied'. Express terms are stated either orally or in writing – the parties are said to have 'expressed themselves' on any particular subject and this is thereby included into their contract. Implied terms arise in other ways.

2.5.1 Express terms

The meaning of the terms is a question of fact where the contract has been made on a wholly oral basis. The court must decide the issue from the evidence which is given to it. The 'parol evidence' rule applies where contracts have been recorded in writing. The rule prevents oral or other extrinsic evidence to be admitted to vary or contradict the terms of the written agreement. There are some exceptions to this rule, among the most important being:

- to prove a trade custom or usage;
- to indicate the existence of a vitiating factor (see Section 2.6);
- where the written contract is not the whole contract.

2.5.2 Implied terms

A contract may contain terms which are not express but which are implied by custom, by the courts or by statute. A term will not be implied by custom if the express wording of the contract shows that the parties had a contrary intention. Where appropriate, a term will be implied by the custom of a particular geographical area as was the case in *Hutton* v *Warren*.[39] The claimant was a tenant of a farm and was given notice by the landlord to quit before the crops (which the claimant had planted) were harvested. Local custom dictated that the tenant was allowed to recover a fair allowance for seeds and labour involved in the planting of a crop where a tenant was prevented from harvesting the crop. A term was therefore implied into the lease to this effect. Other than by custom, terms can be implied in one of two ways: by the courts or by statute.

The courts

The courts will imply a term in the following sets of circumstances.

WHERE SUCH A TERM IS A NECESSARY INCIDENT OF THE TYPE OF CONTRACT IN QUESTION

The reasonableness of the term and the nature of the subject matter are important in this context. In contracts to build where a builder constructs a house, there is a term implied that the dwelling will be reasonably fit for habitation when completed. Similar terms are implied in respect of the quality of work and the time in which the contract will be completed. Likewise on the employer's part, it is implied that the contractor will be given possession of the site on which the building works are to take place and that a reasonable sum shall be payable for the work. These terms may be rebutted where there are express terms dealing with these matters to the contrary.

TO GIVE THE CONTRACT 'BUSINESS EFFICACY'

Business efficacy involves the notion of making a contract into a workable agreement in such a manner as the parties would clearly have done if they had applied their minds to the matter arising. From this, the court creates the common implied intention of the parties. It may be necessary to fill a gap in the contract and the term(s) must be formulated with a sufficient degree of precision. There must be no express term dealing with the same matter, and the stipulations to be implied must be so obvious that they go without saying. The principle comes from *The Moorcock*[40] where a vessel was damaged unloading a cargo owing to the unevenness of the riverbed, which became exposed at low tide and which had not been maintained properly. It was decided that there was an implied term in the contract in that the owners of the wharf had impliedly contracted to take reasonable care to see that ships unloading their cargos were safe and would not be damaged on the riverbed.

WHERE A TERM IS IMPLIED BY THE PREVIOUS COURSE OF DEALINGS BETWEEN THE PARTIES

Terms will be implied into a contract of a similar kind where the parties have previously entered into contracts of the same particular variety containing express terms. This will only occur in the absence of express stipulation to the contrary.

Statute

In some instances, statute will imply terms into a contract. Examples are the Sale of Goods Acts 1979 and the Supply of Goods and Services Act 1982 (as amended by the Sale and Supply of Goods to Consumer Regulations 2002).

THE SALE OF GOODS ACT 1979

TERMINOLOGY

A contract for the sale of goods is one whereby the seller transfers or agrees to transfer the property in goods for a money consideration called the price. 'Goods' includes all personal chattels (basically a person's belongings other than money) and choses in action (intangible personal property conferring a right to sue on the owner – for example, a policy of insurance or a bond).

TERMS IMPLIED BY THE ACT

The Sale of Goods Act 1979 implies terms into contracts for the sale of goods as follows.

TITLE (SECTION 12)   In every such contract, there is an implied condition on the part of the seller that he has the right to sell the goods or that he will have the right at the time when the property is to pass. If this condition is broken, the buyer may treat the contract as at an end even though he has done some act which would otherwise have amounted to acceptance. Section 12 also provides that there is an implied warranty that the goods are free and will remain free from any encumbrances not known to the buyer before the contract is made.

DESCRIPTION (SECTION 13)   The vast majority of sales are by description (i.e. the goods are not physically examined by the buyer before purchase), and in such a case, there is an implied condition that the goods will correspond with that description. This provision also applies to private sales. It extends to matters such as measurements, quantity and methods of packing as in the case of *Moore & Co v Landauer & Co*.[41] The claimants were entitled to reject imported goods merely because they were packed in boxes of 24 rather than 30 as agreed.

SATISFACTORY QUALITY (SECTION 14)   The following provisions only apply where the goods are sold in the course of a business. By section 14(2), there is an implied condition of satisfactory quality except as regards defects specifically drawn to the buyer's attention before the contract is made or where the buyer examines the goods before the contract as regards defects which the examination ought to reveal. 'Satisfactory quality' means that the goods meet the standard that a reasonable person would regard as satisfactory having regard, *inter alia*, to their description and price. The Sale and Supply of Goods to Consumers Regulations 2002 have improved the position of the consumer further by adjusting the burden of proof. Goods returned by the consumer within six months of the date of sale with request for repair are assumed to have been faulty at the date of purchase unless the retailer can establish otherwise.

Section 14(3) relates to fitness. The implied condition is that the goods are reasonably fit for any purpose which the buyer makes known to the seller. This purpose will normally be implied, but if goods have a number of purposes, the buyer must indicate the one required.

SAMPLE (SECTION 15)   Certain items such as tiles, carpets and wallpaper are bought on the basis of a sample inspected before the sale takes place. In such circumstances, the following conditions are implied:[42]

- that the bulk will correspond with the sample in quality;
- that the buyer will have a reasonable opportunity of comparing the bulk with the sample; and
- that the goods shall be free from any defects rendering them unsatisfactory which would not be apparent on reasonable examination of the sample.

THE SUPPLY OF GOODS AND SERVICES ACT 1982

A contract for the sale of goods must be distinguished from a contract for the supply of services where goods are also being sold under the contract. A contract for work done and materials supplied comes into this category. The sale aspect of the contract is governed by sections 2–5 of the 1982 Act which contains implied terms similar to the Sale of Goods Act 1979 (sections 12–15). The works aspects are governed by Part 2 of the Act, which imply terms relating to reasonable skill and care, reasonable time for completion, and reasonable charges for services where no price has been fixed at the outset. The supplier must be acting in the course of a business for the terms to apply.

THE LATE PAYMENT OF COMMERCIAL DEBTS (INTEREST) ACT 1998

This Act was introduced to improve the position of small businesses being kept out of money owing to them by larger businesses. The Act implies terms into contracts between qualifying businesses that simple interest calculated at the appropriate rate can be claimed in the event of late payment.

Other statutes introduced specifically for the construction industry are discussed in Section 6.6.

2.5.3 Conditions and warranties

Conditions and warranties are two different types of contractual term and the rules governing them apply equally whether the term is express or implied. This distinction is important because there are different remedies if there is a breach of contract and things go wrong. There is no special test to determine into which category a term falls. Each situation has to be decided on its own merits.

Conditions

These are the more important terms of contract. These are terms of the contract but for which the injured party would not have entered the contract. A breach of condition (known as a repudiatory breach) allows the injured party to treat the contract as finished, whereupon he will be discharged from his obligations under the contract. In addition, the other party can be sued for damages. The contract is voidable, which means the innocent party has a choice between treating the contract as having been irreparably damaged by the breach of condition (this is known as rescission) or continuing with the contract notwithstanding the breach. The choice is based around the fact that the innocent party should be entitled to keep the benefit of the contract if he so wishes. In either case, it is open to the innocent party to pursue a claim for damages.

Warranties

These are terms which are secondary or ancillary to the main purpose of the contract. A breach of warranty allows the injured party to claim damages but not to repudiate the contract. The contract is not voidable. The contract survives the breach of a minor term whilst preserving the right to damages for the innocent party. It is for a court and not for the parties to determine the category of term.

The difference between conditions and warranties and the practical consequences of the difference are evident from a comparison of the cases of *Poussard* v *Spiers and Pond*[43] and *Bettini* v *Gye*.[44] It was a coincidence that two cases involving singers should be reported in the press in the same year. Both cases involved singers who were late for their performances at theatres and were told that the agreements to hire them had been fundamentally breached by reason of their lateness. In Poussard's case, the court agreed. The singer had missed the all-important opening night, and her absence amounted to a breach of condition. Bettini, however, had only missed rehearsals, and the producers were not entitled to terminate the contract but they were able to sue in damages for the breach of warranty.

A problem which has arisen is that a stipulation in a contract may be classified as a term but is neither a condition nor a warranty. The difficulty that arises is then to determine whether or not the injured party can rescind (avoid) the contract. Such provisions have become known as intermediate or innominate terms. The consequences of the breach are important here. Unless the contract makes it clear that the parties intended that no breach of the contract should entitle them to terminate the contract, the term will be classified as an intermediate term. The question is whether or not the nature and effect of the breach are to deprive the injured party of substantially the benefit which it is intended they should obtain under the contract. If so, the injured party is entitled to terminate the contract as well as claiming damages. This was the question considered in the case of *The Hansa Nord*.[45] The question here was whether an innominate term of the contract, that 'shipment be made in good condition', should be treated as a condition or a warranty. The claimant sought to reject the contract based on a small percentage of the goods being substandard. The defendant suspected that the real reason for the rejection was a reduction in the market value which rendered the cargo much less profitable. The court did not view the partial unsoundness of the cargo to be grounds for total rejection and treated the breach as a warranty entitling the claimant to damages only.

2.6   Vitiating factors

These are factors which affect the validity of a contract. A contract may be challenged notwithstanding the appearance that it has been properly formed with the essential three ingredients of offer/acceptance, consideration and intention to create legal relations being present. A state of affairs may have arisen which may make the contract no longer enforceable. Depending upon the circumstances, a vitiating factor will make the contract:

- void (no contract);
- voidable (the contract exists but can be avoided at the election of the innocent party);
- binding (an enforceable contract exists).

2.6.1 Capacity of the parties

The first challenge to an agreement is based around the notion of incapacity. The central premise is that, regardless of any assertion to the contrary, a person may lack capacity to enter into a contract. A distinction is made in English law between natural persons and artificial persons. Natural persons have full capacity to enter contracts. There are exceptions to this rule in the case of minors (those under the age of 18) and, to a certain extent, in respect of those suffering from a mental disorder or those under the influence of alcohol. The latter lack capacity to enter into contracts whilst so intoxicated as to not know what they are doing and without their intoxication being obvious to the other party. In these circumstances, a contract is voidable but may be ratified by the individual on becoming sober.

As a general rule, contracts are not binding on minors. A practical exception to this rule is to be found in respect of contracts of employment which are benefical to the minor. An example of this would be a contract with a newsagent to deliver newspapers to the local neighbourhood. Another exception involves contracts for necessaries. The notion here is that minors are obliged to pay for goods or services they need. The rather bizarre corollary is that minors do not need to pay for goods or services they do not need. This was demonstrated in the case of *Nash v Inman*.[46] A Savile Row tailor attempted to recover £122 16s 6d for items, including 11 fancy waistcoats, supplied to an undergraduate. The action failed as the minor could demontrate that he was already adequately supplied with clothes and those supplied by the tailor did not amount to necessaries.

Artificial persons are corporations, local authorities and companies registered under the Companies Acts. The ability of artificial persons to enter into contracts may, in certain circumstances, be limited by their incorporation and internal adminsitration documentation.

2.6.2 Mistake

A mistake arises where the parties make an error as to a term of the contract such as to prevent there being *consensus ad idem* (or meeting of the minds). Mistake does not affect the validity of a contract unless the mistake is a fundamental one. This is known as an operative mistake. Such a mistake will render the contract void. In a limited set of circumstances, the contract will be voidable (this may be the position where there is a mistake as to the identity of the person contracted with). Special rules apply to documents. The contract is binding where a person signs a document and by so doing enters a contract. This is so even if the signatory has not read the document nor understands it. The principle of *caveat emptor* (let the buyer beware) applies here. There is a very limited defence based on *Non Est Factum* (it is not my deed) where a person is induced by a fraud or a trick to sign a document. In practice, the defence is difficult to rely on because the signatory must show that:

- all reasonable precautions were taken when signing the document; and
- the document signed was radically different from the one the person thought they were signing; or
- the person signing the document was under some legal disability, such as being illiterate, blind or senile.

The rule does not apply if there is simply a mistake by one of the parties in expressing intention. In *W Higgins Ltd* v *Northampton Corporation*,[47] the claimant stated an incorrect price for work in connection with a tender relating to the building of houses. The error was due to incorrect estimating on the claimant's part. The tender was accepted by the local authority without realising a mistake had been made, but the contract could not be rectified. In this instance, the principle was *caveat venditor* (let the seller beware). The claimant found itself in the position that it could not now avoid a contract where it had legitimately (but erroneously) offered to carry out the work for the price stated.

2.6.3 Misrepresentation

Misrepresentation is concerned with the effect of statements which are made before a contract is formed. A representation is a statement of fact and not a statement of opinion which is made by one person to another with the object of persuading the other party to enter into a contract. Where that party relies upon the statement, a contract is formed, and the statement turns out to be incorrect, a remedy may be provided by the law of misrepresentation. An example of an opinion rather than a fact comes from the case of *Bisset* v *Wilkinson*[48] where the vendor of a farm in New Zealand expressed an opinion that the farm would support 2,000 sheep. In fact, the farm had never held any sheep. The claimant's sheep farming business failed but they were unable to sue for misrepresentation based on what was an honestly given opinion.

Remedies are governed by the Misrepresentation Act 1967 and the tort of deceit. A misrepresentation can be fraudulent, innocent or negligent. A claimant in such circumstances will be able to claim damages and/or rescind the contract. In the case of a non-fraudulent misrepresentation, the court has power to refuse rescission and to award damages instead. A delay in applying for rescission may also be fatal to such a claim. This occurred in the case of *Zanzibar* v *British Aerospace (Lancaster House) Ltd*[49] where problems experienced with a private jet aeroplane were not reported until several years after the alleged problems first arose.

2.6.4 Duress, undue influence and inequality of bargaining power

There are grounds upon which to challenge an agreement where the parties have not given their free consent to be bound by the terms of the agreement. Coercion or actual threats or undue pressure can vitiate a contract. Mere inequality of bargaining power is, as a general rule, insufficient in seeking to set aside agreements. Statutory interventions such as the Unfair Contracts Terms Act 1977 and the Unfair Terms in Consumer Contracts Regulations 1999 go some way towards protecting consumers in commercially vulnerable positions. The Consumer Credit Act 2006 allows the court to adjust a credit agreement if the court decides that the relationship between lender and borrower is unfair.

2.6.5 Illegality

A contract containing an illegal element is void. The illegal aspect may be the contravention of a statute, or it may be against public policy to allow certain types of contract to be enforced. Examples include defrauding the Inland Revenue or contracts involving the commission of a crime. The contract price will not be recoverable where a building contractor enters a contract knowingly in contravention of a law. In *Stevens* v *Gourley*,[50] statutory provisions required buildings to be made of incombustible materials. The contract was found to be illegal when the builder erected a wooden building upon wooden foundations.

2.7   Privity of contract

One fundamental aspect of the common law is the principle that no person can sue or be sued on a contract unless a party to it. This can cause particular problems in the construction industry where many parties are involved in a building project. Some of the participants will have privity of contract between themselves while others will not. The rule means that a stranger (or third party) to a contract cannot sue under a contract to which they are not a party even if the terms of the contract express that the third party should so benefit. The general aspects of the principle are illustrated by *Tweddle* v *Atkinson*[51] where a couple were about to marry. The bridegroom's father and the bride's father agreed between themselves that each would make payments to the couple. The husband sought to enforce the contract when the bride's father refused to pay. It was held that there was no privity of contract between the parties and that, therefore, the claim could not be enforced.

It has become standard practice in large-scale construction projects for those who finance the transactions to insist that all the parties involved enter into contractual warranties with each other. The reason for this is because of the difficulties that claimants have in being able to successfully sue a defendant in the tort of negligence for financial losses. The collateral warranty is a way of ensuring that in the event of default, an injured party will have a potential defendant to sue and claim damages.

This tendency to insist on the use of collateral warranties has continued notwithstanding the change in the law brought about by the Contracts (Rights of Third Parties) Act 1999.[52] This non-mandatory Act allows a third party to sue on a contract to which he is not a party but which purports to transfer a benefit on him. The effect of these provisions, when used, is to circumvent the privity of contract rule. Standard forms of building contract were initially slow to embrace this statutory change. However, the most recent edition of the Joint Contracts Tribunal standard form of building contract gives parties the option of having a third party schedule of rights either alongside or as an alternative to collateral warranties.

2.8   Exemption clauses

2.8.1   Common law

A common practice in contract formation is for a party to insert a clause which, in the event of their breach, either excludes liability completely under the contract or limits that liability to a specific amount of money. The former type of clause is an exemption clause whilst the latter is known as a limitation clause. Such clauses are used by suppliers of goods and services in respect of potential problems such as defective materials and unsatisfactory work. The courts have always disliked exemption/limitiation clauses and tend to decide against them if there is any ambiguity either in what the clause may mean or whether it has been incorporated properly in any given situation.

At common law, an exemption clause is valid so long as certain basic requirements are satisfied. In *L'Estrange* v *Graucob*,[53] such a clause was upheld notwithstanding that it was written in 'legible but regrettably small print' in a sales agreement. The claimant failed to read the document before signing it but had the opportunity of so doing. A person seeking to rely on such a clause must show that it is a term of the contract and that, as a matter of construction, it covers the damage in question. The case of *Chappleton* v *Barry UDC*[54] involved a claimant injured by a hired deckchair which collapsed on him. At the time of hiring the deckchair, no notice was given of the exclusion clause on which the council sought to rely. The clause was contained in a receipt given to the claimant at which he only glanced. The judge held it was unreasonable to communicate conditions by means of a mere receipt. The exclusion will not therefore be effective if the clause or the notice is contained in a document which the average person would not assume to contain contractual terms.

The clause will only be a contractual term if adequate notice is given of it or such a term is implied by a clause having been incorporated into documents where there has been a previous course of dealings between the parties. A common practice is to print exemption clauses on order forms or on a notice. Notice of the clause must be given before or at the time of the contract. Belated notice of an exclusion clause was given in the case of *Olley* v *Marlborough Court*[55] where a contract for the hire of a hotel room was made over the telephone. Mrs Olley's fur coat was stolen from the hotel room during the guests' absence. The hotel sought to rely on a notice avoiding liability from theft displayed in the hotel's lobby. The judge deemed that this notice was not adequate as the contract for the hire of the room had been made over the telephone, and this unilateral attempt to introduce new terms once the contract had been completed would not be successful. Where exclusion clauses are properly incorporated, the position is that liability can only be limited or excluded by clear words. The purpose of the exemption clause is another factor for the court to consider in deciding whether or not it is reasonable, as demonstrated by the case of *O'Brien* v *Mirror Group Newspaper Limited*.[56] This case involved scratch cards distributed with a British newspaper. Due to an error on the cards, 1,472 people were each told they had won the sum of £50,000. The claimants attempted to hold the defendant to their promise to pay. The latter relied on its terms and conditions in seeking to avoid liability. The judge allowed the defendant to rely on its exclusion clause notwithstanding the fact that the terms and conditions were not printed in the edition of the paper the claimants purchased. The effect of the exclusion clause here was not to exclude liability altogether but merely to prevent a windfall for the claimants.

2.8.2 Statute

The principal statutory provisions regulating unfair reliance on exemption clauses are the Unfair Contract Terms Act 1977 (UCTA) and the Unfair Terms in Consumer Contracts Regulations 1999. The Act covers liability in respect of a breach of contract and in respect of the tort of negligence. The Act only has application in respect of 'business liability' where the other party deals as a consumer. 'Business liability' relates to a transaction which emanates from business activities while the other party makes the contract in a domestic capacity. Any provision in a contract excluding or restricting liability for death or personal injury is void. In other cases based on negligence, liability can only be excluded or limited if it is fair and reasonable to do so. The reasonableness provision also applies in contractual cases. The fair and reasonable provision will also apply where a party to a contract deals as a consumer or contracts on the other party's standard business terms. The Act does not apply to contracts of insurance and to contracts which contain a foreign element. This provision was considered by the House of Lords in *Smith* v *Eric S Bush*,[57] where a firm of general practice surveyors inserted an exemption clause into a valuation report for mortgage purposes which had been prepared for a building society. The report was shown to the purchasers. The Law Lords concluded that it would not, in the circumstances, be fair and reasonable to rely on the disclaimer. The factors to be considered when deciding whether a purported exclusion is fair and reasonable include:

- the practical consequences of exclusion – in particular, the cost and availability of insurance cover;
- equality of bargaining power between the parties;
- whether it is practicable for the relevant party (or parties) to seek advice from an alternative source;
- that an exclusion could well be reasonable if the task is unusual or dangerous.

In the case of *Rees-Hough Limited* v *Redland Reinforced Plastics Limited*,[58] the judge decided it was not fair and reasonable for the defendant to rely on an exclusion clause in their terms and conditions of sale. They sold pipes to the claimant that were not fit for the purpose for which the defendant knew they were required. Following this case, it is difficult to see how a party can escape liability for supplying defective goods. Similarly in the case of *Feldaroll Foundry plc* v *Hermes Leasing (London) Ltd*,[59] liability for a Lamborghini Diablo motor car bought for £64,995 but sold with a serious and dangerous steering defect could not avoided by the defendants.

One feature of construction law which is discussed in Part 3 (Section 13.4) relates to the meaning and exclusion of consequential loss. The general position appears to be that provided the contract is specific enough in the categories of loss which are being excluded then there is a realistic chance that some exclusions are possible.

2.9 Discharge of a contract

Where a contract is at an end, it is said to be discharged. The rule is that on discharge the parties are freed from their obligations. Contracts do not end automatically and discharge must come about by an act of the parties. It can come about in one of four ways:

- performance;
- agreement;
- frustration;
- breach of an appropriate term of the contract.

2.9.1 Performance

Each party must perform precisely and completely what he has bargained to do. A person who claims to be discharged from obligations on the basis of performance must show that the work has been completed in full, the goods supplied or the land transferred as appropriate. A building or engineering contract will be discharged by performance when the contractor has completed all the work and the employer has paid all sums due. The contract has not been performed if there are hidden defects. This common law rule is enshrined in *Cutter* v *Powell*[60] where the defendant agreed to pay Cutter a sum of money for performing duties as a seaman on a ship during a ten-week voyage. Cutter died seven weeks after the commencement of the voyage. His widow attempted to recover wages under the contract. The claim failed because the ten-week contract had not been performed. Similarly in *Sumpter* v *Hedges*,[61] the claimant builder contracted to construct two buildings on the defendant's land for an agreed price. The buildings were half-finished when the project was abandoned because of lack of funds. The claimant was unable to recover any money in respect of the work carried out. In order to avoid possible injustice and to prevent unfair manipulation of the common law rule, the courts have recognised certain exceptions where a person may claim a reasonable sum for the work carried out on the basis of a *quantum meruit* application. *Quantum meruit* means 'as much as it is worth'.

The exceptions are as follows.

Severable contracts

Payment can be claimed for work done on completion of each stage where a contract is severable, sometimes described as divisible. On the other hand, if the contract is an entire one then the work must be completed in full before the other party becomes liable to pay. The terms of the contract are important in this context and much will depend on the intention of the parties. The courts tend to lean against finding a contract entire. A severable contract will arise where payments are due from time to time (such as interim or stage payments in construction contracts) and the other party's obligations are rendered in return.

Substantial performance

This arises where performance under a contract, although not complete, is virtually exact. In such a case, a claim may be made for the work completed subject to any counterclaim in respect of work remaining unperformed. The contract price can be claimed subject to a deduction equal to the cost of remedying any omissions or defects. In essence, the defects must be of a trifling nature before the exception can be claimed. In *Hoening* v *Isaacs*,[62] the doctrine was applied where the cost of remedying the defects amounted to £56 on a contract worth £750. In *Dakin* v *Lee Ltd*,[63] the court stated that if a builder has done work under an entire contract but the work has not been completed in accordance with the contract, a claim can be made on a *quantum meruit* basis unless the customer received no benefit from the work, the work was entirely different from that which was contracted for, or the contractor had abandoned the work. The exception will not apply if the contract has not been substantially completed. The actual cost of remedying the defects in relation to the contract price is a relevant factor. In *Bolton* v *Mahedeva*,[64] a contract was entered into to install a central heating system for £560. The system performed very poorly and the remedial work would have involved a further cost of £174. The judge held that there had not been substantial performance and that the claimant was entitled to nothing for the work done.

Prevention of performance

A claim for damages can be made on a *quantum meruit* basis for work carried out if a party to a contract is prevented from performing what he has agreed to do by the other party. In *Roberts* v *Bury Improvement Commissioners*,[65] a contractor was not in default where an architect neglected to supply him with necessary plans, which prevented the contractor from completing the work by the contract date. The contractor was entitled to be paid for the work performed.

2.9.2 Agreement

It is logical that where contracts are created by agreement, they may also be discharged in the same manner. Another contract must therefore be entered into if the parties agree to waive their obligations under an existing contract. This will be appropriate where neither of the parties has completed their obligations due under the contract. Where one of the parties has completed obligations due under the contract but the other party has not, the party to whom the obligation is owed may agree with the other party to accept something different in place of the former obligation. This is known as the doctrine of accord and satisfaction. The subsequent agreement represents the accord, and the new consideration which has been given is termed the satisfaction. Alternatively, an obligation under a contract may be discharged by the operation of one of the terms of the contract itself. Contracts of employment often contain provisions whereby the contract can be determined by either party giving a period of notice to the other. In building and engineering contracts, it is common to find provisions whereby, on the happening of specific events, the contract can be determined before completion. The serious default or insolvency of the other party is a common example.

2.9.3 Frustration

'Frustration' in the context of contract law means impossibility. As a rule, contractual obligations are absolute in that a contract is not discharged simply because it is more difficult or expensive than expected to carry it out. Modern-day examples of building contracts that have been frustrated are rare, and to succeed, a claimant must show that the circumstances have changed to such an extent that the performance of the contractual obligation has become fundamentally different.

2.9.4 Breach of an appropriate term of the contract

Any failure by a party who contracts to fulfil obligations under the contract amounts to a breach. Whenever a party to a contract is in breach, this gives rise to an obligation on his part to pay damages to the person who has suffered loss, subject to a valid exemption clause to the contrary. The obligations of the parties to perform the contract remain unchanged unless the breach can be classified as a repudiatory breach. In each case, the issue as to whether the breach is to be taken as a repudiation depends upon the importance of the breach in relation to the contract as a whole. Repudiation will assert itself in a building contract where execution of the works is so unsatisfactory as to affect the very basis of the contract. The injured party has the option either to terminate the contract or to affirm it in the case of a repudiatory breach. The appropriate test to ascertain whether or not a party has repudiated the contract is whether or not a reasonable person would believe that the other party did not intend to be bound by the contract. If the test is satisfied, the injured party may treat the contract as at an end and is released from further performance. The claimant may rescind the contract but must act without delay to do so. Unjustified rescission of a contract does not always amount to a repudiation.

The injured party need not wait for the date set for performance where he knows that a breach will take place. A right of action exists immediately as if there were a breach, and the claimant may sue for breach of contract. In the case of *Lovelock* v *Franklyn*,[66] the defendant committed such a breach when he sold an item he had contracted to sell to the claimant to a third party instead.

The claimant knew of the sale to the third party and was able to seek damages before the envisaged sale between the parties was due to take place.

2.10 Damages

2.10.1 Unliquidated damages

The claim is for unliquidated damages where the parties to a contract make no pre-assessment of any damages payable in the event of a breach. The object of such a claim is to put the injured party in the same position as if the contract had been performed. Any loss suffered as a consequence of the breach, whether physical or financial in nature, can be claimed by a successful claimant. It is against public policy to allow compensation for every consequence which might logically result from the defendant's breach as the potential liability of the defendant in this situation could be unlimited.

The rule in *Hadley* v *Baxendale*[67] states that the only losses that are recoverable are those which may fairly and reasonably be considered as arising naturally from the breach of the contract or losses which may reasonably be supposed to have been in the contemplation of both parties at the time they made the contract in the event of a breach. The facts in the case involve the repair of a shaft used in a mill. An example of the rules as to the recoverability of damages under what became known as the *Hadley* v *Baxendale* tests came from the decision in *Victoria Laundry Ltd* v *Newman Industries Ltd*.[68] The claimants ordered a new boiler for their business and this arrived late. The claimants were entitled to recover damages for normal loss of profits by applying the first part of the *Hadley* v *Baxendale* test. The boiler supplier should have anticipated the lost profits. The contentious item was whether the claimant was entitled to recover further losses due to the demise of a profitable government contract. The defendants were unaware of the existence of the government contract, and it could not therefore be said to be within the parties' reasonable contemplation as envisaged in the second part of the *Hadley* v *Baxendale* test.

This narrow approach to the remoteness of contractual damages was upheld by the House of Lords in the case of *Transfield Shipping Inc* v *Mercator Shipping Inc* (*The Achilleas*).[69] The claimant chartered a ship from the defendants for a seven-month period. The claimant was late in returning the ship, causing losses to the defendant who had arranged for a third party to re-charter the ship on its return. The claimant was content to pay damages based on the normal market rate of charter rates but the defendant insisted on additional damages caused by the volatility in the market being experienced at that time. Their Lordships decided that only the damages based on a normal market, rather than an exceptionally high market rate, could be claimed from the defaulting party.

The measure of damages awarded is usually the actual monetary loss. The cost of reinstatement will constitute general damages in cases of defective building work. If this formula is incapable of representing the actual loss to the claimant then the damages will amount to the decrease in the value of the property.

Damages for mental distress

The traditional rule was that damages for breach of contract should not take account of any mental distress and inconvenience suffered by the claimant. This rule is illustrated in the case of *Addis v Gramophone Co Ltd*[70] where a company wrongfully dismissed its manager in a way that was 'harsh and humiliating'. The claimant received damages for loss of salary and commission but nothing for his injured feelings. Deviation was made from that position in *Jarvis v Swan Tours Ltd*[71] where the Court of Appeal decided that, in an appropriate case, such damages could be an additional element in the sum awarded. This case involved a holiday to which the claimant was looking forward to immensely. In the event, he had a thoroughly miserable time due mainly to events advertised in the holiday brochure either not occurring at all or being substandard. The claimant was awarded damages for loss of enjoyment on the holiday. The damages awarded amounted to twice the original price of the holiday itself.

From this last case, the principle emerged that damages may be awarded where the object of the contract is to give pleasure, relaxation, peace of mind or freedom from molestation. The case of *Farley v Skinner*[72] extended this principle further to apply to contracts which were not solely for pleasure, relaxation and peace of mind. The claimant had asked his surveyor to consider in his report the issue of aircraft noise above a property he was conisdering buying near Gatwick airport in London. The surveyor reported that the property was unlikely to suffer greatly from such noise. Unbeknown to the surveyor, the property was close to a navigation beacon over which the aircraft 'stacked up' in busy periods. The claimant was entitled to recover damages from the surveyor for loss of enjoyment due to regular noise pollution suffered.

Damages for loss of enjoyment are not usually permitted on commercial contracts. However, damages were awarded in the case of *Watts v Morrow*[73] in respect of the physical discomfort and mental distress resulting to the claimant from the repairs necessitated to remedy defects a surveyor had failed to notice during his work.

2.10.2 Liquidated damages

In some cases, the parties to a contract make an attempt in the contract to assess in advance the damages which will be payable in the event of a breach. This provision for liquidated damages will be valid if it is a genuine attempt to pre-estimate the likely loss suffered as a consequence of the breach. There can be no inquiry into the actual loss suffered. If the clause is a valid liquidated damages clause, the sum stipulated is recoverable upon that breach even though the loss is less or even nil. A liquidated damage clause must be distinguished from a penalty clause. The latter is a stipulation which is inserted to frighten the potential defaulter and to compel performance of the contract. It is not a genuine pre-estimate of loss and is therefore unenforceable. A penalty clause is invalid but the injured party may, if the liquidated damage clause is not an exclusive remedy, recover any actual loss on the Hadley and Baxendale principles set out above in relation to unliquidated damages.

It is common practice for construction and engineering contracts to contain a liquidated damages clause for delay alongside an extension of time clause giving the contractor the right to extra time if he is delayed by matters outside the control of either party or due to the default of the employer and/or his agents. Relevant matters would include industrial action, bad weather, or the employer delaying the execution of the works by failing to supply necessary plans or to give possession of the site. Any claim to liquidated damages will be lost if any delay in the performance of the contract is caused by the employer. The question as to whether a clause is a penalty or liquidated damages depends upon its construction and the surrounding circumstances at the time of making the contract.

The leading case in this area is *Dunlop Pneumatic Tyre Co Ltd v New Garage Ltd.*[74] This case involved a garage and its contract to sell Dunlop tyres. The contract between the parties provided that the defendant would have to pay £5 for every tyre it sold under the list price provided by the claimant. In this instance, the courts were satisfied that this was not a penalty and was an honest attempt to quantify the effects of a breach of the contract, and it was therefore upheld. A number of propositions were put foward to determine whether a provision is a liquidated damages clause or a penalty.

- The name which the parties give to the clause is not conclusive and it is the task of the court to decide the category.
- The essence of liquidated damages is that the sum stated is a genuine pre-estimate of the probable loss while the essence of a penalty is that it is a threat to carry out the contract.
- It is presumed to be a penalty clause if the sum stated is extravagant compared with the greatest possible loss.
- It is a penalty if the breach consists only in not paying a sum of money by a certain date and the sum fixed is greater than the sum which was originally to be paid. For example, X agrees to pay Y £5,000 on January 1, and if he fails to pay at the contract time, X must pay £10,000 as 'liquidated damages'.
- There is a presumption that where a single sum is made payable on the happening of one or more events, some of which are serious and some of which are of little consequence, the sum is likely to be a penalty. There is a likelihood of greater losses emanating from a major breach than from a minor one. In *Law v Redditch Local Board*,[75] the sum stated in the relevant clause was to be paid on the happening of a single event only. Therefore, it was deemed to be a valid pre-estimate of likely loss and was considered to be a liquidated damages clause.

An interesting take on liquidated damages is the point that occasionally the level will be set lower than the anticipated loss where the operation is intended to be a limitation clause. A number of industries, such as haulage and lift installers, have negotiated standard terms whereby their liability is limited to a low percentage of the contract sum. In the case of *M J Gleeson (Contractors) v London Borough of Hillingdon*[76] the low figure was challenged for failure to reflect a genuine pre-estimate of loss. The judge rejected this argument on the basis that, pursuant to the overriding freedom to contract, the parties were free to agree a limitation of liablity in this manner.

Liquidated damages are discussed further in Section 10.7.2 as one of the distinctive features of construction law.

Mitigation of loss

The injured party has a duty to mitigate any losses suffered. This means that reasonable steps must be taken to minimise damage. An employee who is wrongly dismissed must attempt to find alternative employment, while a buyer of goods which are not delivered must try to buy as cheaply elsewhere. Failure to do so will be taken into account when assessing damages. Only reasonable steps to mitigate need be taken. The claimant is not expected to incur great expenditure or undertake great risks. The law around mitigation equates to the common-sense actions that would objectively be taken in any situation to limit the effects of the breach of contract.

2.11 Limitation of actions

Limitation periods are the time periods within which a claimant must commence an action for damages. It is recognised that a claimant must pursue a claim with due diligence while potential defendants should not have the threat of litigation pending indefinitely. Specific time periods are laid down by statute within which actions for damages must be brought. The reliability of evidence diminshes over time as memories fade. Claims that have been in circulation for a considerable period of time make bad law and the eventual outcome of a case unreliable.

Limitation in contract

The law is statute based and governed by the Limitation Act 1980. Section 5 of the Act states that if a claim for breach of contract is based on a simple contract, the action must be brought before the expiration of six years from the date on which the cause of action accrued. Section 8 deals with the case of a deed, stipulating that the action must be brought within 12 years from the date when the cause of action accrued. The claim is statute barred and cannot be pursued if it is not commenced within these time limits. The date when the cause of action accrued is the date when the breach of contract occurred.

If the breach is a continuing one, a claimant can choose the last date that the defendant was in breach before the time period begins to run.

2.12 Equitable remedies

These may be available to an injured party where the common law remedy of damages is inadequate or inappropriate. They comprise the remedies which were only available in the courts of equity before the Judicature Acts 1873–75. The courts of equity and common law were fused by these Acts, and the resultant position is that any civil court is able to grant these remedies, if thought appropriate. They are discretionary and not automatically available in the event of winning a case. Certain principles apply generally to the grant of equitable remedies while others apply, in addition, to individual remedies. In all cases, the conduct of the parties is taken into account. No equitable remedy will be available if damages are considered to be an adequate remedy in the circumstances of the case. Limitation periods are not relevant to equitable remedies. Time is of the essence in such circumstances and each equitable remedy must be sought with reasonable promptness.

2.12.1 Specific performance

Specific performance is a court order directing the defendant to carry out a promise according to the terms of the contract. It is appropriate where the subject matter of the contract is not freely available, such as a rare painting or antique. In such cases, the payment of a sum of money would not be adequate compensation in the event of a breach of contract. It is mainly used as a threatened remedy where contracts for the sale or lettings of land are broken. The normal conditions required before equitable remedies will be granted apply. Specific performance will not be awarded to enforce a contract for personal services or where the contract requires supervision to ensure that it is being enforced properly. This was the case in *Wolverhampton Corpn v Emmons*[77] where the defendant owned the land on which the claimant sought an order compelling specific performance of a building contract. The court refused to grant the order in these circumstances. A claim for damages was more appropriate in the circumstances. Specific performance of a building contract may be awarded where:

- the building work to be carried out is clearly specified in the contract;
- the claimant has a special interest in having the work done which cannot be satisfied by an award of damages;
- the builder is in possession of the land, so it is not practicable to employ another builder.

2.12.2 Injunction

This equitable remedy restrains a person from continuing a wrong, or alternatively it can be used to order a party to do something to remedy a breach of contract. Unlike specific performance, it can apply to contracts containing a personal services element. The usual rules as to equitable remedies apply and an injunction will not be granted if damages are an adequate remedy. It has been held to be appropriate to restrain a contractor from continuing work after a breach of contract has taken place. Similarly in the case of *Bath and NE Somerset District Council* v *Mowlem plc*,[78] it was decided that contractual provision for liquidated and ascertained damages would not prevent the council from obtaining an injunction to restrain the contractor's refusal of access to the site to replacement contractors.

Failure to obey an injunction is a contempt of court and may result in the imposition of a fine or imprisonment.

2.13 Conclusion

This chapter has looked at the law of contract in a level of detail not contained in the rest of this book. The rationale for this approach is that construction law is overwhelmingly based on contract law, and the use of this chapter as a reference tool is encouraged.

Contracts are used in everyday situations in order to give the parties some certainty of the outcome of transactions they enter into. Everybody wants to know that if they hand over money or some other commodity, such as their time, then in return they will receive what the other person has undertaken to provide, such as goods or a salary.

Contracts are governed by law so that any breach of contract can be taken through the courts and a remedy, usually in the form of damages, can be sought. This chapter has looked at the special rules that govern:

1 the creation of contracts involving offer and acceptance;
2 who can make and who can sue under a contract;
3 the categories of terms contained in a contract and their relative importance; and
4 the damages payable and time limits in which you can sue for breach of contract.

2.14 Further reading

Wood, D., Chynoweth, P., Adshead, J. and Mason, J. (2011) *Law and the Built Environment*, Second Edition, London: Wiley-Blackwell, Chapter 2.

Wild, C. and Weinstein, S. (2013) *Smith and Keenan's English Law*, Seventeenth Edition, Harlow: Pearson, Part 2.

McKendrick, E. (2014) *Contract Law: Text, Cases and Materials*, Oxford: Oxford University Press.

Taylor, R. (2013) *Contract Law*, Oxford: Oxford University Press.

Notes

1 Section 2 of the Law of Property (Miscellaneous Provisions) Act 1989.
2 The Law of Property (Miscellaneous Provisions) Act 1989 section 2 repealed the Law of Property Act 1925 section 40, which required such contracts merely to be evidenced in writing.
3 [1993] 1 Lloyd's Rep 25.
4 [1961] 1QB 394.
5 [1952] 2 All ER 456.
6 [1979] 1 All ER 972.
7 [1952] 2 All ER 456.
8 [1893] 1 QB 256.
9 (1880) C P D 344.
10 [1972] 2 Lloyd's Rep 234.
11 [2003] EWCA Civ 1741.
12 (1840) 3 Beav 334.
13 [1979] 1 All ER 965.
14 (1880) 5 QB 346.
15 (1863) 1 New Rep 401.
16 [1969] 3 All ER 1593.
17 See *Carlill* v *Carbolic Smokeball Co.*
18 [1877] 2 App Cas 666.
19 See *Trentham (G Percy) Ltd* v *Archital Luxfer Ltd.*
20 [1973] 2 All ER 476.
21 (1818) 1 B & Ald 681.
22 [1955] 2 All ER 493.
23 [1995] CLC 1011.
24 [2001] EWCA Civ 274.
25 (1875) LR 10 Ex 153.
26 [1975] 1 All ER 198.
27 [1987] AC 87.
28 (1809) 2 Camp 317.
29 (1857) 7 E & B 872.
30 [1990] 1 All ER 512.
31 (1884) 9 App Cas 605.
32 (1602) 5 Co. Rep 117a.
33 [1965] 3 All ER 837.
34 [2007] EWCA Civ 1329.
35 (1877) 2 App Cas 439.
36 [1947] KB 130.
37 [1966] 2 All ER 284.
38 [1955] 3 All ER 10.
39 (1836) 1 M&W 466.
40 (1889) 14 PD 164.
41 [1921] 2 KB 519.
42 See Section 15(2).
43 (1876) 1 QBD 410.
44 (1876) 1 QBD 183.
45 [1975] 3 All ER 739.

46 [1908] 2 KB 1.
47 [1927] 1 Ch 128.
48 [1927] AC 177.
49 (2000) The Times, 23 March.
50 (1859) 7 CB NS 99.
51 (1861) 1 Best & Smith 393.
52 See Section 6.6.4.
53 [1934] 2 KB 394.
54 [1949] 1 KB 532.
55 [1949] 1 All ER 127.
56 [2001] EWCA Civ 1279.
57 [1990] 1 AC 83; see also Section 3.6.1.
58 [1984] CILL 84.
59 [2004] EWCA Civ 747.
60 (1795) 6 Term 320.
61 [1895] 1 QB 673.
62 [1952] 2 All ER 176.
63 [1916] 1 KB 566.
64 [1972] 2 All ER 1322.
65 (1870) LR 5 CP 10.
66 (1846) 8 KB 371.
67 [1854] EWHC J70.
68 [1949] 1 All ER 997.
69 [2008] 2 Lloyd's Rep 275.
70 [1909] AC 488.
71 [1973] 1 All ER 71.
72 [2001] All ER 801.
73 [1991] 4 All ER 937.
74 [1915] AC 79.
75 (1892) 1 QB 127.
76 (1970) EGD 495.
77 [1901] KB 515.
78 [2004] BLR 153.

3 Aspects of commercial law

3.1 Introduction

The first two chapters set out the basis on which legal systems operate and gave an account of the law of contract. The importance of contract law was discussed in terms of the requirement for certainty in commercial dealings. A contract is the means by which trading partners can have confidence that the promises exchanged in a contract will be kept or that the party in default can be made to face the consequences, usually in the form of damages.

The law of contract does not operate on its own to regulate trade and commerce. There are different sectors of the law, such as banking, finance, insurance and commercial law, which are all relevant to a wider discussion. For example, a business needs to know more about the types of organisation which they may encounter and the implications of the other's association on the bargain made. Having rights to take action under contract law is one thing, but if the other contracting party is not financially sound or has no assets from which to settle any action taken then there is little point in seeking a remedy through the courts.

This chapter examines some of the background to commercial law in which construction contracts operate. Commercial security involves a consideration of the investigations that can be made and steps taken to minimise exposure where a contracting party defaults.

Commercial law, also known as business law, is the body of law that applies to the rights, relations and conduct of persons and businesses engaged in commerce, trade and sales. It is a branch of the civil law and deals with issues of both public and private law. The field also comprises company law and regulation applying to other forms of business organisations such as partnerships. Commercial law may be viewed as the collection of norms and procedures which allow transactions to occur with an appropriate level of safeguards in place to ensure against default.

This chapter covers certain aspects of commercial law to complete the framework required before the reader can start to form an appreciation of the operations of construction law. The chapter examines the types of business arrangements recognised by the law and their key features before discussing issues around insolvency of companies. The chapter concludes by looking at the law around the different types of security available for lenders and judicial rules of interpretation often used in commercial contract cases.

In this chapter and the remainder of this book, reference is made to employer – meaning the client or building owner – and contractor – meaning the builder or provider of services appointed by the employer.

3.2 Business organisations

Carrying on business usually involves taking steps to set up a separate entity to the person(s) involved. Most people see the benefit in having their personal life kept separate from their business affairs. This is not universally true and sole traders using their own name for business are common, having varying degrees of separation between business and personal affairs.

There are three main methods of carrying on a business in England and Wales. These are as:

- a sole trader/practitioner;
- a partnership; or
- a limited company.

Figure 3.1 represents the differences between the three forms of organisation in terms of the creation of a separate entity at law. The following section examines the three main types of business organisation before giving more coverage to some aspects of company law. The additional detail on company formation reflects the point that most employers and contractors organise themselves in business as limited companies or plcs.

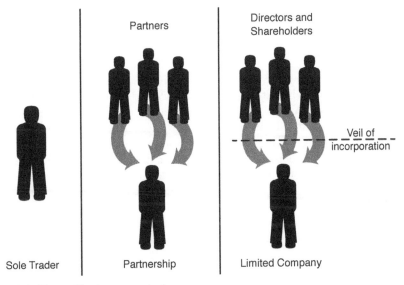

Figure 3.1 Types of business organisation

3.2.1 Sole trader/practitioner

A sole trading situation exists where a person sets up in business without creating a separate 'vehicle' or legal person through which the business is conducted. As such, the business owner is entitled to receive the profits from a business organisation. This may arise in an enterprise run entirely by one person or where the proprietor of a business employs a number of people. Those who carry on business in this manner will normally do so because of personal preference. The construction industry has a higher incidence of sole traders than most other industries. Subcontractors and tradesmen will often start out in business and remain as a sole trader because it suits them.

The principal advantages are control over the organisation and fewer legal formalities. The sole trader is entitled to the profits from their business and to a considerable degree of privacy as there is no need to file separate business accounts subject to public scrutiny. This is not the same as saying that there is not a requirement to complete tax returns, which even a small sole trading organisation ought to seek help from an accountant in compiling. The disadvantage is that if the business incurs losses then the sole trader will be personally liable for all the debts of the business. Further, it can be difficult for this form of business organisation to raise capital for investment in the business due to a lack of assets in the name of the business.

In the context of professional practices (e.g. surveyors, solicitors and accountants), it is difficult for a sole practitioner to combine working with the management of a practice whilst remaing up to date with developments in their profession. The high premiums required for professional indemnity insurance also act as a barrier to sole practice. As a consequence, many sole practitioners have tried to build niche practices in their respective areas of expertise. Sole practitioners sometimes amalgamate with others to form a partnership.

3.2.2 Partnerships

A partnership exists where a number of individuals agree to work together in circumstances where they collectively own the business and any profits generated. The individuals come together to create a partnership. This does not involve the incorporation of a separate company. The partnership agreement records the terms of their association.

Traditionally, partnerships were chosen as a method of carrying on business for professionals where there was a restriction on forming a company. The central notion is that professionals should be prepared to stand by their advice and be accountable for services rendered. Thus, by trading as a partnership, the partners remain personally liable for any debts and errors without any limit on their potential liability. This was considered to be more appropriate than allowing professional advisors to hide behind the so-called 'veil of incorporation' of a limited company whereby the liability of individual owners (shareholders) is limited to the value of their shareholding. In recent years, these restrictions on how professionals organise their business arrangements have been largely removed, yet many professional practices appear to prefer to keep their partnership status.

The law is governed by the Partnership Act 1890. The major characteristic of a partnership is that there is in existence the common intention between the partners to carry on a business with a view to making a profit. A partnership is valid even though the agreement is entered into by word of mouth with no formalities. However, for practical reasons, partners most often choose to enter into a written partnership agreement. This will deal with matters such as the nature of the business, capital, profit and losses, partnership property and the frequency and form of meetings. The most important aspect of the agreement is that it overrides the Partnership Act 1890 which would otherwise operate to imply certain basic terms. A partnership agreement also has the flexibility of being able to cover the position of each individual partner. Neither the agreement nor the accounts of the firm are open to public inspection.

Each partner owes a duty of good faith to the other partners. Partnerships operate on an agency basis. Each partner therefore has implied authority to bind the firm by transactions entered into during the ordinary course of business. Each partner is entitled to take part in the management of the firm and is entitled to inspect the partnership accounts. Decisions are made on a simple majority basis, but unanimity can be required in the case of major changes such as the admission of a new partner. Partners are said to be *jointly and severally liable* for partnership debts. This means that the firm may be sued *en bloc* or individual partners may have proceedings brought against them for the entire amount owed by the firm (rather than simply their share of the debt). Joint and several liability does not allow for an account to be taken of what share of the liability would be just and equitable to attribute to each partner. Each and every one is liable for the whole amount.

Retired partners (and the estates of deceased partners) are also liable for debts incurred by the firm during the period when they were partners. The partnership relationship can be brought to an end by dissolution of the firm. This may be effected without a court order and is usually done by mutual agreement.

Limited liability partnerships

The advent of limited liability partnerships (LLPs) has blurred the distinction between partnerships and companies. The Limited Liability Partnership Act 2000 created this new form of organisation for businesses which is essentially a hybrid between the partnership and the limited liability company. An LLP has a separate legal identity to that of its members in the same way as a company. The LLP is therefore capable of entering into contracts and owning property. The formalities required to set up and run an LLP are less onerous that those associated with setting up and running a company. In the absence of a partnership agreement, the standing arrangement will be that profits, losses and capital are shared equally and all members have an equal right to manage the LLP. The LLP has more in common with a limited company in terms of managing its affairs.

The key feature of an LLP is that the maximum liability of the firm is limited to a predetermined figure. The figure is usually several million pounds or a greater sum for larger LLPs. The liability of the partnership is therefore limited but at a level sufficiently high to give its trading partners and clients comfort that the LLP has sufficient standing and backing to meet its obligations should the need arise. Therefore there is little risk, from the creditors' point of view, that the LLP will be able to walk away from the liability, and the chances are that the LLP will honour any debt it admits. The risk of debts being left unsettled is much more pronounced when dealing with limited companies.

3.2.3 Limited companies

The question of standing and financial resources is frequently less clear where limited companies are concerned. A company is an incorporated body which exists as a legal 'person' that is quite distinct from its members. The key difference from partnerships and sole traders is that the assets belong to the company and not to its members. Companies are regulated by statute and there has been a long list of company law statutes with the current law to be found in the Companies Act 2006. The vast majority of companies are limited liability companies. The company's liability is said to be limited by shares, which means that the liability of its shareholders is limited to the extent of their individual shareholding. The shareholders cannot therefore be pursued by the creditors for anything more than the value of their shareholding.

For example, John has 100 issued shares in John Limited which he subscribed to at the price of £100 per share. He has paid for 50 shares but has 50 shares that are allocated to him but for which he has not yet paid. The business enters into a situation where it owes money to a third party and has no assets from which to settle the debt. The maximum amount that John would owe to the company's creditors in the event that the company could not pay its debts would be the balance on the shares not yet paid for – 50 × £100, being £5,000. The first call on settling this sum would be from existing share capital held by the firm or from a positive trading account.

The benefit to the shareholder of this arrangement is that their maximum financial exposure through the business is predetermined. This is not a protection enjoyed by partners or sole traders. Private companies can be identified by the word 'Limited' or the abbreviation 'Ltd' at the end of the company name. The other suffix seen with companies is 'plc' which denotes a public limited company. These are larger, frequently multinational companies in which the public can apply for shares and trade the same on the stock exchange. Non-profit-making companies sometimes form themselves as companies limited by guarantee, while a handful of companies operate on the basis of unlimited liability. This occurs primarily because unlimited companies are not obliged to file accounts.

The downside for a limited company is the filing requirements and placement of company affairs on the public record. A potential trading partner can therefore carry out credit checks. In the event that they are not satisifed that sufficient protection is given to them, they can seek additional security.

The capacity of the company

The question of capacity in terms of the freedom to enter into contractual arrangements has already been discussed.[1] Trading partners may justifiably be concerned about the ability of business organisations to enter into contracts. This is not an issue with sole traders or partnerships but was, at one time, more complicated in relation to companies. Before the enactment of the Companies Act 2006, companies were required to define the extent of their powers to enter into contracts by setting out the objects of the company in their memorandum of association. If a company subsequently entered into a contract that fell outside these objects, it was said to be acting *ultra vires* (beyond the powers) and the contract could be void.

The leading case in this area is *Ashbury Railway Carriage Ltd v Riche*[2] where a company had been formed with the object of carrying on the business of making and selling rolling stock. The company entered into a contract with Riche to construct a railway in Belgium but defaulted on their obligations. When Riche sued for breach of contract, the claim failed as the construction of a railway (rather than rolling stock) was *ultra vires* and the contract was therefore void and the debt could not be pursued.

The fiction around the decision in the above case was that before entering into a business transaction, every company should enquire into the authority of the other to enter into the proposed contract by examining its memorandum of association. Such a state of affairs was clearly not in the interests of open trade, and revisions were made to the legislation to rectify the situation. The Companies Act 2006 (and the proceeding Companies Acts) completely removes the *ultra vires* rule. Section 39(1) now provides that the validity of an act done by the company shall not be called into question on the grounds of lack of capacity by reason of anything in the company's constitution.

3.3 Operating a business

Statistically, most businesses operate as private companies. Upon formation of the company, the business owners become its first members by subscribing for shares in it. Private companies tend to have a nominal share capital which is issued by dividing the sum collectively invested into shares of £1 each or similar. This nominal capital may have little relationship to the actual financial state of the business. Private companies are frequently family businesses. They tend to be small to medium-sized organisations, although there are instances of large family businesses remaining as private (as opposed to public) companies to enable the owners to retain as much control as possible. The members of a private company will also generally be its managers. If the articles (discussed below) so provide, there may be restrictions on the rights of members of private companies to transfer shares and on the number of members that the company can have. A public company has no such limitations on its membership and transfer of shares but must have a minimum share capital of £50,000. Shares of public companies are bought and sold on the stock exchange and are freely transferable. Professional managers are invariably recruited to act as managers and directors of public companies.

As with partnerships, the operation of the business is based on majority rule. However, the separation of roles between the shareholders and the board of directors can complicate matters. Basically, the board has responsibility for the day-to-day running of the company whilst the shareholders must be consulted on the bigger decisions. Different actions in companies are taken by shareholder resolution, which are either ordinary – requiring a 51 per cent vote in favour – or special – requiring 76 per cent in favour of the proposed action. One downside of company organisations is that the minority shareholders can frequently consider that their views are not being addressed in the company's policies. This can lead, in smaller companies, to the creation of a shareholders' agreement which contains express provisions about which decisions require unanimous support. This is also a means of ensuring that any change in the ownership of the shares can be controlled. This may be achieved by requiring that anyone seeeking to sell their shares in the business must offer them to the existing shareholders in the first instance.

A company is formed by lodging documents with the Registrar of Companies, including the Memorandum of Association and the Articles of Association. The Memorandum lists the original members of the company and records their agreement to take at least one share in it. The Articles represent the company's internal constitution. They describe the company's management structure and set out the rules that all members will be subject to as a matter of contract law. Once formed, a Certificate of Incorporation is issued to the company, signifying its creation. After this has been created, it will continue to exist, despite changes in the organisation, until the organisation is wound up. A company may borrow money, like any other person, in addition to raising capital by issuing more shares. In the case of small companies, this is often in the form of loans made in return for personal guarantees given by the major shareholders. A loan may be secured by a debenture, which is a charge over the company's property. This is more likely to be an option for companies with substantial fixed assets.

Companies operate through the decisions of their boards of directors and through the wishes of their members at general meetings. Companies are required to keep annual accounts and to file a copy with Companies House which is open to public inspection.

A credit check is frequently the first action undertaken when one business is contemplating entering into a contractual relationship with another. Credit agencies commonly give a weighted score to businesses based on their assets, trading history, any satisfied or outstanding court judgements and reputation. A poor score should indicate to a would-be investor or trading partner that they may wish to avoid this business or take additional steps to protect themselves from any likelihood of insolvency. This is the next topic considered in this chapter.

3.4 Insolvency issues

The limited liability status of private limited companies gives rise, all too often, to situations where the company runs out of money. The construction industry has the highest rate of corporate failure within any sector. When this occurs, a set of procedures are needed to govern what happens next. The company has a separate legal identity at law and, in the first instance, it is the company – not the directors or shareholders – that is liable for the firm's debts. Most companies do not plan for insolvency, and a once-viable business model can unravel in the face of tough economic conditions or unforseen events. Consider the position of a steelworking subcontractor who has to pay for his materials up front from the steel fabricator and yet is not paid by the main contractor for 60 days from the date of his invoice. It would not, depending on the credit terms extended to such a business, take much to destabilise this business model if the cash flow is delayed or challenged.

The test for solvency of a company is a straightforward one. The Insolvency Act 1986 does not define 'insolvency' itself but uses the phrase 'unable to pay its debts'. Section 123 of the Insolvency Act 1986 sets out when a company is deemed unable to pay its debts; this includes failure to comply with a statutory demand for a debt over £750, failure to satisfy enforcement of a judgement debt or proof that the value of the company's assets is less than the amount of its liabilities. A company that finds itself behind with its payments is often in the position of being in arrears of paying its taxes. Her Majesty's Revenue and Customs often bring or support statutory demand procedures against debtor companies leading to their winding up.

A company may be wound up either at the petition of its creditors or at the election of its members. Sometimes the business has run its course and the members perceive no need to continue with it. Liquidation is used therefore when there is little or no prospect of the business continuing. Governments and policymakers have been sensitive to taking a heavy-handed approach with companies in difficulties, and other less severe forms of company procedures than liquidation are possible, some of which are aimed at corporate recovery. The different forms of insolvency events include the following.

Company Voluntary Arrangement

A Company Voluntary Arrangement is where the company and its creditors come to an agreement which is then implemented and supervised by an insolvency practitioner under Part 1 of the Insolvency Act 1986. It is an arrangement whereby the company intends to trade through its solvency issues and asks its creditors to give it some breathing space to repay a proportion of the debts owed. This procedure may be supported by a court-ordered moratorium.

Administrative receiver

An administrative receivership is a procedure whereby a secured creditor installs a receiver to run the company pending the realisation of company assets subject to security. This form of insolvency is less about corporate recovery and more about the major creditors wishing to continue with the business until such a time as it can repay their secured lending. If there are other unsecured creditors then the winding up of the company can follow once the liability to the major creditor is repaid. Alternatively, the business may be sold at this stage if it is still a viable going concern.

Administration

This form of insolvency occurs where a company may be rescued or reorganised under the protection of a statutory moratorium. The company is put into administration and an administrator appointed. A relatively recent phenomenon is the emergence of the 'pre-pack administration'. This refers to a situation where a business is put into adminsitration having already negotiated a deal for the sale of the business to a third party. Once the administration is ordered, the sale is triggered and the new owners take over the business. This can leave the unsecured creditors of the old business somewhat disgruntled at having their debt extinguished whilst ostensibly the same business continues for the benefit of the select few creditors involved in the pre-pack deal.

Where liquidation of the company becomes necessary, the liquidator takes control of the company's property. The powers of the board are taken over and the liquidator is required to call in all assets, realise them, and distribute the proceeds to the appropriate creditors so far as is possible. The liquidator will be particularly concerned with transactions of the company made immediately before liquidation to try to ensure that any disposals made with the intent of defeating creditors are traced. If so, they could well be rendered void. Once the liquidator has paid the costs and expenses of the liquidation, he must pay off the creditors who have submitted formal proof of their debt. If there are surplus funds after distribution to the creditors, these are distributed to the members of the company according to their interests and rights.

The order in which debts are settled from the recovered assets of the company is subject to a strict order of priority. Those creditors who have 'secured' their debt over company assets, either by way of mortgage or some other charge such as a debenture or bond, are in a much better position than the unsecured creditor, who is unlikely to see any of the owed money repaid. Sometimes a partial distribution of assets recovered is possible, which is usually on the 'penny in the pound' basis. This arrangement operates to pay out a fraction of each pound owed, and this is divided up between the unsecured creditors. This is a common outcome in a creditors' voluntary arrangement where there is a predetermined level of payout and the company trades on until the agreed level of debt is settled.

Individuals (and therefore sole traders) are subject to a process of personal insolvency known as bankruptcy. A debtor is automatically discharged from bankruptcy on the first anniverary of his becoming bankrupt.[3] Once discharged from bankruptcy, a debtor has no further liability for his bankruptcy debts although he will emerge with a chequered credit history.

Insolvency law contains some additional court powers which are sometimes applied for. Freezing orders by way of injunctions can prevent the disposal of assets from bank accounts. Director disqualification proceedings can be brought by the Secretary of State to prevent further abuses of the position by individuals shown to have fallen below the standards required. Attempts to reclaim assets wrongly dissipated from businesses are also a fertile area of court proceedings and include cancelling preferences and transactions at under value. These two transgressions involve looking into the recent history of the company before its demise and challenging any transaction which appears dubious.

3.5 Security

Insolvency is a regrettably frequent event in the construction industry, mostly at the subcontractor level where cash flow is more difficult to manage effectively. The effects of insolvency are felt not just by the organisation and its staff entering into an insolvent event but also by its direct and indirect trading partners. A main contractor entering insolvency could leave hundreds of subcontractors and suppliers exposed and jeopardise their own credit terms extended to the business. Similarly, an insolvent client developer causes potentially serious repercussions for the professional team and supply chain.

Managing the risk of insolvency on a construction project involves a number of steps, including taking frequent credit checks on a business' creditworthiness and ensuring as little work is performed at risk as possible. Interim payments assist with this last point given that they involve being paid for work done at the intervals set in the contract, usually monthly. This effectively limits the exposure of the payer and the payee to the potentially disastrous effects of the other's insolvency.

The risk of insolvency is frequently managed in construction by the adoption of the following procedures which are used in domestic and international contracts as a means of protection against non-performance. The issuer of the bond or guarantee undertakes to be responsible for the fulfillment of a contractual obligation owed by one person to another if the first person defaults.

The following security instruments are now considered:

- parent company guarantees;
- retention;
- bonds.

3.5.1 Parent company guarantees

Parent company guarantees can be useful where a credit check may reveal that a potential trading partner has a short trading history or a small amount of share capital and assets. It is common in construction for companies to be created specifically for a project. On occasion, the ownership of the company formed can be shared between the stakeholders on the project, which can include the employer and contractor. The Private Finance Initiative was well known for its SPVs or Special Purpose Vehicles which were companies set up for the delivery of the project undertaken. Many subcontractors working for SPVs raised concerns about their liquidity and may have sought parent company guarantees.

This is straightforward if the company is a member of a group in which one or more linked companies has more share capital or assets than the linked company and may be approached to guarantee the performance of the trading company. Clearly, this can only happen where such a group exists or another company is willing to stand as guarantor.

3.5.2 Retention

The purpose of retention payments is not primarily associated with insolvency but effectively covers the same risk. It is common practice in the UK industry, and worldwide, for a percentage of the contractor's earned value on an interim payment to be retained by the client pending the completion of the works and the associated remedying of defects. The percentage is typically 3 to 5 per cent but can be as high as 10 per cent on civil engineering and international projects. The first moiety of the retention is released on practical completion of the works and the rest on final certificate following the completion of the defects liability period. The rationale for the practice is that if the contractor is unable or unwilling to fix the defects that appear after he has left site then the employer has a fund from which to finance the same.

Retention is seen by many in the industry as being a backward-looking practice which does little to promote harmonious relations between the participants on a building project. Retention is conspicuous by its absence from some of the more forward-looking contractual arrangements, such as partnering. Instead of a retention, the contractor may be able to negotiate the release of interim payments for work done in full by providing a retention bond. This bond guarantees to the employer that any cash retention from the price which the employer would have withheld but which is released to the contractor will be repaid by the bondsman if the work is not completed. This type of bond is relatively uncommon as surety companies have been slow to offer this product. Where they are available from banks, they are frequently treated as being part of the contractor's overdraft facility and are therefore not popular with contractors themselves.

3.5.3 Bonds

Retention bonds are relatively rare in the construction industry. A very common occurrence in construction projects is for the employer to seek a performance bond from the contractor. This involves the contractor providing a surety company to 'put up' a bond for him. The idea is a simple one – in the event that the contractor does not perform its contractual obligations then a call may be made on the bondsman who then pays the employer directly. The cost of securing the bond is part of the contractor's overheads, indirectly charged to the client in the tender price. The range of bonds available includes the following.

Performance bond

The performance bond is the most common and usually entitles the employer to call for a fixed percentage of the contract sum (usually 10 per cent) in the event that the contractor does not 'perform' its obligations. The surety company usually has strict requirements of the circumstances in terms of notice and time limits in which the bond can be called, and the employer needs to pay close attention to these.

Advance payment bond/re-tender bond

In the unusual circumstance that the employer pays the contractor up front, the repayment of the advance can be secured by way of a bond. Similarly, having secured a tender to build, the employer may seek a bond against the cost and losses that would be incurred in recommencing the tender exercise if the contractor selected is not able or unwilling to proceed.

On-demand bond

The on-demand bond is a curious category of bond which is rarely found in the UK but is found internationally. This bond allows the employer to make a call upon it at any time and without the need to show any lack of performance on the contractor's behalf. The employer may make a call 'on demand' and the contractor, in the absence of being able to show bad faith or fraud, must arrange for the payment to be made. The rationale for the bond is hard to fathom when viewed from the standard practices of the UK. Contractors who operate in markets where these bonds are prevalent need to tread carefully at the risk of being caught out. Lord Denning, in the case of *Edward Owen Engineering Ltd* v *Barclays Bank International Ltd*,[4] had evident sympathy with the contractor's position but was unable to assist and duly ordered payment commenting that *so long as the… customers make an honest demand, the banks are bound to pay.*

The judgement is a salutary tale for businesses operating in a different geographical market to the one they are used to trading in and what happens when they are caught out by the unexpected. The judge sympathised with the plight of the contractor but was powerless to intervene in the absence of bad faith. This is a useful reminder of the sanctity of the contract and the certainty which contracts can deliver. The case demonstrates that, in the face of clear and unambiguous wording, a judge has little scope for intervention.

The picture which emerges from this discussion so far about business organisations and security is one of understandable caution on the part of would-be trading parties. Businesses set out to be successful and must navigate as best they can between the threats of bad bargains and the risk of bad debts they are unable to recoup. The reader should now have formed an appreciation of these risks and the strategies adopted to mitigate them.

This chapter has discussed aspects of commercial law based on the premise that businesses require as much certainty as possible in their dealings as provided through the medium of contract law and security arrangements. What if the wording of these documents is in itself unclear? This involves a consideration of the main rules of interpretation that judges may use in seeking to interpret how commercially worded agreements should be construed. It should be noted that these rules normally only operate in the event of an ambiguity or inconsistency in contract terms.

3.6 Rules of interpretation

Interpretation is the task of ascertaining the meaning that a contractual document would convey to a reasonable person and involves broad principles rather than strict rules. The words in the document are the starting point, and commonly the end point, for questions of interpretation. Reference has already been made to judges' frustration at having to give effect to what the contract states or what the statutory draftsmen wrote. Neither are they permitted to undo a bad bargain. The law will not help a fool. The broad principles which apply may vary on the facts of the case and the nature of the ambiguity alleged. The consequent main principles of interpretation are as follows.

3.6.1 *The mischief rule*

The mischief rule of interpretation applies to statutes and involves asking the question: what was the mischief this statute was intended to avoid? For example, if the purpose of a statute was to ban hunting with dogs then the interpretation of a section in the statute should be the one that is consistent with this outcome rather than, say, reading a clause as permitting a loophole around allowing hunting otters with dogs. This is a particular way at looking at contractual interpretation and represents the common-sense approach. Where there are two or more readings of a provision then the interpretation most consistent with common sense will be preferred.

3.6.2 *Contra proferentem*

In terms of contractual interpretation where there is any ambiguity or lack of clarity in relation to a clause then the *contra proferentem* rule may have effect. Translated from the Latin, this means 'against the offeror', also known as 'interpretation against the draftsman'. This provides that where a promise, agreement or term is ambiguous, the preferred meaning should be the one that works against the interests of the party who provided the wording. The doctrine is often applied to situations involving standardised contracts or where the parties are of unequal bargaining power, but it is applicable to other cases. If a contract is co-drafted by the parties then this rule will not apply. The rule is embodied in statutory form for business-to-consumer transactions under regulation 7 of the Unfair Terms in Consumer Contracts Regulations 1999.[5] This provision may be seen as going further than the rule itself as it requires that the interpretation most favourable to the consumer be adopted.

3.6.3 *Other approaches*

The literal or plain meaning rule requires a literal application of the language used in the statute. The words used should be given their usual meaning and the statute applied according to this and no other rule of interpretation. Another approach is the 'whole contract' principle where the document must be construed as a whole, in its context. Under this approach, it is inappropriate to focus excessively on a particular word, phrase, sentence or clause.

The rules of interpretation can be seen as being contradictory in parts and disputing parties may choose the approach which most suits their case. This appears unhelpful but is simply an example of the uncertainty which surrounds legal submission and the scope for legal argument around any given scenario.

3.7 Conclusion

The reader should now be able to reflect on the existence of commercial risk and the attempts made to protect interests against the worst effects. This chapter has reviewed some of the basic commercial law required for an understanding of the background in which construction law operates. The types of business organisations have been reviewed and a discussion entered into around capacity and security of commercial dealings. The on-demand bond serves as an example of different traditions existing in construction law and a note of caution that different legal traditions have different approaches.

3.8 Further reading

Adams, A. (2014) *Law for Business Students*, Eighth Edition, Harlow: Pearson.
Dobson, A. and Stokes, R. (2012) *Commercial Law*, Eighth Edition, London: Sweet & Maxwell.

Notes

1   See Section 2.6.1.
2   (1875) LR 7 HL 653.
3   Section 279 of the Insolvency Act 1986.
4   [1978] 1QB.
5   SI 1999/2083.

4 Aspects from the law of torts

4.1 Introduction

The aim of this chapter is to introduce the aspects from the law of torts which apply directly to construction law. By the end of the chapter, the reader will gain an understanding of their importance, interrelation and application. Construction professionals need to gain this appreciation of the operation of tort law in order to practise in this area.

The word 'tort' comes from the French, meaning 'wrong'. A tort is a *civil wrong* in the sense that a tort is committed against an individual rather than seen as an indictment against the state. A wrong committed against the state is a crime and is subject to the criminal law and the courts. Tort is the name given to the branch of law that imposes civil liability for breach of obligations imposed by the law.

It was established in Chapter 1 that one of the most important functions of the law is to protect people in certain defined situations. Tort pays a crucial role in this protection. The protection offered by tort law is more complete than that extended by the other two main areas of law included under the umbrella of construction law – contract and property law. Contract law is primarily concerned with protecting our financial interests in situations where we enter into dealings with one another. Property law protects our right of enjoyment over our land and our ability to own the land in a variety of different ways. These two subjects have a particular application in that they govern the arrangements between individuals who have already decided upon a course of action.

Tort law offers a wider range of protections than financial and proprietorial. Tort protects our basic well-being insofar as attacks on the person are protected by the torts of assault and battery. We have the right not to fear (assault) or be occasioned bodily harm (battery), and anyone infringing these rights could be guilty of a tort. The crossover here between tort and criminal law is apparent, but the law draws no distinction in that one set of facts gives rise to both criminal and civil proceedings.

Tort law also protects our dignitary interests. Our reputation and public good name can be thought of as our property, and these are protected by the torts of libel and slander. Anyone attacking our reputation unjustly can be guilty of these torts and proceedings could be commenced.

The most relevant application for the law of tort in the field of construction law concerns accountability. Professional negligence refers to the area of law where someone carrying out services can be liable for damage caused as a result of their actions. Contract law is often the more obvious route for someone seeking recourse against another for a breach of duty. However, contracts are not always present and in their absence, residual liability to third parties is owed under the law of tort.

Tort law protects physical, financial, dignitary and property rights. This usually involves asking a court to award damages for infringement of a protected interest. The general rule of compensation is to place the innocent party in the same position as if the tort had not been committed. Damages are usually awarded in the form of a lump sum. The basic recipe for a successful tort action is an act or omission by the defendant that causes damage to the claimant.

This approach can be simplified to the simple equation that a party suing for tort will need to demonstrate a duty, a breach of that duty and the loss which results from it.

There are more than 70 torts in the laws of England and Wales. This chapter considers only those directly relevant to the built environment and omits many of the minor torts. The most important tort is negligence and this is covered first.

Act (or omission) + causation + fault + protected interest + damage = liability

---

Figure 4.1 The components of a successful tort action

4.2 Negligence-based torts

4.2.1 Introduction to negligence

The essence of the tort of negligence is a duty to take reasonable care. Negligence has two meanings for the purposes of this chapter:

- It can refer to a tort of negligence. This is the most common of all the torts and accounts for most of the cases being tried in the courts. This covers such circumstances as injuries suffered following road traffic accidents or allegations that people have not done their job properly.
- Negligence is also used to describe careless behaviour. Acting in a negligent manner is a component of several different torts. In order to decide whether someone has been negligent or not, it is necessary to examine their conduct

and ask: Would a reasonable man have acted in such a manner? Sometimes the question is: Would a reasonably competent fellow professional have acted in such a manner?

The key to understanding negligence is to appreciate that it is about making a person (tortfeasor) responsible for their actions whether they intended the consequences of their actions or not. The tortfeasor has to demonstrably fall below the 'reasonable man' standard in order for the negligence to be established. The purpose of the tort of negligence is to make people accountable for their actions or inactions.

The duty to take care is not restricted to the law of negligence and appears in statues and contracts alike. For example, section 13 of the Supply of Goods and Services Act 1982 states: 'In a contract for the supply of a service … there is an implied term that the supplier will carry out the service with reasonable care and skill'.

Statues and contracts have in effect borrowed the reasonable man test from the law of torts. This is probably a case of imitation being the best form of flattery in that the reasonable man test is a succinct and accurate portrayal of where the legal standard should be pitched in this area.

4.2.2 Donohue v Stevenson

The development of the tort of negligence is directly attributable to one case. This is without doubt the most important case in the English legal system. The case is the archetypal common law example of a seemingly trivial set of facts creating a whole sector of the law. The case is *Donoghue* v *Stevenson*.[1] The facts of the case are these.

A boyfriend bought his girlfriend (May Donoghue) an ice cream sundae from a café in Paisley, Scotland, and gave it to her. The ice cream sundae involved the customer pouring ginger beer from an opaque bottle over the ice cream. May Donoghue was enjoying her ice cream until the decomposed remains of a snail came out of the bottle whilst her boyfriend was replenishing the sundae. She became very ill and wanted to sue for the distress and injury caused.

Miss Donoghue had not bought the drink so could not sue the café owner under the law of contract. The common law doctrine of privity of contract dictates that a third party cannot enforce a contract between A and B. She therefore sued the manufacturer of the drink (Stevenson) for failing to take adequate care when preparing the drink for onward sale to consumers. In a landmark decision in the House of Lords, she won. The case single-handedly created the law of negligence. It did not matter that there was no contract between claimant and defendant, nor that Stevenson had no way of knowing who the eventual consumer of his ginger beer would be. The key to the decision was the establishment of the neighbour principle and the seminal judgement of Lord Atkin: *The rule that you are to love your neighbour becomes in law, you must not injure your neighbour. ... You must take reasonable care to avoid acts or omissions which you can reasonably foresee would be likely to injure your neighbour.*

It is telling in this passage which created a new branch of law that the judge traced the origin back to religious teaching. This is an example of a trait picked up earlier (in Chapter 1 on the sources of law) recognising the indirect influence of religion on forming laws.

The importance and ramifications of this case on construction law cannot be overstated. Construction projects are well known for having multiple parties and for the potential of liabilities being created well beyond the confines of the building contract itself. Consider the following example.

The architect's design is not only for the benefit of the employer but will also be relied upon by a future owner. The future owner is not in contract with the architect and, should an issue arise, would be left without a direct contractual remedy; in those circumstances, he would look to the law of tort to provide cover. If the architect could be shown to be negligent then the future owner might fall into the class of 'neighbour' in which case action can be taken. In anticipation of any claim that may be made by the first or subsequent owner, the architect would maintain insurance, called professional indemnity insurance, to cover this risk.

The case of *Donoghue* v *Stevenson* provides us with a three-stage test to establish whether negligence has occurred.

1   Was there a duty of care?
2   Was the duty breached?
3   Has loss been suffered as a result?

4.2.3 Duty of care

Prior to 1932, there was no such thing as a general duty of care. Liability existed within defined relationships such as doctor–patient, ferryman–passenger, blacksmith–customer, but it was not universal. *Donoghue* v *Stevenson* changed that. One of the judges, Lord Atkin, defined 'neighbour':

> *Who, then, in law is my neighbour? Persons who are so closely and directly affected by my act that I ought reasonably to have them in contemplation as being so affected when I am directing my mind to the acts or omissions which are called into question.*

The case extends a duty of care to anyone with whom you either come into contact or should have in mind. This is a duty *not to* injure or make a situation worse for anyone you come into contact with. As such, this is a negative duty – there is no compulsion to do the opposite (i.e. to help people). For instance, ignoring the cries for help from a drowning man in a canal whilst you walked past on the towpath would not be breaking the law even though you have 'come into contact' with them. There may be a moral case for helping but there is no legal case. The architect, in the example given above, ought to have in mind the future owner of the building even if the identity of this person could not be known at the time of designing the building. The architect owes a duty to take reasonable skill and care in relation to this subsequent owner as well as his original client.

4.2.4 Breach

Once a duty of care has been established, the next question is whether or not that duty has been breached. This is a question for the judge in any given case using the 'reasonable man' test: would the reasonable man have acted differently (and therefore not negligently) in the same situation?

Where a person presents themselves as having a particular skill, they are required to show the skill normally possessed by persons doing that work (e.g. solicitor, doctor or plumber). The question then becomes: would the reasonably competent solicitor, plumber, surveyor have acted in the same manner or would they have acted differently (and therefore not negligently)?

This can be a difficult test and a judge is often helped by expert evidence in this regard. The party alleging the negligence and the party defending will both seek to call an expert from the profession in question to guide the judge as to the adequacy of the action or inaction being called into question in terms of meeting professional standards.

A case demonstrating the difficulty of ascertaining whether a breach has occurred is *Barnett v Chelsea & Kensington Hospital Management Committee*.[2] The casualty officer failed to check a patient in an emergency department and sent him home. Unbeknownst to both of them, the man had been poisoned, and he later died. His wife brought the action against the hospital alleging negligence. The test therefore would have been for the judge to decide whether or not the casualty officer had fallen below the standard of a reasonably competent fellow professional in failing to diagnose the poisoning. It would be difficult to establish exactly what a casualty officer is able to ascertain from a short consultation and whether this amounted to breach of duty. In the event, it was decided that any discussion around breach of duty was irrelevant as the poison had already taken hold and it would have been extremely unlikely that any antidote could have been administered in time.

The case therefore failed because, regardless of the position on breach, the third component of a successful tort action could not be established. No loss had occurred and therefore neither the defendant nor his employer who was vicariously liable for the actions of the casualty officer had aggravated the situation as a result of the alleged breach of duty. This position can be contrasted with the rescuer cases where a well-intentioned person can make the situation worse for the person needing assistance. If, for example, the passer-by tried but ultimately failed to assist a drowning person and frustrated a professional rescue then the passer-by could be liable.

4.2.5 Loss

The claimant must prove that their damage was caused by the defendant's breach of duty and that the damage was not too remote. The casualty officer in the Barnett case was not found to have been negligent irrespective of the breach point as the patient would have, in all probability, still have died if the antidote was administered there and then. In this sense, no loss has occurred – the claimant was no worse off by reason of the failure to act.

The other element of the loss test is that the chances of the loss occurring were not too improbable or too far removed from what the parties might have reasonably thought would happen. This is known as the 'foreseeability' test. The leading case on the remoteness of damage is *Overseas Tankship (UK) Limited* v *Morts Dock & Engineering Co*,[3] known as the *Wagon Mound* case.

A ship in Sydney harbour called the *Wagon Mound* was in the process of refuelling when some diesel was spilt into the harbour. Nearby, a dockyard was carrying out some welding repairs to another ship. At the time, it was a commonly held belief that oil on water would not ignite if in contact with burning material. The belief proved unfounded as a fire broke out at the welding site and damage was caused to the dockyard. The dockyard tried to sue the *Wagon Mound's* owners for causing the fire. However, it was held that the resulting damage was 'too remote' from the oil spill and damages were refused. Basically, the state-of-the-art defence was available to the defendants. It was not known at the time that the oil was combustible so no liability could be established. The cost of cleaning up a neighbouring jetty covered with the oil was a foreseeable consequence of the accident and this head of damage was recoverable in a separate case.

The remoteness of a consequence of an action or inaction can involve some esoteric discussions about what may or may not have happened. The issue of causation is considered in Section 13.2 with specific reference to construction law issues. Negligence is the most common and important tort. To establish tort, you need to prove all three of the components – duty, breach and loss. The risk of negligence actions is a real consideration for all professionals operating in the built environment, and the ensuing cost of insurance is a major overhead. The preference remains for contracts notwithstanding the availability of tortious remedies. This is for two main reasons:

- There is an element of uncertainty in relying on tortious liability. Most people would prefer to have the obligations written into a contract, which can be specific and relied upon should the need arise.
- Certain types of financial claims are not recoverable under tort law. This is known as the 'economic loss' exception.

4.2.6 Economic loss distinguished

The initial distinction made between contract and tort was that the former regulated financial dealings and the latter, protection. This is manifested in the economic loss exception. The rule is that pure money claims are not recoverable under the law of negligence. There needs to be an element of 'damage' in tort claims in the sense of physical harm being caused to either person or property. If no damage has occurred then a tort remedy may not be available.

The case of *Spartan Steel & Alloys Ltd v Martin & Co (Contractors) Ltd*[4] illustrates this point. A road worker was conducting repairs in the vicinity of an iron works. The road worker cut through a cable, thereby interrupting the power supply to the works. During the time it took to repair the cable and restore power, the works had missed out on three 'smelts' – the operation by which the steel was poured and rolled. Only one of the three had been in process when the power went out; the remaining two had been scheduled but had not yet commenced. The works sued the road worker for the cost of all three as they doubtless would have been carried out in the normal course of events. However, the judge only allowed recovery of the monetary value of the one that was in progress as only this was 'damaged'; the other two were strictly pure money claims and therefore fell outside the economic loss exception.

The potential protection afforded to construction professionals by the economic loss rule is curtailed by the law around negligent misstatement. Negligent misstatement is a subcategory of professional negligence and, somewhat confusingly, does permit the recovery of pure money claims.

The law was set down in the lead case of *Hedley Byrne & Co Ltd v Heller & Partners Ltd*[5] where the judge found a 'special relationship' existed between an accountant giving a credit reference to a financier. The credit reference did not represent a true position on the solvency of the company and the financier lost the money advanced. This special relationship was sufficient to allow the economic loss rule to be bypassed. Professionals found to be in breach of the principles of negligent misstatement can be pursued for damages in the widest sense. Happily for the accountant in Hedley Byrne, he was able to rely on an exclusion clause in his terms of appointment which extricated him from any liability found. The judgement nevertheless formed the principle of law and also provides a good example of the interplay possible between tort and contract arrangements.

It can seem illogical to draw a distinction between types of damages when ultimately it all comes down to money as the means of recompense. Typical negligence claims are for the cost of remedying such things as cracks in the foundations of a house. However, the complexities of the subject can be simplified by returning to first principles – tort is the means by which damage is compensated and contract is the means by which financial losses are compensated. It is often possible to contemplate an action lying in both and this is supported by case law. It is apparent, however, that judges have taken a dim view of overreliance on tort law where a perfectly good remedy exists in contract. The case of *Simaan General Contracting Co v Pilkington Glass Ltd*[6] illustrates this point.

The claimant was the main contractor on a building project in Abu Dhabi. It was a term of the contract with the building owner that the curtain walling be a particular shade of green (as green is the colour of peace in Islam). The claimant engaged a firm to obtain and erect the glass. The firm ordered the glass from the defendant. The glass was the wrong colour and this caused extra expense to the claimant in terms of his performance of the contract with the building owner. The glass erector went into liquidation, which prevented a contract action against them. The claimant sued the defendant in negligence. The action failed as the claimant was unable to show that the defendant had assumed any responsibility to them under the Hedley Byrne principles discussed above. In the absence of a negligent misstatement case, any claim brought in tort would fail because of the economic loss exception – the glass was not damaged and had not caused damage to any other structure. It was simply the wrong colour, and therefore this was not a claim that would succeed in negligence.

4.3 Property-based torts

Torts arising from the enjoyment of land are important to the built environment. The protection of rights offered to owners and visitors needs to be carefully considered before, during and after construction. Building works are often poorly received by neighbouring landowners, who can feel aggrieved at any interference, real or perceived, with their enjoyment of their land. Any inconsiderate employer or contractor may find themselves facing a claim in tort for damage to the neighbouring land.

4.3.1 Trespass to land

Trespass is defined as an unjustifiable interference with the possession of land. Trespass occurs where a person directly enters upon another's land without permission. The tort is actionable without the need to prove damage. The tort is committed against the possession rather than the ownership of the land, which means that a tenant can bring an action as well as a landlord. The tort of trespass to land is committed by:

- uninvited entry onto land, in which case the slightest crossing of a boundary is sufficient;
- remaining on land after any right of entry has ceased;
- placing objects on land.

It is relatively straightforward to ascertain whether or not a trespass has occurred. Trespass can occur through uninvited occupancy, such as political protestors staging a sit-in on a building site. Trespass to airspace above the land was found to have occurred in the case of *Kelsen v Imperial Tobacco Co.*[7] The defendant had an advertising board which trespassed into the neighbouring land by eight inches. The judge granted a mandatory injunction for the removal of the sign.

A construction example of trespass is tower cranes infringing neighbouring airspace during a construction project. The risk of trespass is usually dealt with during the pre-construction stage and can involve procuring oversailing licences from the neighbour at a cost.

Once the trespass itself has been identified, the next issue for the occupier of the land is whether or not they wish to sue the trespasser or seek an order from the court excluding the person(s) or thing(s) from their land. Damages awarded for minor trespassers are likely to be nominal only. This often leads the complainant to a situation where it is not worth the time and expense needed to action a one-off breach of their rights through trespass. If the trespass is likely to be repeated then the complainant could consider obtaining an injunction as in the above case. The threat of injunctive proceedings and the consequent delay caused to a construction project mean that these issues are taken very seriously in the pre-construction phase.

4.3.2 Private nuisance

There are two types of nuisance known as private and public nuisance. A private nuisance is the unreasonable, unwarranted or unlawful use of one's property in a manner that substantially interferes with the enjoyment or use of another individual's property without an actual trespass or physical invasion of the land. This can be contrasted with a public nuisance, which is an act or omission obstructing, damaging or inconveniencing the rights of the community. Some nuisances can be both public and private where the public nuisance substantially interferes with the use of an individual's adjoining land. For example, pollution of a river might constitute both a public and a private nuisance.

The law of private nuisance involves disputes between adjacent landowners. The law seeks to draw a balance between the right of one person to use their land in whatever way they wish and the right of the neighbour not to have their own property rights in any way compromised. The key question to ask is: is this an unreasonable interference with a person's use or enjoyment of their land?

The degree to which an infringement is unreasonable depends on circumstances, such as the character of a particular location or the types of activities that are carried out there. Activities that constitute nuisance in one location might be acceptable in another. This was the point made in the case of *Sturges v Bridgman*[8] where the judge drew a distinction between practice in a residential and industrial area: *What would be a nuisance in Belgrave Square would not necessarily be so in Bermondsey.*

Most nuisance actions involve noise, smells, vibrations or dust, but they can include other things such as fire. The leading case in this area involved noise. In

Hollywood Silver Fox Farm v *Emmett*,[9] the claimants were breeders of silver foxes (using pelts for fur coats) and erected a noticeboard on their land inscribed: 'Hollywood Silver Fox Farm'. The defendant owned a neighbouring field, which he was about to develop as a housing estate, and he regarded the notice-board as detrimental to such development. He asked the claimants to take the sign down. They would not. He therefore sent his son out to shoot off a shotgun close to the claimant's land with the object of frightening the vixens during breeding. This worked, with many litters of foxes being stillborn. The claimant brought this action alleging nuisance, and the defence was the defendant had a right to shoot as he pleased on his own land. The judge held that defendant did not have such a right as he was wilfully interfering with his neighbour's enjoyment of their land. He was ordered to stop the shooting.

Actions can also arise from construction 'nuisances', and developers and contractors are well advised to ensure satisfactory relationships with their neighbours and to implement steps to avoid, inasmuch as possible, any disgruntlement that could result in legal action. Many contractors run 'considerate builder' schemes to minimise these issues. Damping fine materials and roadways, minimising demolition and keeping neighbours informed are all sensible strategies to deal with potential issues.

4.3.3 The rule in Rylands v Fletcher

The rule in *Rylands* v *Fletcher*[10] is described as a species of private nuisance that imposes strict liability on a defendant for damages caused by their non-natural use of land. The facts of the case are that Rylands employed contractors to build a reservoir. The contractors discovered a series of old coal shafts, which they chose not to block up. The result was that when the new reservoir was filled for the first time, it burst and flooded a neighbouring mine run by Fletcher. The case was originally brought in negligence but would have failed as it is arguable whether a duty of care had been breached as a result of a loss, which may not have been foreseeable. A new subspecies of nuisance was created to cover this scenario to give the innocent party a remedy. The rule in *Rylands* v *Fletcher* was that: *the person who for his own purposes brings on his lands and collects and keeps there anything likely to do mischief if it escapes, must keep it in at his peril, and, if he does not do so, is* prima facie *answerable for all the damage which is the natural consequence of its escape.*

Liability in this case is strict in that there is no need to prove a duty of care or negligence. This departure from the normal fault-based torts has exposed the case to criticism, and other jurisdictions have rejected the approach taken. Feelings ran particularly highly in Scotland where a judge described the case as: *a heresy that ought to be extirpated.*[11] The facts necessary to give rise to this action between landowners are relatively rare, and there have been very few incidences where the case has been pleaded successfully.

4.3.4 Public nuisance

The term 'public nuisance' covers a wide variety of minor crimes that threaten the health, comfort, convenience and welfare of a community. Violations of public nuisance can result in a criminal sentence or a fine or both. A defendant may be ordered to remove a nuisance and/or pay the costs of a clean-up. A potential issue for construction operations is the blocking of the public highway with deliveries and construction works. Any interruption of public rights of way needs pre-agreement with the relevant authorities to avoid any issues arising.

4.3.5 Statutory nuisance

Government has, over the years, shown increasing concern for public health and the environment. This has led to the introduction of legislation concerned with noise, run-down premises, clean air and accumulations. Part III of the Environmental Protection Act 1990 classes certain matters as statutory nuisance including noise, artificial light, odour, smoke, dust, fumes or gases and accumulation.

Any breach of these myriad regulations can be classed as a 'statutory nuisance' inasmuch as they are enforced by a public body (the council or the government), not an individual. This saves the individual the cost and time of having to go through the courts themselves. The first call for a disgruntled neighbour is therefore usually made to the local council with a request for the statutory nuisance powers to be exercised. The most common outcome of a successful action here is an abatement notice against the offender and a fine or stop order for a repeat offender. Statutory nuisances are a risk for a construction operator. Environmental protection and the impact of construction on the ecology are governed by laws arising from the same principles as public nuisance. Environmental protection is of increasing importance to construction stakeholders.

4.3.6 Occupiers' liability

Occupiers can protect themselves from interference with their property by using the laws of trespass and nuisance. However, owning or occupying land carries responsibilities to look after people coming on to your land. This is the area of law is known as occupiers' liability. Occupiers' liability is an offshoot of the law of negligence where the common law arising out of case law has been substituted by Acts of Parliament defining exactly the standard duty of care applying to the situations covered.

The law here involves the person controlling the premises rather than the physical occupier. The action lies against the person(s) responsible for the general upkeep of the land. This applies not only to land but also fixed and moveable structures such as vehicles and aircraft.

The law involves two Acts of Parliament that set out the duties which are owed to visitors and trespassers. They are:

- the Occupiers' Liability Act 1957 (relevant to visitors);
- the Occupiers' Liability Act 1984 (relevant to trespassers).

Visitors

A visitor is someone with actual or implied permission to enter premises. A visitor may sue for damages if injured or suffering damages whilst visiting the property. The duty on the occupier (section 2(2) of the 1957 Act) is to 'take such care as in all the circumstances of the case is reasonable to see that the visitor will be reasonably safe in using the premises for the purpose for which he is invited or permitted by the occupier to be there'. An occupier may be discharged of liability if the visitor was given a *sufficient* warning of a danger (section 2(4) a of the 1957 Act).

The Act allows the occupier to set limits on where the visitor is allowed to go or how long they are allowed to be there. This qualification was in the mind of Lord Justice Scrutton[12] when he said: *when you invite a person into your house to use the staircase, you do not invite him to slide down the bannisters, you invite him to use the staircase in the ordinary way in which it is used.*

Exceptions are made for children. Occupiers must be prepared for children to be less careful than adults. A warning notice, for example, would normally be good enough to alert adults to a potential danger but not to alert children. Thus in *Glasgow Corporation* v *Taylor,*[13] a seven-year-old child died after eating poisonous berries from a bush in the botanic gardens of Glasgow. The berries, which looked like cherries or blackcurrants, were not fenced off and no warning signs were present as to the danger they represented to health. The House of Lords found that the berries constituted an 'allurement' to the child and found Glasgow Corporation, which owned the park, liable.

Trespassers

A trespasser is someone entering land without an invitation of any sort and whose presence is unknown to the occupier or, if known, is objected to. The common law was traditionally hostile to trespassers but this was changed by the 1984 Act. A trespasser can sue an occupier for injury or personal property damage suffered on the property. The occupier has to take reasonable care for the safety of trespassers *if* the occupier is aware of a danger and the risk is one that they should have protected against. Essentially, this is a lower subjective duty than that owed to visitors, but it is a duty nonetheless.

The trespasser's acceptance of a risk, known as the defence of *volenti non fit injuria*, is covered in section 1(6) of the Act, which provides that 'no duty is owed … to any person in respect of risks willingly accepted as his by that person'. In *Ratcliffe* v *McConnell,*[14] the claimant, who had been drinking, dived into the shallow end of a swimming pool marked with warning signs, suffering serious injuries after hitting the bottom. The Court of Appeal held that because of the circumstances (jumping into an obviously shallow pool with warning signs during the winter), the claimant should have known of the risk and had, by acting, accepted the risk.

There is a real risk of trespassers bringing actions against the occupying contractor for the many hazards encountered on a building site. Health and safety reviews should be conducted regularly to manage this risk and there should be careful attention to maintain perimeter security on building sites.

4.4 Employer's vicarious liability

The meaning of vicarious liability is that one person can be liable for the tort of another person. It is most common in an employment situation where the employer is responsible for the actions of the employee. This was evident in the *Barnett* v *Chelsea & Kensington Hospital* case mentioned above. Vicarious liability requires a contract of employment and does not cover independent contractor cases. Where the tort is committed by an independent contractor then liability may be avoided by the primary party sued. The tort must be committed during the course of employment. This finding would have been a necessary component in the case of *Donoghue* v *Stevenson* to make Mr Stevenson responsible for the action or inaction of his contracted bottle washer.

Defining what is 'during the course of employment' has been the subject of a good deal of case law. Essentially, the employer is understandably reluctant to admit that the tortfeasor was 'at work' when the negligent act occurred. The lead case is *Century Insurance Co* v *Northern Ireland Road Transport*.[15] A tanker belonging to the Northern Ireland Road Transport company, and driven by one of their employees, was delivering petrol to a garage in Belfast. While the tanker was discharging petrol at the garage, the driver lit a cigarette and threw away the lighted match. The resulting explosion caused considerable damage. Century Insurance attempted to avoid paying for the damage, claiming that the driver was not acting 'during the course of employment' whilst smoking the cigarette; they argued that he was acting in breach of his instructions and was 'on a frolic of his own'. Century Insurance should not therefore have to pay out for the damage. The court disagreed and held that the driver was within the course of his employment even though he had been extremely negligent in his actions.

The liability here is strict – it is not dependent on the fault of the employer. The thinking behind the principle is that the employer has deeper pockets and, crucially, insurance, which is required by the Employers' Liability (Compulsory Insurance) Act 1969 to meet any claim brought.

4.5 Builder's liability

The effect of the *Donoghue* v *Stevenson* case was to introduce the concept of a duty of care being owed by individuals to all those people with whom they come into contact or should have in mind when acting. This liability applies equally to retailers and road users as it does to road cleaners and professionals such as accountants and surveyors.

If a building proves to be of defective construction then the owner can, in the absence of a contractual relationship, sue the builder in negligence. The owner will need to satisfy the tests of establishing loss resulting from the builder's breach of duty in failing to meet the standard of the reasonable builder in providing his services and/or materials. However, what happens when it is not the original owner, but a subsequent owner, who notices the problem and therefore suffers the loss? The builder may dispute that a duty is owed to someone they never knew about when the negligence occurred. This may allow them to escape liability to a subsequent purchaser.

Fortunately, as far as the subsequent purchaser is concerned, there are two saving legal considerations that apply *but only to a residential owner.* They are:

- the NHBC scheme;
- the Defective Premises Act 1972.

4.5.1 NHBC scheme

The NHBC scheme provides protection for new-build house owners. The scheme applies to all registered builders and developers. Other similar schemes are also available from rival organisations for the protection of new-build houses. The purchaser of a new home is given a ten-year guarantee that the house has been built in an efficient and workmanlike manner with proper materials. The builder/developer agrees to make good defects at their own expense during the first two years of the guarantee. The scheme will satisfy any claims for defects arising in the next eight years. Importantly, the unexpired term of the guarantee can be transferred to subsequent purchasers without requiring the consent of the developer/builder.

A similar level of protection is conferred to subsequent purchasers by the Defective Premises Act 1972.

4.5.2 Defective Premises Act 1972

Section 1(1) of the Defective Premises Act 1972 imposes a three-part duty on builders, subcontractors and designers involved in the construction of a dwelling (i.e. not commercial or industrial property). The Act refers to persons 'taking on work for or in connection with the provision of a dwelling'. The duty is to perform work: (1) in a workmanlike manner; (2) with proper materials so that the house will be; (3) fit for human habitation.

The duty is owed to the person to whose order the building is provided and to every person who acquires an interest in the dwelling up until a period six years from the date the building was completed. 'Dwelling' is not defined but includes both houses and flats. Examples of cases where dwellings were judged unfit for human habitation include:

- the construction of a dwelling on an inherently unstable hillside;[16]
- the omission of a damp-proof course;[17]
- having inadequate foundations.[18]

The clarification provided by the case laws to what constitutes 'fit for human habitation' is helpful given the divergent views that could exist here. This is a good example of where case law supplements the statutory law.

4.6 Limitation

The issue of how long someone should remain liable is clearly important in any consideration of the law. A potential defendant ought not to have the threat of litigation hanging over them indefinitely. There are statutory limitation periods within which a claimant must bring their claim or lose their remedy. The law is contained in the Limitation Act 1980.

The limitation period starts to run when the cause of action accrues. The cause of action accrues when the damage is sustained. The time limits are:

- Section 2 – six years from accrual of cause of action;
- Section 11(4) – three years from the date of any personal injury suffered.

The situation is more complicated where the damage goes undetected and remains latent because the question of when the damage has been sustained arises. This is an area in which the difference between tort and contract law can be appreciated. It ought to be relatively straightforward to ascertain when a contract case is actionable. The issue is when did the breach of the contract in question occur, and then the time limit is six years from that date. The key date for pursuing contract actions under building contracts is usually six or twelve years from the date of practical completion.

However, with tort claims, the breach is not actionable until the loss is known or should have been known. For example, should time start running on limitation from the date a crack in foundations happened or from when it was observable? The situation here was regularised by the Latent Damage Act 1986. This Act introduces a three-stage procedure:

1 The usual six-year cut-off date applies unless no damage was observable until after the six-year deadline has passed.
2 A three-year period commences from the date the claimant knew or ought to have known that a cause of action existed.
3 A long stop period for any discoverability applies 15 years after the original breach of duty.

Potential claimants need to pay particular attention to the second stage as suspecting a problem exists and then doing nothing for three years will see the right to sue become time-barred. This occurred in the case of *Renwick* v *Simon and Michael Brooke Architects*[19] where the defendants successfully applied to have an action brought against them for the inadequate design of a subterranean garden room struck out on the grounds of limitation having expired.

4.7 Remedies

4.7.1 Damages

The rationale of any award of damages is to put the claimant back in the position they would have been if the tort had not occurred. This is easier to quantify in some torts than it is in others. The approach taken in tort is slightly different to the contractual position on damages. Contractual damages can be said to have more of a subjective element than in tort – the second head of the *Hadley* v *Baxendale* test involves an examination of what was in the parties' reasonable contemplation at the time the contract was made. Tort damages are more objective in that the consideration is centred on what was foreseeable in the circumstances, whether or not the parties had the opportunity of turning their minds to consider the issue. The *Wagon Mound* case has demonstrated that if the loss occurring is too remote a consequence of the breach of duty then no loss can result.

In construction disputes, it is common for the parties to seek to dispute the issue of causation. The challenge for a would-be successful party is to demonstrate that, in each case, the breach complained of caused the resulting loss. This can often involve a complicated account of exactly how events led to the delays or disruption experienced. The issue of causation in construction disputes is covered again in Section 13.2.

4.7.2 Injunction

In the vast majority of tort cases, as in contract cases, the remedy will be damages. Occasionally – for example, in the case of a repeated trespass or nuisances – an alternative remedy is required. There may be a need to apply for an injunction – a court order requiring the defendant to do some act or refrain from doing some act. Injunctions are either *mandatory* – the defendant must do something – or *prohibitory* – the defendant must stop doing something. They can be either interlocutory (temporary) or final in nature. The injunction will not be granted if damages are an adequate remedy. This requires the claimant to establish some urgency and pressing need for the injunction.[20]

4.8 Defences to negligence

Three of the defences that can be put forward to an action brought in tort are briefly reviewed below. The focus of this chapter has been on the requirements of establishing liability for a tort. The defendant, in addition to being in a position where he can require that the case against him is proven, may seek to avail himself of a defence and/or counterclaim. If a court action is seen in terms of being a sword fight then the other combatant is able to both use his shield (defend) and strike with his own sword (counterclaim). Construction cases usually provide the ingredients for a contested claim and counterclaim. This can be seen in an example where a claim for extra money on a final account is denied and an entitlement to liquidated damages alleged. The employer is saying: 'Not only do I owe you nothing, you owe me something'. This can quite often come as a surprise to the claimant who, thinking he was on to a good thing in terms of making a claim, now finds he has more than he bargained for. In such a situation, the claimant may regret bringing the case in the first place.

4.8.1 Consent

The defence of consent has already been mentioned in the case of *Ratcliffe* v *McConnell* above.[21] A party who knowingly takes the risk of an event occurring cannot complain if they are subsequently injured as a result. There are limits to this defence as there is clearly a difference between recklessly taking a risk, as in the above case, and taking a risk within certain parameters, such as playing contact sport. The lead case of *Condon* v *Basi*[22] explored this point.

This case concerns a football match. The defendant made a late and reckless slide tackle upon the claimant, resulting in the latter sustaining a broken right leg. The defendant was sent off from the field of play. The claimant sued the defendant for negligence. The defendant argued that football was a contact sport and the claimant knew the risks when he took the field. The court held that although the claimant knew there was a risk, this was not the same thing as consenting to the possibility of being injured where another player falls below the acceptable standard of behaviour. The claimant's case was therefore successful and damages were awarded.

Students often find it bizarre to learn about a book by Kemp and Kemp: *Quantum of Damages*.[23] The book lists out in meticulous detail the value placed on injuries by judges in personal injury trials.

4.8.2 Contributory negligence

Contributory negligence can be thought of as a watered-down version of the consent defence. The defence is along the lines of: if you were partially the author of your own misfortune then you may be unable to claim a full indemnity for your damages. The wording of the Law Reform (Contributory Negligence) Act 1945 section 1(1) states:

where any person suffers damage as the result partly of his own fault and partly of the fault of any other person(s), a claim in respect of that damage shall not be defeated by reason of the fault of the person suffering the damage, but the damages recoverable in respect thereof shall be reduced to such an extent as the court thinks just and equitable having regard to the claimant's share in the responsibility for the damage.

Put simply, if you are partially at fault yourself, your claim will be discounted accordingly. In general, the question of whether the claimant was partly at fault involves consideration of whether the claimant acted as a reasonable person would to protect himself against the damage suffered; for example, by wearing a seat belt to limit injury in a car accident. If the court concludes that the claimant's own culpable act or omission caused or contributed to the damage in respect of which he claims compensation then it will go on to consider what would be a just and equitable reduction to make to his damages. The judge made a finding of contributory negligence in the case of *Lindenberg* v *Canning and Others*.[24] A builder on a project involving a basement conversion was handed a design prepared by the structural engineer on which the walls for demolition were marked (Figure 4.2). The builder acted upon the design and knocked down a wall which was actually structural and had been marked by the engineer by mistake, thereby causing considerable property damage. The building owner sued both the builder and the engineer.

The negligence of the engineer was evident in that marking the wrong wall for demolition represented a breach of his duty, resulting in loss. The negligence of the builder was found to have contributed to the damage and he was found partially liable (liability was split 75/25 between engineer/builder). It did not assist the builder to point out that 'he had done his job'. All professionals can potentially owe a duty to warn others of things within their specialist knowledge and to correct for them any oversight which should have reasonably come to their attention. This case stopped short of creating a general duty to warn, and clarification of the law in this area is still required. This last point is especially true given the platforms currently available for shared data and digital building models.

4.8.3 Duty to warn

The defence involved in the duty to warn is self-explanatory. It involves the defendant seeking to prevent a claim from a claimant where the latter had an opportunity to warn the defendant about the situation and did nothing to prevent the alleged negligence occurring. The allegation is that the defendant could be said to be under a duty to have prevented the situation had they taken positive action themselves. The above case of *Lindenberg* v *Canning and Others* could have generated satellite litigation between the defendants. The engineer could have pursued the builder to reclaim the damages paid out to

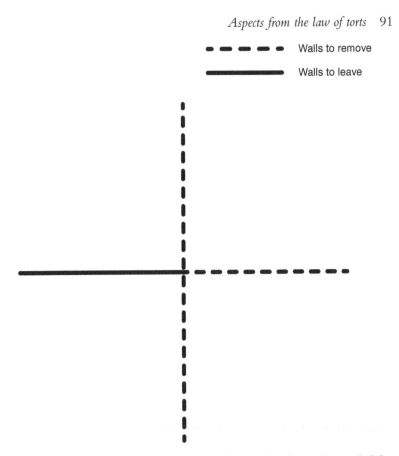

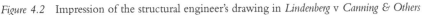

Figure 4.2 Impression of the structural engineer's drawing in *Lindenberg* v *Canning & Others*

the employer on the basis that the failure of the builder to raise his concerns about the drawing led to the loss.

The problem here is that the law is usually expressed in the negative – do not be negligent; do not trespass; do not cause a nuisance. Expressing law in terms of positive obligations – do warn and do act in good faith – is much harder for the law to regulate and set standards for compliant behaviour. This is an area where developments in the law may materialise to meet the requirements of the construction industry in an age of collaborative practice and three-dimensional design and construction. This may lead to more being expected of stakeholders in their legal relations with regard to each other.

4.9 Conclusion

Tort law is a very wide-ranging subject, and this chapter has only examined those torts with a direct relevance to construction law. Tort law is a difficult subject to grasp by those not familiar with the common law approach. The basic recipe for a tort action is similar to a contractual breach claim, involving, as it does, duty, breach and loss. The essential difference is that tort actions can occur between parties who did not envisage that they would necessarily come into contact with one another. In this scenario, the basic question is whether it is reasonably foreseeable that the injured party should receive compensation from the defendant given all the circumstances of the case. The rationale of tort is to make people responsible for their actions whether or not they were intended to have the consequences that they did.

4.10 Further reading

Wood, D., Chynoweth, P., Adshead, J. and Mason, J. (2011) *Law and the Built Environment,* Second Edition, London: Wiley-Blackwell, Chapter 3.

Wild, C. and Weinstein, S. (2013) *Smith and Keenan's English Law,* Seventeenth Edition, Harlow: Pearson, Part 3.

Cooke, J. (2012) *Law of Tort,* Twelfth Edition, Harlow: Pearson.

Notes

1   [1932] AC 562.
2   [1968] 2 WLR 422.
3   (1961) AC 388.
4   [1973] 1 QB 27.
5   [1964] AC 465.
6   [1987] APP.L.R.
7   [1957] 2 QB 334.
8   (1879) LR11 Ch D 852.
9   [1936] 2 KB 468.
10 [1868] All ER 1.
11 Cameron, G. (2005) Scots and English Nuisance … Much the Same Thing? *Edinburgh Law Review* 9(1): 98–121.
12 *The Calgarth* [1927] P 93.
13 [1922] 1 AC 44.
14 [1998] All ER (D) 654.
15 [1942] AC 509.
16 *Batty v Metropolitan Property Realisations* [1978] Q.B. 554.
17 *Andrews v Schooling* [1991] 1 W.L.R. 783.
18 *Bole v Huntsbuild Limited* [2009] EWHC 483 (TCC).
19 [2011] EWHC 874 (TCC).
20 See Section 2.12.2.
21 [1998] All ER(D) 654.
22 [1985] 2 All ER 453.
23 David and Sylvia Kemp's *Quantum of Damages in Personal Injury Claims,* known as 'Kemp & Kemp', was published originally in 1975 by Sweet and Maxwell, and updates are available by subscription.
24 [1992] 62 BLR 147.

5   Aspects of the law of property

5.1 Introduction to property law

This is the last chapter in this part and represents the remaining aspects of background law around which the construction law professional needs to form an appreciation. This topic involves a brief introduction to the origins of property law and identifies some key concepts. The relevance to construction law is mostly felt on the employer side in terms of:

- the acquisition and disposal of interests in land either side of the development project;
- the obtaining of planning permission permitting development;
- the rights existing over the use of land such as rights of way.

The contractor has rights akin to property interests during his occupancy of the site, which are most likely to be in the nature of a licence. The licence is created by the building contract, which establishes the contractor's right to occupy the site, the obligations incumbent on him during this occupation and how the licence can be brought to an end.

Property law is intertwined with the other background law covered in this part of the book. Leases and conveyances of land are forms of contract, albeit in the form of deeds. Obligations contained in leases such as restrictive covenants comprise express terms of contract. Property law is also concerned with responsibility for things emanating from land, as discussed in the property-based torts.[1] Some of the property torts concern obligations on the occupier whilst others (most notably the rule in *Rylands* v *Fletcher*)[2] concern obligations owed between landowners. Forming an appreciation of how land is owned and transferred is therefore of importance to the study of construction law. The interplay between commercial law and property law is also evident in the ability of organisations to buy and sell land and developments.

5.2 The development of property law

In former times, a person's wealth and standing were directly related to the amount of land owned. This was before such modern trappings as bank accounts and stock market shares existed. Wars were fought and marriages sought with the sole aim of increasing stakes in land. Then, and to a lesser extent now, landowners were the most powerful and richest people in society.

The historical importance of land stemmed not only from the status gained from owning the land. The potential to generate income from tenants on the land was of fundamental importance. The landowner could expect a rent or proportion of any money/goods/crops generated using the land, whether used for agriculture or some other form of industry.

Who, then, owns the land? In Britain, the monarch ultimately owns all land. Consider the position if someone dies intestate (without a will) and with no living traceable relatives. The land is said to be *bona vacantia* (meaning 'vacant goods') – the name given to ownerless property which by law passes to the Crown. In a sense therefore, ultimately, all land is merely borrowed from the monarch. This is demonstrated by such exercises as compulsory purchase orders.[3] The state may, upon paying reasonable compensation, force a landowner to sell a property to it. This may be seen in the context of the building of a ring road around a city where the householder is unlucky enough to be in the way.[4]

The rule that the monarch owns all land stretches back to the Norman Conquest of 1066. William of Normandy defeated King Harold at the Battle of Hastings. The new King William I was quick to make sure he and his issue enjoyed the fruits of their victory. The Normans realised that they could not simultaneously run the whole country. They therefore gave what are called 'estates' in land to the people who were loyal to them and in their service. In this manner, barons and knights were given castles and estates for as long as they remained loyal and in service.

The rules developed that an estate owner could:

- enjoy possession of the estate until the right was revoked;
- bequeath the estate to anyone of their choosing (invariably the eldest male heir, if available);
- grant an estate (or sub-estate) to another person.

For example, the local knight enjoying ownership of a fortified farm on the land of the local baron (himself enjoying royal patronage) could give an estate to the resident blacksmith for a smithy. In exchange, the blacksmith would shoe all the knight's horses at no cost.

The next step was for the incumbent invested with an estate, or sub-estate, to want to sell on the valuable right of possession to someone else. The right to sell estates between individuals therefore developed. These rules seem simple but soon resulted in unforeseen complexities. The problem became that there was a confusing mix of estates, sub-estates and inherited rights. Disputes between people claiming the right of ownership over pieces of land became commonplace. The absence of a central record was keenly felt. For this reason, a separate court was set up in London known as the Court of Chancery. The judges of the Chancery court had expert knowledge of land law matters and were able to resolve related disputes that arose. However, the lawmakers had to address another function in relation to property, namely the regulation and administration of all dealings in land.

5.3 The Law of Property Act 1925

By the early twentieth century, the state of confusion surrounding estates and ownership had reached a critical point. Problems included:

- non-uniform approaches to documentation;
- different types of estate being recognised;
- no copy documents being kept in a single place;
- local laws varying quite dramatically up and down the country;
- false and dubious claims being made by some people to rights over other people's land.

An additional problem to the general confusion was the increasing privatisation of what had been public land, known as common land. Large swathes of common land had been converted to private ownership during the late nineteenth and early twentieth centuries. For large numbers of people, this meant that their established way of life and attitude to the space they lived in had to change. The protection of people's interests became increasingly important to those who had been affected by these changes. In a separate development, this gave rise to the need for planning law.[5] The enclosure of land used commonly is currently an issue with nomadic peoples in the developing world. The suspicion of any attempt to make people register their interest in land can lead to problems for policies aimed at ensuring that rights are protected and entered onto the land registers concerned.

A key role for an Act of Parliament is to create new laws and/or to regulate existing situations. The positive effect of an Act of Parliament is writ large in the Law of Property Act 1925.

The Act introduced the following measures.

- A central registry was created where the details of every transaction relating to land in England and Wales are recorded and made available for the public.
- The types of estates available were simplified to two: an estate in fee simple absolute in possession; and a term of years absolute.
- No local variations were allowed to the rules.
- The rights of non-owners over other people's land had to be registered to be binding on third parties.

In one series of measures, the problems surrounding property law were largely swept away. However, the effect of the 1925 Act was not immediate. The changes to how property law was managed were gradual because it was only when a property was sold that it had to be registered for the first time. Property not changing hands for many years was therefore left unaffected by the new system. It is still possible to purchase a property today that has yet to be registered. Unregistered property requires an epitome of title to be created. This is in effect a bundle of documents pulled together to prove the current legal owner's title. The documents delve back into the history of the land and establish what is possible about the transactions appertaining to its ownership.

5.4 Extent of ownership

An estate owner's (simplified to owner's) land has six parts which can be remembered thanks to the following acronym: MAGPIE. The component parts of landownership are set out below. The overall effect of ownership is to give wide-ranging rights over the land owned. The Latin phrase *de profundus ad astra* is a colourful way to remember the extent of ownership. It means 'from the depths to the stars'.

5.5 Restrictions on ownership

The extent of ownership is wide-ranging. Table 5.1 establishes that the owner of a piece of land owns not only the ground itself but the airspace above, as well as the plant life and the minerals below the surface. However, it does not necessarily follow that the owner can do as they please with the land. With ownership comes some responsibility.

Historically, the owner was free to enjoy their land without any outside interference. This was a time before the needs and comfort of other people

Table 5.1   The extent of ownership

Minerals	Under the surface of the ground; ownership means that the owner has a right to mine any minerals (e.g. coal, tin, gold) located under the surface
Airspace	Above the surface; the limit of ownership is somewhere between the height of a tower crane, which constitutes a trespass, and the height where commercial aircraft fly, which does not[*]
Ground	The surface itself
Plants	Vegetation growing on the land's surface
Incorporeal hereditaments	Rights over other people's land and their rights over your land, including covenants and easements
Erections	Buildings and structures on the surface of the ground

[*] The Civil Aviation Act 1982 section 76(1) states that no action shall lie in respect of trespass or nuisance, by reason only of the flight of an aircraft over any property at a height which is reasonable.

had to be taken into consideration in the same way as they do today. The Victorians in particular were very keen on using land to suit their needs. It was during this era that monuments and towers were put on hilltops and stately homes and factories were constructed in the rush to cash in on the wealth brought by industrialisation and colonial trades.

5.5.1 *Planning law*

Towards the end of the Victorian period, the clamour for regulation of development became intense. Serious concerns had to be addressed about the living standards and air quality in industrial areas. This gave rise to the first laws in the now well-developed area of planning law. The decision on whether or not to allow a development was taken away from the parties with the deepest pockets or loudest voices and given to government. The government operates a devolved regime of local laws with decisions on planning being given back, in the first instance, to the communities affected by the proposal.

Permission to build, known as planning permission, is the *sine qua non*[6] of construction projects. The granting of permission over a piece of land can have a dramatic effect on its value. Planning consultants are a key appointment that a developer needs to ensure a proposed project becomes a reality. The following features of planning law have been selected for their relevance to construction.

- Planning permission is not required for all projects. Smaller projects such as small extensions, porches, garages and temporary structures do not usually need permission.
- Planning permission can be full, conditional or outline. Outline permission enables the developer to receive an indication about whether or not the local authority is amenable in general terms to the development.
- Planning permission is not restricted to new builds and is also necessary for changes of use. For example, permission would be needed to alter a derelict mill into housing units.
- Each local authority must maintain a local plan which can be consulted by the general public to gain an appreciation of where development is being encouraged and discouraged.
- Where, for example, a new housing estate is being constructed, a key planning consideration revolves around the adoption, post construction, of the roads and sewers. The relevant statutes are section 38 of the Highways Act 1980 and section 104 of the Water Industry Act 1991.
- The system of listed buildings, conservation areas and Areas of Outstanding Natural Beauty bring with them additional hurdles to overcome in order to be allowed to develop in these areas. The presumption is that building impinging on the amenity or pleasantness of the area will not be allowed.

5.5.2 Environmental law

Many aspects of environmental law are based around the notion of sustainable use of the Earth's resources and the prevention of pollution. In the area of construction law, environmental considerations are often translated into Building Standards[7] that require environmentally friendly materials and technologies to be used. In addition, such initiatives as the Code for Sustainable Homes[8] ensure that new developments achieve high performance in terms of energy efficiency in both construction and maintenance of property. The financial savings possible over the life of a building are considerable and act as an incentive towards early adoption of these measures.

The key principle in environmental law is *the polluter pays*.[9] This is an area where the European Union has been very active in promoting environmental performance, and a huge number of regulations and directives have been passed in this area. Government targets continue to challenge the construction industry to perform to higher standards. The laws of nuisance provide a means whereby individuals can seek action for any environmental impact on their land.

5.6 Types of ownership

Following the Law of Property Act 1925, the recognised ways of 'owning' property were restricted to two, known as freehold and leasehold ownership. A brief look at the implications and meaning of both is helpful for construction professionals.

Freehold property

The freehold interest in land (referred to above in legal terms as the fee simple absolute in possession) is the most complete interest that can be held in England and Wales. In practice, this means the outright ownership of land or property for an unlimited period held subject to the Crown. The property is bought and the new owners take the title 'absolutely' – there is no question of the ownership reverting to someone else at any time provided the beneficiaries for the property exist either under a person's will or through the law of intestacy. A freehold property will maintain its value when the property market remains even and will increase/decrease in value when prices rise/drop.

Leasehold property

Leasehold property is held in a situation where the owner (lessor) leases the property to a paying tenant (lessee) for a specified length of time. The lease will be a lengthy legal document based on property and contractual law that sets out the rights and obligations of both parties. A lease is for a fixed term.

Although the lease on any property will be a complex document, the basic principle is that it will give the lessee the right to occupy the property for the period specified in the lease. The lease will normally specify that the lessor will allow the lessee to have 'quiet enjoyment' of the property provided that the lessee in turn observes their duties under the lease. The principal duties are normally the requirement to pay the rent specified in the lease and to maintain the premises in a good condition. Commercial leases are sometimes known as maintenance and repair leases. The value of the dilapidations is a key area of practice for building surveyors to advise on. Commercial leases also frequently involve negotiations about rent increases and break clauses. The latter refers to the frequency of occasions on which the lease can be brought to an end without penalty during its term.

The rent payable under a long residential lease will normally be quite modest. When the lease comes to an end, the lessee must usually give up possession. The value of leasehold property is a diminishing asset. At the beginning of a long lease, there will be little difference in value between a freehold and leasehold property. As the lease gets shorter, the value begins to fall more rapidly. For example, a flat with a 50-year lease will only be worth about 70 per cent of what an identical flat with a 99-year lease would be worth.

Lease arrangements are quite common; for example:

- a university will lease a room to a student for residential accommodation for a year;
- a petrol company might lease a petrol station to a station manager for a renewable five-year term;
- Mr and Mrs Jones sell the leasehold interest in their home, an unexpired term of 999 years, to Mr and Mrs Smith. The Duke of Dursely owns the freehold reversionary interest (worth only a nominal sum).

In each case, the tenant is said to be the leasehold owner of the property. These three leases are clearly for different periods of time. The third example with a particularly long lease of 999 years is as valuable as a freehold interest in the property. The longer the term of a lease, the more valuable it is. The value is in the transfer value – if *x* wanted to sell the lease with *y* to a third party, what would the third party be willing to pay for it?

Some short leases have restrictions on whether they are capable of being sold or not and are known as full repairing leases. However, long leases (such as 99-year leases and above) are more akin to freehold ownership, and the leaseholder may resell and possibly create subleases.

For example, the Duke of Stinchcome is the freeholder (owner) of 1, Wotton Place in Barchester. The property is leased to Hardy Property Limited (the tenant) for 99 years; they sublet the top floor to Fiona (the subtenant) for five years; she rents out a room to student Janet (the sub-subtenant) on a weekly tenancy.

5.7 Rights over the land of others

The 'I' from the MAGPIE acronym stands for incorporeal hereditaments, which are intangible rights attached to property that are inheritable. The central notion is that land can be bound over to other users irrespective of the wishes of the landowner themselves.

Mostly, but not exclusively, a person's right to use someone else's land is as a result of either a long-standing arrangement or permission given by a previous owner that therefore binds the land. These rights usually appear on the title for the land kept at the Land Registry. These rights are also created when one parcel of land is divided into two. Certain rights can be kept back for the benefit of one part of the land over the other. Consider the following.

1 A marked footpath allows ramblers to walk through the garden.
2 Electricity cables and sewerage pipes linking the house to the mains services pass through a neighbour's garden.
3 A farmer grazes his cows on the local landowner's fields.

All of the above are capable of binding the land irrespective of the views of the current owners. The importance of rights over land is particularly important in construction law. Obstruction of property rights can have serious repercussions and lead to injunctions halting developments and even demolition orders in extreme cases.

The categories of rights include:

- easements;
- covenants.

5.7.1 Easements

An easement is the right to use the land of another or the right to restrict that other person from using his land in a particular way. Some examples follow:

- a right of way (vehicular or pedestrian);
- a right to light (not to block or obscure);
- a right to a view (not to block or obscure);
- a right to support a building;
- a right to take water.

Easements are created in a number of ways. These vary from an express grant or reservation to being established through prescription, meaning long enjoyment of a right without protest. In any multi-unit construction project, the developer must be careful to ensure that the necessary easements are granted and reserved in the deeds to each of the units to ensure that no subsequent issues will arise over ownership, enjoyment and maintenance of the properties. This involves ensuring that the titles of both the dominant (beneficiary of the easement) and servient (grantor of the rights) tenement are duly recorded with an easement.

Interference with an easement gives rise to an action for private nuisance. The party claiming interference with an easement must show that they are entitled to the benefit of the easement and that the interference is substantial in nature. The remedies for interference include injunctions, damages and the abatement of a nuisance order.

5.7.2   Covenants

A covenant is an agreement made by deed that binds the land. Covenants are usually expressed in terms of agreements to act (positive) or refrain from (restrictive) doing something. Covenants are often contained in leases between landlords and tenants. This is particularly the case where the tenancy is one of many in, for example, a block of flats where the interests of other tenants need to be considered. Examples include:

- positive covenant – to maintain the leased property in a fully repaired state;
- restrictive covenant – not to play music after 9 p.m.

Restrictive covenants are also fairly common between landowners, particularly where a plot of land has been divided for separate use. Covenants not to build or carry on certain activities on land are examples of this. Clearly, a developer must take careful note as, irrespective of the planning regime, a covenant can prevent development given that it 'runs with' the land and binds later purchasers of the land who did not sign the original deed. Covenants are also usually subject to registration requirements.

At common law, the remedy for breach of a restrictive covenant is damages. However, a person with the benefit of a restrictive covenant is likely to prefer to have the breach stopped by injunction, and this equitable relief is therefore sought in most cases. The damages are calculated on the basis of the sum that would be reached in negotiations between the parties, assuming that each made reasonable use of their respective bargaining positions without holding out for unreasonable amounts.[10] This case involved a consideration of the size of hotel that would have been built if the restrictive covenant had not been breached. The judge awarded damages based on the extra income received by virtue of being able to have built a larger hotel, which he assessed at £375,000.

5.8   Conclusion

Property law is a wide-ranging and important area of law that has only been very superficially dealt with in this chapter. The subjects chosen have sought to raise awareness of important issues and can form the basis for further study.

Those property rights affecting development are of particular interest to a study of construction law. This chapter has examined:

- the development and consolidation of property law;
- the extent of ownership in terms of physical property and the restrictions made by planning authorities;
- the types of ownership; and
- the creation and operation of rights to use the property of others.

5.9 Further reading

Wood, D., Chynoweth, P., Adshead, J. and Mason, J. (2011) *Law and the Built Environment*, Second Edition, London: Wiley-Blackwell, Chapters 4 and 5.

Wild, C. and Weinstein, S. (2013) *Smith and Keenan's English Law*, Seventeenth Edition, Harlow: Pearson, Part 4.

Notes

1   See Section 4.3.
2   See Section 4.3.3.
3   Compulsory Purchase Act 1965.
4   This idea is taken to its extreme in the book *The Hitchhiker's Guide to the Galaxy* by Douglas Adams where planet Earth is deemed to be in the way of a galactic highway. Unfortunately for the residents, the planet is blown up without any compensation being paid.
5   See Section 5.5.1.
6   Meaning 'without which, nothing'.
7   Part L of the Building Regulations.
8   The Code works by awarding new homes a rating from Level 1 to Level 6, based on their performance against nine sustainability criteria which are combined to assess the overall environmental impact.
9   It is mentioned in Principle 16 of the Rio Declaration on Environment and Development 1992.
10 *Amec Developments Limited* v *Jury Hotel Management* (UK) Limited [2000] EWHC Ch 454.

Part 2

The processes of construction law

6 Parameters of construction industry practice

6.1 Introduction

Construction law is the term used to represent a mixture of different laws examined in Part 1. The principal ingredient is contract law with smaller measures of commercial, tort and property law. Construction law developed from these principles and established its own norms and processes to service the needs of the AEC industry. These are manifested in the devices and legal instruments that are prevalent in the industry but not elsewhere. The following are examples, many of which are covered in this book.

- *Contract law* – standard forms of building contract such as JCT, NEC and FIDIC,[1] letters of intent, collateral warranties and development agreements.
- *Commercial law* – performance bonds and parent company guarantees.
- *Tort law* – professional indemnity insurance and oversailing licences.
- *Property law* – options to purchase completed projects, asset-backed trusts and concessions to operate from built premises.

The processes of construction law are also supported by statutes and case law created specifically with construction practice in mind. More has to be known about the landscape of the AEC industry before construction law documents can be considered in any depth. The purpose of this chapter is to supply the additional detail required to fully appreciate the parameters in which the law operates.

This landscape of construction law departs from the background law relating to regular commerce in some subtle and unusual ways. Chapter 2 examined how contracts seek to ensure that the parties' rights and remedies are dealt with inside the contract rather than having to rely on any background law such as implied terms or tortious relationships. Contracts used in the AEC industry are no different in this but are characterised by their length and comprehensive nature. Certainty retains its position as the most highly prized contractual commodity, and every attempt is made by contract writers (known as draftsmen) to provide it.

This chapter examines the following influences on construction contracts:

- the design function;
- the influence of multiple parties;
- the role of the overseer;
- the proliferation of standard form contracts; and
- construction law statutes.

6.2 The design function

A major distinction between construction law and contract law generally is the separation of the design function from the physical act of building. This feature has developed through custom and the allocation of roles and responsibilities in construction law. This allocation of roles and responsibilities can be termed 'procurement', using the simple sense of the word. To understand construction is to appreciate that, originally, someone is responsible for designing the building and someone else for the construction itself. Only later were these twin responsibilities carried out by a single firm in the form of procurement known as design and build.

Where design and construction are separate, the issues involved include the standard to which the design must perform. The design obligation is in part defined in common with standard contract law owing to the application of the Sale of Goods and Services Act 1982. This Act provides[2] for the implication, in certain circumstances, of a fitness for purpose obligation into contracts for the sale of goods. However, contracts for the supply of services are treated differently. The standard of care for the supply of services is the reasonable skill and care standard discussed in Chapter 3. Construction activity, whether represented by design and/or build, is treated as the supply of services, unless agreed otherwise.

Fitness for purpose is not therefore the usual test in respect of construction operations. The key question is: what must the claimant establish to prove that the problem experienced was the defendant's fault? In order to establish a fitness for purpose warranty, the claimant need only show that the building does not perform as intended. It is more onerous on the claimant to establish that the defendant fell below the standard of the reasonably competent fellow professional. The defendant can adduce evidence to show that reasonable care was taken notwithstanding any damage accruing. The higher test of reasonable skill and care is the default position for the standard to which services are to be supplied. Construction projects are not viewed as simple goods and services contracts as they are more complex and subject to the test of: what would the reasonably competent fellow professional have done?

Thus a designer's liability is usually dependent upon a finding of 'fault' in the sense of professional negligence. The test of professional negligence was well stated by Lord Justice Denning in *Greaves & Co (Contractors) Ltd v Baynham Meikle & Partners*:[3]

> *The law does not usually imply a warranty that [the professional man] will achieve the desired result, but only a term that he will use reasonable skill and care. The*

surgeon does not warrant that he will cure the patient. Nor does the solicitor warrant that he will win the case.

The case involved a design and construct package arrangement where the building contractor took on the project of building a new factory, warehouse and offices. The warehouse was to be used for the storage of barrels of oil, and it was a requirement that the floor had to take the weight of forklift trucks carrying the barrels of oil. The floors failed due to the vibrations caused by the use of the trucks and the builder was pursued in respect of the failure. The decision at first instance suggested that the fitness for purpose warranty applied to the project, but this was later corrected by Lord Justice Denning in his speech (quoted above). In any event, the failure of the floor was found to be in breach of the duty to take reasonable care as well as its unfitness for purpose.

The point of the distinction between the two hurdles is that with the higher hurdle, a party can make mistakes and not be liable. This is demonstrated in the state-of-the-art defence, as shown in the case of *199 Knightsbridge Development Ltd v WSP UK Ltd*.[4] The case held that while an engineer breached its duty of care when designing a new high-rise building's water system, the claimant property developer's evidence did not demonstrate that the breach caused it to suffer a loss. The court had to consider whether an engineer was liable for not addressing a risk that was perhaps obvious with hindsight in circumstances where there was no body of knowledge about that issue and no common best practice to address it. The engineer was found not liable.

Standard form contracts in the UK typically provide for reasonable skill and care and fitness for purpose depending on the subject matter. Reasonable skill and care is much more common. The JCT design and build contract states: 'The contractor shall in respect of any inadequacy in such design have the like liability to the employer as would an architect or, as the case may be, other appropriate professional designer'.[5]

Using reasonable skill and care is not adequate when the subject matter of the build is more specialist and regulated. Large engineering projects, particularly on the international stage, are procured using FIDIC contracts, which typically use fitness for purpose. FIDIC contracts state: 'When completed, the Works shall be fit for the purposes for which the Works are intended as defined in the Contract'.[6]

Why should the majority of standard contracts in the UK have embraced the reasonable skill and care approach and not the client-friendly fitness for purpose warranty? One answer involves the issue and cost of insurance. Insurance companies require that their clients seek to limit their potential liability, and professional indemnity insurance is not readily available for this higher standard of care. Another explanation as to why the reasonable skill and care test is used requires looking at the background of construction projects. Construction projects tend to be complex, involving a multitude of parties and designers who are all brought together for these largely one-off projects. The convention has arisen that parties must be able to limit their potential liability and have scope to defend themselves.

Those contracts requiring a specialist designer where the role is separate from construction require a form of professional service agreement. These typically contain provisions for the timing of services, payment, co-ordination, limits on liability and intellectual property rights. Statutory appointment such as Principal Designer under the Construction (Design and Management) Regulations 2015 is also likely to feature in the professional appointment of a designer.

6.3 Multiple parties

The ramifications on construction law of the multitude of parties with interests in a building project need to be considered further. It is quite easy to list a dozen people and organisations involved in even a modest project. Take, for example, the construction of a new shopping unit on a small retail park. The building contract between the employer and contractor is only one aspect in a chain of agreements emanating from it. There can be contracts for funding, lending, professional appointments and security arrangements for the employer. In addition, the needs and requirements of subsequent owners of the build and their rights vis-à-vis the designers and constructors need to be considered. For his part, the contractor will have subcontract appointments, supply agreements, bonds and insurances to arrange.

The contractual groupings required are shown in Table 6.1. The contracts put in place need to complement each other and form a network capable of protecting individual party interests and conferring as much certainty as possible to all those involved. This is no small challenge and involves additional consideration of bargaining power and each participant's place in the supply chain.

A common feature of the traditional method of executing building works is for the main contractor to appoint independent subcontractors to carry out specialist tasks or designs or to supply materials. The subcontractor is usually liable to the main contractor in the event of any default.[7]

The contractual links set out in Table 6.1 have to be considered alongside the operation links that also apply. The position quickly becomes complicated if different members of contractual teams deal with third parties outside of their direct arrangements. Keeping contractual and operational links separate can be a major headache for those concerned.

6.4 The role of the overseer

Another departure from standard contract law arrangements is the need for the overseer on construction projects. In JCT parlance this role is known as the architect/contract administrator (A/CA) whilst the NEC uses the term 'project

Table 6.1   Contractual teams of the parties to a construction project

Employer's contractual arrangements	Contractor's contractual arrangements
Architect/contract administrator	Site manager
Subsequent purchaser	Subcontractor
Engineers	Supplier
Quantity surveyor	Bondsman
Project manager	Insurer
Funder	Specialist designer

manager'. Regardless of which contract is used, there is the requirement of an overseer to plan and run the project once construction work has begun.

The need for a person to be in control of the development is self-evident given the complexity and large amounts of time and money potentially involved. Investors in a company would not expect to see the operation run without a managing director. Some construction projects involve more money being spent than a multinational company might turn over in years. This necessary function is usually delegated from the employer to a third party. This has long been the province of the architect – there to lead the employer from inception of the project to completion. Who is better placed to be the overseer of the site than the person whose design is being built? There is certainly a compelling case for the architect to be the overseer and, if willing, they are well placed and can achieve the best results.

The temptation in recent times has been to view the role as being better handled by a project manager. This may or may not be an architect. This is a major point of divergence for the two most popular standard forms in the UK construction industry, which is investigated in the following sections. Essentially, the JCT stands by the dual A/CA role whilst the NEC recognises that a person with a construction management background is more likely to run a better project and give the role to a project manager. This does not exclude architects from the role, but the change in emphasis is apparent and understandable given the pedigree of the two forms. In the FIDIC contract, the key role is that of the engineer. The FIDIC approach is more akin to the JCT view of what the person specification should entail.

One image which is helpful to use in this regard stems from the speech by Cassius in Shakespeare's play *Julius Caesar*. In it, Caesar is decried as being like a colossus bestriding the earth whilst other mere mortals rush between his legs. The allegory follows that architects have been regarded in the same way by others in the construction industry in a rather churlish display of envy. To continue the metaphor further, one can see that the pedestal on which the statute of the colossus stood has now been chipped away by other professionals proving themselves at least the equal of architects in performing specific roles. Hence, for example, the quantity surveyor takes the accounting side of the overseer role away from the architect; the architectural technologist takes away the detailed parts of the design role; and the project manager, the co-ordination of the trades and the flow of information roles.

This silo approach to the relative merits and demerits of each professional in the industry detracts from the collaboration needed to make a project a success. Architects mostly appreciate the specialist help on which they are able to rely whilst retaining their role as lead designer and overseer.

The perspective that counts more than most in terms of the role of the overseer is that of the contractor. The contractor is the entity that must 'buy in' to the role and be prepared to act in accordance with the overseer's instructions. The safeguard for the contractor is the knowledge that when performing the overseer role, the professional concerned must act as an independent certifier between the parties. In other words, in most instances, the overseer must have regard to the interests of both parties in reaching any decision and act accordingly. There is a conundrum at the heart of this role as the overseer is evidently likely to arrive at an interpretation of the contractual issue that favours their client. The saying 'he who pays the piper calls the tune' is relevant here. Nevertheless, from the contractor's point of view, relying on the professionalism of the overseer is better than nothing.

The key to knowing when the overseer must exercise their professionalism is, under the JCT regime, in the granting of certificates. The requirement to certify – whether an interim payment certificate or a practical completion certificate – signifies that the A/CA must act with impartiality in deciding how much to pay or when an event has occurred.

Judges have recognised that an architect is predisposed in some measure to favour a client-friendly version of events. It has been established, in one of the few reported cases involving the NEC, that the project manager is in a similar position to act fairly between the parties when certifying. This occurred in the case of *Costain Ltd* v *Betchel Ltd*[8] where the obligation to act fairly was implied and the client's direction to disallow some legitimate costs of the contractor should not have been followed. The duty to the contractor stops short of a formal duty of care as found in the law of tort. No contractor has ever successfully sued an employer for under-certifying. However, employers have sued overseers for over-certifying, as in the case of *Sutcliffe* v *Thackrah*[9] where the employer, on the insolvency of the contractor, was left exposed to the overpaid costs. The employer was able to recover the amount it had overpaid the contractor from the architect. The latter was found not to have exercised reasonable skill and care in satisfying himself that the contractor's claims were legitimately made. The contractor had inflated the amount claimed, no doubt reflecting his financial problems.

Many contractors recognise the limitations in the system and the risk to themselves of having a partisan overseer. The remedy for a disgruntled contractor is to seek to overturn the certificate, which they are able to do in dispute resolution procedures such as adjudication. The professionalism of the overseer and the realisation that their action may be subject to scrutiny act as encouragements to certify correctly.

6.5 The proliferation of standard form contracts

It is a widespread practice in the AEC industry to use standard forms of contract. Contracts involving the sale of land and buildings are notable examples. A question frequently posed in the context of standard form contracts is: why reinvent the wheel? This rhetorical question is frequently used to extol the benefits of using contracts already in existence rather than starting with a blank piece of paper every time a contract is required. In the construction and engineering context, there are a plethora of standard forms which reflect the divergent interests of the parties involved. In 1964, the Banwell Committee recommended the use of one single standard form of contract in the construction industry.[10] This call for a single standard form has been repeated on several occasions through the decades, most notably by Sir Michael Latham in 1994.[11] These calls for industry improvement are discussed further in Part 5.

The existing situation is that there are a variety of forms of contract used both domestically and internationally. Although not essential, it is more convenient and safer to use a standard form where the project is complex and has a high value. Using a standard form gives the stakeholders guidance as to how they should proceed with their likely obligations under the contract. Repeated use in practice means that its users become familiar in the form's application and how it distributes risk between the contracting parties. Moreover, it is open to the parties to agree among themselves to modify the forms where appropriate. Simply because a standard form of contract is used does not mean that the ordinary rules of contract law can be avoided. All of the standard forms of contract discussed here and used in the UK have been amended to ensure compliance with those statutes containing mandatory provisions.

The popularity of the various forms of contract is analysed regularly, and statistics based on varying sample sizes are published.[12] The JCT forms have remained the most popular forms over the course of the surveys. However, the later surveys have been characterised by the rise of the NEC which has mounted a serious challenge to the dominance of the JCT. Other standard form contracts, which represent other views on the possible approaches to contract writing and strategy, are the Project Partnering Contract (PPC2000) and the FIDIC Red Book.

The different forms of construction contract can appear to the first-time user to be overwhelmingly complex. In fact, construction contracts are strikingly similar when paired down to the bare essentials they seek to address. The loyalty and preference shown by clients and contractors to the form of contract they know and trust can be striking. Parties are reluctant to try alternative types of contract for fear of being outside their comfort zone and in unknown contractual territory. This factor is a major obstacle to the establishment of new contracts in the market. A new contract usually requires public sector assistance to make a name for itself. The different standard forms and their strategies are discussed in Chapter 9.

The approach taken by contract draftsmen and their parent bodies does not represent the whole story when deciding what to include in a contract. The next section addresses another major factor in construction contracts, namely the need to comply with those statutes that specifically target construction industry practice.

6.6 Construction law statutes

6.6.1 Introduction

Part 1 covered the background law on which a study of construction law is based. The lawmaking function of the executive and legislature was discussed in relation to Acts of Parliament or statutes. Freedom to contract is a fundamental principle in English law and something that lawmakers are reluctant to depart from. However, if the case is made for intervention then Parliament can turn its attention to a specific issue and legislate accordingly. Statute law has played its part in shaping the AEC industry's laws, and examples of these laws are considered in this section.

The AEC industry has been impacted by two statutes in particular in the last 20 years. The importance of these statutes is writ large on the standard forms. The mandatory nature of the statutory regimes imposed meant that the contract writers had to make sure that their contracts were compliant with the Act, failing which the non-compliant clause would be replaced with default scheme provisions. Such a result was clearly unpalatable for the contract writers who reissued their complaint contracts in 1998 and again in 2011. More recent amendments to standard forms are similarly motivated in part by ensuring the contracts are compliant with newly introduced laws.[13]

6.6.2 Housing Grants, Construction and Regeneration Act 1996 (HGCRA)

This is the first of two statutes of great importance to the construction industry and worthy of special mention. Statutory intervention into the practices of any industry is a risky undertaking for the legislature and something not undertaken lightly. Usually, under English law, the parties are left to their own freedom to contract. Provided the contract is enforceable and complies with the Unfair Contract Terms Act 1977, the contract will stand. The Unfair Contract Terms Act imposes limits on the extent to which liability for breach of contract, negligence or other breaches of duty can be avoided by means of contractual provisions such as exclusion clauses. HGCRA signalled a departure from this principle of freedom to contract. In the words of one commentator:

> the Act constitutes a remarkable (and possibly unique) intervention … whereby the ordinary freedom of contract between commercial parties to regulate their relationships has been overridden. Apart from consumer legislation, there had never before been legislation (other than in a state of emergency) which intervened to regulate the

freedom of contract in a sector of the economy that is generally buoyant and which should be capable of looking after itself.[14]

The purpose of the intervention was simple – to improve cash flow, memorably described in 1974 by Lord Denning as the *lifeblood of industry*.[15]

The benefits of the new system were primarily targeted at the main contractors and specialist contractors who quite often found themselves in the dark in relation to reasons why payments were not being released to them for work performed.

The results have been viewed as a success by many in the industry notwithstanding the drastic nature of the changes brought. Construction industry practice was turned on its head by 13 clauses buried inside this statute, dealing mainly with unconnected provisions. The clauses impacted greatly on payment provisions and the resolution of construction contract disputes. The effect of the terms set out below is to empower those individuals and firms within the construction industry who hitherto found themselves at the mercy of their contractual partners when a dispute situation arose. Armed with the provisions of the HGCRA, a disgruntled party may now know why they have not been paid and, if they do not approve of the reasons given, have the power to do something about it through the dispute resolution procedure of adjudication. Adjudication has replaced litigation and arbitration as the industry's preferred form of dispute resolution.[16] At a stroke, HGCRA ensured compliant procedures were introduced across the industry.

Background

This Act was the result of a number of recommendations made in the Latham report and subsequent Department of the Environment consultation papers. It is also consistent with the desire of the Woolf report to promote alternative dispute resolution (ADR) as a means of resolving disputes. It received the Royal Assent in July 1996 and applies to all construction contracts entered into after 1 May 1998. Several small amendments were made to the Act following a long consultation with construction industry stakeholders.

The Act can be divided into three main areas:

- definitions and scope;
- adjudication;
- payment.

Definitions and scope

The Act applies to 'construction contracts' which are defined as agreements with any person for the carrying out of construction operations. It extends to subcontracts and management contracts. It also covers agreements for the provision of architectural design or surveying work and advice on building, engineering, interior work and exterior decoration or landscaping work in relation to construction operations. The contract in question must be in writing or evidenced in writing to receive the protection of the Act. There have been many instances where the courts have been asked to decide whether or not a contract is in, or is evidenced in, writing sufficient for a reference to adjudication to be valid.

Exceptions

The following are not covered by the Act but may, by agreement between the parties, import the terms of the Act:

- contracts with residential occupiers;
- contracts relating to drilling for or the extraction of oil or natural gas;
- contracts relating to the extraction of minerals and certain nuclear work;
- contracts relating to the manufacture or delivery of building or engineering components, materials or plant (if there is an installation of these items, that aspect will be covered); and
- purely artistic work.

Adjudication

The discussion around adjudication appears in Chapter 20.

Payment

The payment sections are contained in sections 109–112 of the Act. A construction contract must deal with payment in a specific way to prevent the provisions of the Act applying by default.

1 If a construction contract is at least of 45 days' duration, any party to the contract is entitled to payments by instalments.
2 The parties to the contract are free to agree the amounts of the payments and the interim periods when payable. If they do not, the provisions of the Scheme for Construction Contracts will apply.
3 An adequate mechanism must be in place to determine what payments are due and the final date for payment must be specified.
4 Every construction contract must contain provisions relating to the serving of notices by the payer and payee in respect of making payments and, where appropriate, notice of an intention to withhold payment. Certain additional protection is given to the payee in the absence of an effective notice of intention to withhold payment.
5 In the event of non-payment by the payer or if there is no effective notice of withholding payment, the payee has the right to suspend performance of

the contract. This may be effected by a notice to suspend the performance of the contract. This right ceases when the amount due is paid in full.

'Pay when paid' clauses

A further initiative introduced by the Act is the outlawing of conditional payment clauses. Provisions making any payment under a construction contract conditional upon the payer receiving payment from a third party are ineffective. The intention here was to outlaw the sometimes nefarious practice of main contractors denying subcontractors money properly owing under their subcontracts because they themselves had not been paid by the employer. The performance of the subcontract was completely removed from any main contract arrangement and these clauses have been consigned to history. The only exceptions arise in situations of insolvency.

6.6.3 Local Democracy, Economic Development and Construction Act 2009 (LDEDCA)

Wide-ranging consultations were held following the introduction of the HGCRA, resulting in some relatively minor changes being made through this follow-up Act of Parliament. LDEDCA received Royal Assent on 12 November 2009. Part 8 deals with construction contracts. The new Act became law in England and Wales from October 2011.

Expressions of concern about the shortcomings of HGCRA were raised by representative bodies of the construction industry to the Chancellor and other ministers. This led to a review of the operation of the adjudication and payment provisions. These were the first steps along the road that led to the introduction of the new Act. The journey was to prove a slow one, involving seemingly endless rounds of industry lobbying, government-backed reports, consultation, analysis, post-consultation events, a change from secondary to primary legislation, and impact assessments. The stakeholders in the construction industry wanted a statue that provided improved cash flow between parties, improved the operation of construction contracts and promoted adjudication as a form of dispute resolution.

The stakeholders consulted recorded that they were generally pleased with how HGCRA was working. Amongst the minor changes made were:

- the application of adjudication to oral or partly oral contracts;
- the extension of the ban on pay when paid clauses to cover any conditional payment terms;
- a ban on any attempt to predetermine responsibility for adjudicator's costs;
- the introduction of a payee's notice in the event that the payer does not provide a payment notice;
- a change in terminology from withholding notice to notice to pay less; and

- an unpaid party may suspend part only of the works if a separable part is affected by the non-payment.

The challenge for LDEDCA was to deliver its improvements in the face of a vastly different industry to the one originally contemplated. The recession spanning the end of the first decade of the 21st century and the start of the next caused suffering in the construction industry. The LDEDCA sped up the flow of payments and widened access to a quick method of dispute resolution. However, LDEDCA arrived too late for many, given the depth and breadth of the downturn experienced.

These two statutes have done a great deal to improve practices across the construction industry. The government has recognised that mandating contract clauses is not going to reform practices on its own, and current efforts at promoting good practice have now taken over. In an industry as diverse and complex as construction, a complementary range of measures is definitely required.

6.6.4 *Contracts (Rights of Third Parties) Act 1999*

This Act of Parliament brings together a number of outstanding threads emerging from contract and tort law. The background to this Act was that it started life as a European directive seeking to harmonise third party rights across the European Union. The English doctrine of privity of contract was at odds with the position elsewhere on the continent whereby third parties have wider access to rights over contracts for their benefit. The Act radically affects the law in relation to privity of contract. The Act confers on a third party 'a right to enforce a term of the contract' where either the contract contains an express term to that effect or where the contract purports to confer a benefit on that third party. In both cases, the third party must be expressly identified in the contract by name, class or description. General references to third parties are not capable of being enforced. The right to enforce a term of the contract means the right to all of the remedies that would have been available to a third party through the courts if it had been a party to the contract.

The third party's rights will be subject to all the defences and set-offs that would have been available to the contracting party had the third party been part of the original contract. The effects of this legislation have been far-reaching. The Act could have obviated the need for separate collateral warranties in favour of funders, purchasers and tenants. The Act has established itself as an alternative to the collateral warranty procedures in the industry (see Section 12.3). The non-mandatory nature of this Act is in sharp contrast to the two construction Acts referred to above. Parties can opt out of this Act and it is a common clause in building contracts and professional appointments to seek to exclude its operation.

This rather negative approach has now been replaced with a more balanced use of the benefits the Act has to offer. JCT standard form contracts contain a third party schedule which accommodates the Act and recognises the problems created by the need to secure the execution of collateral warranties up and down the supply chain. Collateral warranties remain the preferred option for many funders largely because of the provision of 'step-in' rights whereby the funder can mitigate against the unwanted effects of insolvency on project delivery. It is apparent that third party schedule arrangements may be necessary for BIM (Building Information Modelling) implementation in an effort to keep abreast of the users of different categories of information. The legal implications of BIM are discussed further in Section 21.7.

The three statutes here have had very different receptions in the construction industry. The HGCRA has left a legacy of change which has transformed the legal landscape. The LDEDCA made some minor changes which required careful account to be taken, but this was largely following up on the earlier Act. The Contracts (Rights of Third Parties) Act has been less successful, and the effort seeking to normalise contract law provisions across Europe has had mixed results. Essentially, the reception given to this Act shows the differences in approach between the common law and civil law systems. Any attempt to graft an essentially alien law on to a domestic application often does not have the desired results. Privity of contract is well entrenched in English law and the industry has been resistant, on the whole, to this attempt to bypass its effects.

6.6.5 Construction (Design and Management)Regulations 2015

Legislators are on more familiar ground when setting standards in relation to safety measures required for construction projects. The CDM Regulations are a series of statutory instruments aimed at improving the construction industry's health and safety record. The Regulations require the appointment of a Principal Designer whose duties are to:

- advise and assist the client with their health and safety duties;
- notify details of projects to the Health and Safety Executive;
- co-ordinate health and safety aspects of design work and co-operate with others involved in the project; and
- prepare and maintain the health and safety file.

The CDM Regulations are enforced by the Health and Safety Executive which has wide-ranging powers to stop non-compliant projects and even prosecute those involved.

6.7 Conclusion

This chapter has moved the discussion on from the background law to introduce the reader to specific detail of construction law. The special features of the construction industry have been considered including the design role, the role of the architect/contract administrator, the prevalence of standard form contracts and the impact of multiple parties. Statutes impacting on construction law have been introduced and discussed in order to familiarise the reader with the concepts in operation. The next chapter seeks to provide more key concepts and processes under the heading of procurement. Procurement choices are regarded as the start of the construction journey, which later chapters then continue to examine.

6.8 Further reading

Ndekugri, I. and Ryecroft, M. (2009) *The JCT 05 Building Contract: Law and Administration*, Second Edition, Oxford: Butterworth Heinemann, Chapter 2.

Uff, J. (2013) *Construction Law*, Eleventh Edition, London: Sweet & Maxwell, Chapter 1.

Notes

1   Discussed in Chapter 9.
2   Section 4 of the Act.
3   [1975] 1 WLR 1095.
4   [2014] EWHC 43 (TCC).
5   JCT Design and Build Contract, 2011 edition (DB11), clause 2.17.1.
6   FIDIC Silver, Yellow and Gold Books, clause 4.1.
7   Subcontracting is discussed in more detail in Section 11.6.
8   [2005] TCLR 6.
9   [1974] AC 727.
10 Banwell, H. (1964) *The Placing and Management of Contracts for Building and Civil Engineering Work*, London: HMSO.
11 Latham, M. (1994) *Constructing the Team: Final Report of the Government/Industry/Review of Procurement and Contractual Arrangements in the UK Construction Industry*, London: Department of the Environment.
12 NBS National Construction Contracts and Law Survey 2013.
13 The latest CDM Regulations (2015) will need addressing in standard forms.
14 HHJ Humphrey Lloyd QC in *Outwing Construction Limited v H Randell & Son Ltd* [1999] 15 Const LJ vol 3.
15 *Danways Ltd v FG Minter* [1971] 2 All ER 1389.
16 See Part 4.

7 Procurement

The remainder of Part 2 examines the construction processes necessary to undertake a building project. A distinction is drawn between legal provisions – essentially rules governing conduct and entitlement – and process – involving making choices and selecting a strategy from options available for the delivery of a construction project.

The choices featured in these chapters highlight the procurement approach, the selection of the best contractor and the choice of a standard form contract to deliver the project.

7.1 Introduction to procurement

Procurement can be succinctly defined as 'buying with care'. Procurement concerns the decisions made by the employer before entering into a building contract about how the project will be commissioned and delivered. A substantial number of clients pay too little regard to the procurement stages and concern themselves with deadlines and progression towards these dates with as much haste as possible. A key observation here is that if more time was spent by the stakeholders in ensuring that a correct and timely procurement strategy had been followed then fewer problems would be encountered and fewer disputes would doubtless arise. Failing to prepare is preparing to fail.

Selecting the most appropriate procurement strategy for a client is very much dependent on tailoring the choice to suit individual characteristics and needs. Three important initial considerations to take into account are: the complexity of the project, the experience the client has at its disposal and the commercial drivers behind the project.

Complexity

The previous chapter introduced the multiple contributors encountered on a construction project. Projects range from the small and straightforward contracts undertaken by local tradesmen for residential occupiers to massive undertakings involving complex issues and funding arrangements. The legal arrangements vary according to project size, and whilst informal arrangements are found on smaller projects, they are not suitable for larger projects. It is a simple rule of thumb that the bigger the project, the more pronounced the need for sophisticated and interrelated appointments and contracts. Larger projects also involve more interested parties, sometimes with competing interests. In this sense, procurement can be thought of as the network of arrangements needed to define responsibilities, protect interests and allocate the risks arising.

Client's experience

This is also true of the client's experience, which ranges from the perspective of the first-time one-off client to that of the repeat client with construction as a core business. The procurement choices will need to match the client's profile in terms of experience and planned involvement in the project. Not all clients with experience will necessarily want to perform the management function, and they may seek to delegate on any given project. Less experienced clients will be attracted by procurement approaches transferring risk and responsibility onto the supply side. Client involvement is closely linked to the issue of experience. How involved does the client wish to be in the project? A 'hands-off' client might prefer the turnkey approach whereby opening the front door of the project on completion is their first proper involvement. Other clients may wish to be involved in every minute decision before, during and after the project.

Commercial drivers

The business case for the construction of a new building will depend on the funding arrangements and envisaged use of the finished product. All clients want the best quality building they can have in the quickest time possible and for the best price. In reality, compromises need to be made between these three criteria in order to arrive at a deliverable procurement strategy. Cheapest and best quality are not usually found in the same product, and building too quickly also risks both the quality of the build and the budget. A discussion around these issues is the usual starting point for deciding on a procurement strategy. For example, the procurement approach taken by an investor looking to maximise its return in budget hotels may be appreciably different from the prospective owner/occupier of a modern-day stately home.

The desire for design changes

It is preferable to develop the design of a project at the early stages to the point where it will not greatly alter further. However, factors such as rapidly changing technology and shifting priorities of the client may alter requirements during the design process. Changes in the scope of the project often result in an increase in cost. These changes are manageable during the detailed design stage, but if they are introduced during construction then they may have a disproportionately high effect on the project in terms of cost, delay and disruption. The design should therefore go through a series of 'freezes' as it develops, and clients should be encouraged to set a final design freeze date after which no significant changes will be allowed.

The role of government

Procurement is an area where the government has been extremely active in promoting best practice. The government is a major client of the construction industry and can clearly dictate policy in relation to public projects. The government's ability to influence what happens in the private sector is more difficult. New initiatives are often slow to catch on in the private sector with the blame often being put at the door of reactive funders and unenlightened clients and their hired professionals. The new directions in construction projects are considered in detail in Chapter 21.

A consideration of these initial factors is the first stage in arriving at a procurement strategy for any particular client on any given building project. The client's initial profile needs to be weighed against the options available.

7.2 Procurement options

The first stage in deciding a construction strategy is to consider the external factors influencing the project, as set out above. Once these have been identified, analysed and prioritised, the next stage is to evaluate the available options and recommend a preferred strategy.

In practice, it is likely that there may be more than one way of achieving the project objectives. Each option should be considered in some detail as each will address the various influencing factors to a different extent. It is also possible for aspects of the different approaches to be 'borrowed' between the main choices.

The common procurement strategies differ from each other in relation to:

- the financial risk to which the employer is exposed – how important is it to the client that they have cost certainty?
- the degree of control the employer has over the design and construction processes – what is the level of involvement desired by the employer?
- the level of information available at the time the construction contracts are let – how far along is the design?
- the distribution of responsibility and accountability – does the employer require single-point responsibility for both design and construction?

In simple terms, the choice for an employer in deciding a headline procurement strategy is between traditional, design and build, and management contracts. The choice between the three approaches places different emphasis on the relative importance of the three main criteria against which the success of the completed project will be judged, namely time, cost and performance.

- *Time* – earlier completion can be achieved if construction is commenced before the design is completed. The greater the overlap between the two, the less time will be required to complete the project.
- *Cost* – with the exception of simple standard buildings and certain design and build contract strategies, the final contract sum cannot be established until the design is complete. Any overlap between design and construction means that construction starts before the cost is fixed. This obviously increases the importance of accurate cost forecasting.
- *Performance* – the quality and performance characteristics required from the completed building or facility determine the project time and cost. Some strategies reduce the ability of the client to control and make changes to the detailed building specification after the contract has been let.

The relative importance of these three criteria needs to be carefully considered because they will inevitably have an impact on the other objectives (see Figure 7.1).

In a well-managed project, the three objectives of time, performance and cost should be in constant tension. The three criteria are interdependent, and decisions affecting one will affect one or both of the others. Usually an improvement in one objective can only be achieved to the detriment of another. Too many projects overrun on time or cost, or they underperform because the project manager fails to keep all the objectives clearly in view. If tight targets are set for all three objectives, the likelihood of meeting them all is small.

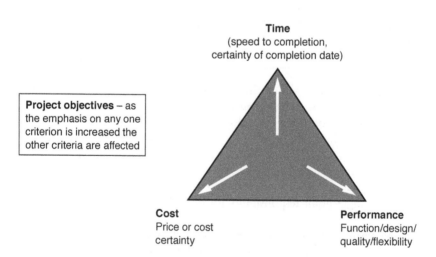

Figure 7.1 The relationship between the primary criteria

The strategy adopted should also reflect the client's technical ability and resources and the amount of control they wish to exert through their overseer. The organisational arrangements of the most common procurement strategies are described below, together with summaries of their respective advantages and disadvantages.

7.3 Traditional contracts

Reference was made earlier to the capture of statistics in relation to the most prevalent forms of construction contracts currently in use.[1] Those same statistics indicate that, by number, traditional contracting is the most popular form of procurement.

Traditional, in this context, means that the design and construction roles are kept separate from one another. The contractor is employed in the 'traditional' sense to simply build to the design he is given and subject to the instruction of the overseer fulfilling the role outlined in the previous chapter (Section 6.4). The overseer is also on hand to run the tender competition for the employer and advise on the selection of the appropriate contractor. The overseer is able to monitor the performance of the contractor and to require any quality control issues to be rectified by the contractor. The overseer certifies payment and can reject any aspects of the work with which issues have been identified.

The popularity of the traditional form is explained by its performance as an all-rounder. The employer's interests are well looked after by having their

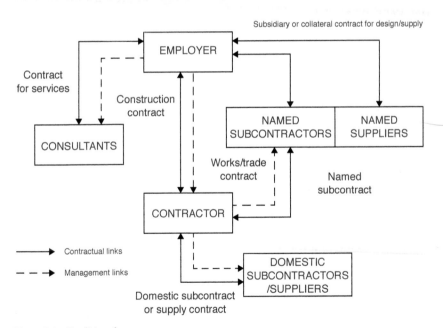

Figure 7.2 Traditional contract arrangements

Table 7.1 Strengths and weaknesses of traditional procurement

Potential strengths	Potential weaknesses
• reasonable price certainty at contract award • competitive fairness • ability to achieve high level of quality in design and construction • direct contractual relationships between client and design team • client can recover costs from contractor in event of failure to meet obligations • changes easy to accommodate and value	• client must have ability and resources to administer consultant and contractor contracts • timely and accurate construction information essential to avoid major claims • limited 'buildability' input by contractor • client shares risk with contractor • encourages adversarial relationships • lack of single-point responsibility for design and construction

overseer in place with the ability to ensure quality and value for money are being achieved. Anecdotally, the weakness in the traditional form stems from its inability to allow an early start on-site. The employer's design team needs to finalise its plans before the project can be let for tender and work started. There are measures that can compensate for this such as the use of provisional sums and/or the use of a contractor's design portion.

7.4 Design and build contracts

Design and build procurement has been around in its present form since the late 1970s when it was championed by the Conservative government as a means by which much-needed improvements and developments in the built environment would be delivered. The industry statistics bear testament that this new method of procurement quickly established itself as pre-eminent in the field. The statistics show that design and build is the leading form of procurement when the value rather than the number of projects is measured.[2]

The advantages of design and build are hard to deny as it represents a 'one-stop shop' for the employer. In a traditional construction contract, the task of design is essentially a matter for the design team comprising the architect/engineer as appropriate. It is common to find situations where the contractor has partial or complete responsibility for design matters. A number of design and build contracts in standard form are now available. Probably the best known of these 'package deal' or 'turnkey' contracts is the JCT With Contractor's Design contract. The procedure for determining the design in such contracts is made up of the 'employer's requirements' and the 'contractor's proposals', which will become incorporated into the contract. An employer entering a design and build project on the basis of lowest costs might find difficulties if he wishes to vary it at a later stage because these types of package deals tend to allow the contractor to object to a variation which will affect his responsibility for design. It is common practice for design and build contracts to provide for subcontracting of design elements.

The negotiation of the contractor's proposals and employer's requirements has echoes of the offer and acceptance ritual necessary for contract formation. At first glance, employer's requirements resembles the invitation to treat against which the contractor's proposals are made, equating to an offer. The employer's acceptance that the proposals meet his requirements completes the agreement. The consensus forms on the proposals and these form the basis of the agreement. However, the employer often takes care when accepting the proposals to ensure that the initial requirements remain the most important in the event that any ambiguity or inconsistency arises between the requirements and proposals. The statement is therefore made by the employer that the proposals 'appear' to meet the requirements. This reserves the employer's position and allows him to argue that he is entitled to what he wants and not what the contractor wants to give him.

The employer does not have the protection of the A/CA in the same way as under the traditional route. Essentially the contractor is in a more powerful position to dictate terms. The employer will employ an employer's agent with a role to represent his interest and sign off on such events as interim payments. In terms of the performance of the design and build procurement route, it is said to perform well with regard to time and money though the quality may be compromised. The reasoning for this is that design and construction can be overlapped to allow a time saving on the project whilst the employer will also benefit by paying for 'one of everything' given that the design and construction sides are no longer divorced from each other. The quality issue is where the sticking point may arise as the employer will essentially be given what the

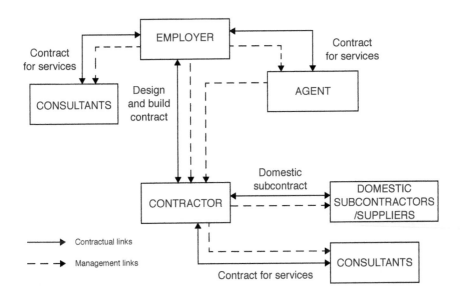

Figure 7.3 Design and build contractual arrangements

contractor wants to provide, which should be, but is not always exactly, what the employer had in mind.

Under design and build procurement the contractor essentially interprets the brief from the client contained in the employer's requirements and makes proposals. The likelihood is that the employer will give the contractor a relatively free hand in how those requirements are fulfilled. The contractor will appreciate the flexibility inherent in doing things his way and will be able to manage the process accordingly. Similarly, the contractor will have a good idea of his monthly entitlements to payment as it is in his own interest to maintain scrutinisable records for the valuation process.

The overseer is not on hand in the same way as on a traditional project to ensure the integration of any client change into the construction operations. As has been noted, should the employer instigate change through the variation procedure then this is likely to incur considerable cost as the contractor can legitimately include for the knock-on effects of having to accommodate a change. The employer may be in a position to give the contractor a free hand in deciding on the length of the build and its programme. The employer usually invites the contractor to state how long they need to deliver the project for the budget set. The contractor is still able to apply for additional time and money on the occurrence of certain events, and it then falls to the employer's agent to issue a statement accordingly.

On a design and build project, the contractor is much more likely to take steps to satisfy himself as to access and site requirements. This will extend to making arrangement for ingress and egress to the site and any considerations involving neighbours and third party access to the site during the construction phase.

7.5 Management contracts

Management contracting has a very small share of the market according to the statistics and is not really a viable alternative to the other two approaches. What was already a dwindling market share was made smaller by the tendency of all

Table 7.2   Strengths and weaknesses of design and build procurement

Potential strengths	Potential weaknesses
• early completion achievable through overlapping of design and construction processes • price certainty is possible before the construction starts (subject to adequate specification and changes not being introduced) • single point of responsibility and contact for client • 'buildability' potential increased	• client changes can be expensive in terms of cost and time implications • unsuitable for very complex projects • client has limited control of design quality • client's requirements must be precisely defined and specified prior to signing contract at the risk of client not getting the desired building

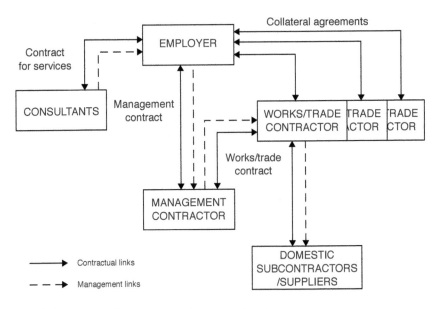

Figure 7.4a Contractual arrangements in construction management

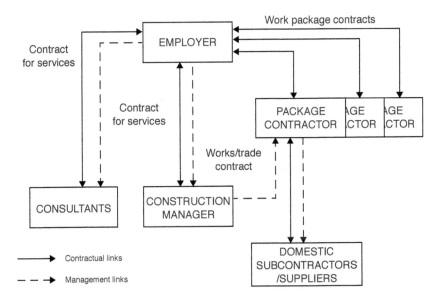

Figure 7.4b Contractual arrangements in management contracting

other procurement approaches to embrace its philosophy. Most of what were the features of management contracting have found their way, in some shape or form, into general good construction practice, obviating the need for this separate procurement route.

There are two types of management contracts: construction management and management contracting. The difference between the two lies in the contractual links between the trades contractors and the contractor/employer. Basically, the employer hires the trades contractors directly in construction management whereas the contractor retains contractual liability in management contracting.

The essence of a management contract is that the main contractor does not phsyically undertake any of the building work involved. Instead, the actual physical work is carried out by works contractors (subcontractors) and the main contractor manages the operations. The advantage to the management contractor is that contractual liability is diminished because, as a general course, the management contractor is only liable if the subcontractor can also be held responsible. There are indirect economies of scale in that the main contractor can reduce operating costs by not having to keep a large workforce, with appropriate reductions in equipment and plant.

Management contracts should not be confused with construction management, where the individual contracts are made directly between the employer and the works contractors. In both cases, the distinctive feature is that the management contractor/construction manager is essentially promoted to the 'top table' in terms of being treated as one of the employer's team of professionals tasked with delivering a successful project. The contractor's knowledge of the buildability of the project can result in some savings in terms of:

- meticulous programming of trades to ensure the build is rapid;
- knowledge of sequencing issues and the ability to condense delivery times and prevent the need to store materials on-site;
- suggestions and enhancements to the build itself in terms of input into the design process.

The client is able to insist on the works packages being let to works contractors who are the leaders in their field to ensure a high-quality build. The downside to this is that the cost of the project may increase. Further, if the contractor is not in control of the supply chain then his ability to keep prices down through negotiation and trade contacts is also reduced.

7.6 The time/cost/performance triangle applied

Figure 7.1 demonstrated the tension between the three corners of the triangle and how each wishes to exert its influence on the project. The three procurement routes can be overlaid on the triangle to give an indication of their core qualities. The diagnosis given in the triangle approach should only

Table 7.3   Strengths and weaknesses of construction management/management
contracting procurement

Potential strengths	Potential weaknesses
• early completion achievable through overlapping of design and construction processes • 'buildability' potential increased • changes in design can be accommodated later without paying a premium • clarity of roles, risks and relationships for all parties • client has direct contracts with package contractors and pays them directly • poorly performing contractors can be more easily removed from the project	• lack of price certainty until last work packages let • client must have resources to administer separate design team members and package contractors • client takes risk for co-ordination of design and construction and bears risk for delays, disruption and associated costs • strong cost, time and information control required

Figure 7.5   The time-cost-quality triangle applied to traditional procurement, design and
build procurement and management procurement

ever be viewed as a starting point for a procurement discussion. Each of the
general assumptions made can be challenged and the advice departed from.

The reputation of traditional procurement for being an all-rounder is shown
by its position on the triangle. 'Traditional' aligns itself with:

- good cost control which can be achieved by the overseer;
- good performance whereby the employer's brief is given to the overseer to deliver;
- weaker time certainty given that the design must be complete before the contract can be let to the contractor.

Design and build provides many advantages for the employer as discussed. Design and build aligns itself with:

- competitive pricing based on the cost savings implicit in having a single organisation structure;
- good time management in that the design and construct period can be co-ordinated by the single organisation;
- weaker performance risk in that the employer does not have the control of an overseer in the same way as occurs in traditional and management contracting.

The management approach completes the series of applications around the triangle. It has been noted that management is not a strong force in terms of procurement choice. Nevertheless, it illustrates some further important points in relation to initial procurement considerations. The management approach aligns itself with:

- good performance standards and the ability for the employer to insist on hiring highly skilled works contractors;
- good time management whereby the construction manager pushes the importance of the programme;
- weaker performance on cost as the contractor's skills in levering the supply chain for best value are underutilised.

7.7 Evaluation of procurement routes

The time/cost/performance triangle has its limitations in terms of being a broad-brush approach. A slightly more detailed evaluative route can be taken by seeking to further unpick what each of these concepts represents for the employer in question. The detailed client brief can then be numerically scored to the extent to which each strategy is able to meet the objectives and requirements. The relative importance of these objectives and requirements is taken into account by assigning a weighting. The total of the weighted scores illustrates how suitable each contract strategy is for the project under consideration. A simplified example of such an evaluation is given in Table 7.4.

This process provides guidance for selecting a preferred option. However, it should be noted that the inability of a particular procurement strategy to satisfy project objectives may exclude that strategy from further consideration. It is unlikely that there will be a clear-cut 'right' procurement strategy as each

Table 7.4 Time, cost and quality expanded

Priority	Criteria	Relative importance Score 1–5
Cost	Lowest possible capital expenditure	
	Certainty over contract price	
	Best value for money overall	
Time	Earliest possible start on-site	
	Certainty over contract duration	
	Shortest possible contract period	
Quality	Top quality, minimum maintenance	
	Sensitive design, control by employer	
	Detailed design not critical and can be left to contractor	

option will have some disadvantages or an element of risk; nonetheless, some will be better suited than others. The final recommendation must therefore include an element of professional judgement.

7.8 Other approaches to procurement

Procurement also involves consideration of wider factors than simply the client's priorities during the build. The interest of third parties in the building project can have a huge effect on the procurement approach chosen. Public clients are subject to direction from the government and its various ministries on how they should procure work. In the private sector, the funders can dictate the strategies on which they are willing to lend money to allow the build to go ahead. Construction procurement routes are not always applied in a rigid way. Projects in different industry sectors have developed bespoke procurement solutions and hybrid construction procurement routes. Several of these approaches are considered in the following sections.

7.8.1 EPC – Engineer, procure and construct

The notion that the employer has the inclination and expertise to design the required facility is not always appropriate. Large undertakings such as power plant and civil engineering and infrastructure projects in the developing world will require the contractor to liaise much more closely with the design team and to assist the employer and/or relevant stakeholders in creating the design solutions required for their project before implementing the same. EPC contracts are a common form of contract used by the private sector to help the client through the procurement process and essentially guide them towards the best choices.

EPC projects have single-point responsibility, a fixed programme, a fixed price, and guaranteed performance and reliability levels. The performance levels can be underwritten by performance damages, which operate in a similar way to liquidated damages clauses.

7.8.2 BOOT – *Build-own-operate-transfer*

Another notion that is not universal is that the employer needs to own the building from completion. The interest in a developing country is geared toward the provision of facilities or generation of power more than the benefit of simply owning the building itself. This departure from the usual importance of ownership can be reflected in the procurement strategy. A facility can be built and run by the contractor for the benefit of the employer who pays for the services rendered provided the performance targets are met by the contractor. The contractor essentially recuperates the profit, build and running costs over time. These types of arrangement can be offered by a consortium which agrees to build, own and operate the facility. At the end of the concession period, typically 25 years, the facility reverts in ownership to the employer.

7.8.3 PFI/ PPP – *Private finance initiative/public private partnerships*

In the UK, BOOT-type procurement is known as PFI (private finance initiative) or PPP (public private partnerships). Essentially the principles are the same as in the BOOT-type arrangement in that the contractor is paid for constructing and operating the facility for the benefit of the employer. The difference is that the building and facilities management work packages are usually kept separate. The PPP angle draws on the arrangement whereby what were public services (such as schools, prisons, road schemes and hospitals) are now supported by the private sector. The benefits of these arrangements include the introduction of competition and market best practices to improve the services offered. Others view the transfer of what were public services to partial ownership by the private sector as a regrettable step. This criticism is levelled in particular at land-backed schemes where the public authority places the ownership of land in private hands in return for the investment made.

This form of procurement has had a mixed press in terms of whether it represents good value for money. Critics point out that the final sum paid by the employer is far higher than it would have been had the employer funded the work from day one. However, many successful public schemes were financed this way during the last 20 years, and it is difficult to see how they would have been built without such a scheme. The popularity of this approach has now waned in the UK just as the international appetite for the schemes has greatly risen.

Projects built under PFI/PPP may be built by any construction procurement method. However, it is common for the project company to use design and build procurement.

7.8.4 Multi-contracting procurement

This approach divides a project into distinct contractual packages. A contractor may lack the expertise or be wary of the risks involved in contracting to deliver an EPC-type project. This has been seen in the UK in relation to offshore wind farm projects where the expertise to build the turbines and ensure transmission onshore has been regarded as too large an undertaking for one firm to undertake the whole project while still meeting output performance. A multi-contracting approach where the turbine construction and cable connections packages are let separately is a potential alternative to EPC for the employer. Multi-contracting procurement is less popular with funders because less risk is passed to the contractors involved, leaving a greater risk remaining with the employer.

7.8.5 Framework approaches

Framework contracts have been popular in recent years with employers who have a portfolio of construction work to perform in any relevant period. Examples of employers using frameworks include:

- a rail company seeking to let maintenance work on its network;
- a local authority requiring reactive maintenance for call-outs on its housing provision;
- a highways authorities requiring a programme of road building;
- a supermarket requiring the building of new stores.

Both public and private sector clients may enter into framework agreements by which they can award to the same contractors a series of specific contracts following a stated procedure. Frameworks usually include several contractors, all of whom have a chance of being awarded the contracts allotted to the planned works. Other frameworks operate by rotating the works around the contractors or even allowing the contractors to decide the work share amongst themselves. The framework contractors are effectively 'ring fenced' in terms of not having to compete through an open tender procedure for the projects envisaged. Steps are usually introduced to ensure that the performance and costs inside the framework remain competitive in a procedure known as benchmarking. The employer will 'benchmark' the performance of the contractors and their supply chains and expect improvement over any given period on the basis of the economies of scale and commercial advantage from keeping the same delivery team together during the projects.

7.8.6 Partnering

A partnering approach to procurement is often found in situations involving frameworks and other medium- to long-term arrangements between project teams. The benefits of working collaboratively over time are emphasised in this approach. The central notion here is that the culture and approach of the parties involved should be more closely aligned to the project's success, and steps are taken to positively encourage this environment. Partnering approaches are known for incorporating such tools and techniques as good faith clauses and incentives for exceeding expectations. Partnering is considered in more depth in Chapter 21.

7.9 Public procurement

The way in which government procures work is regulated by the rules mandated by the European Union. The single market in Europe extends the potential consumers for construction projects to some 492 million people. The founding principles of the European Union are that business in each of the 28 member states should have an open market place. The free movement of people, services and capital and the adoption of common policies are corner-stones of European practice. Moves continue to bring about the harmonisation of laws and technical standards to facilitate the fundamental objectives of free trade.

The opening up of procurement by government bodies and utilities to EU-wide competition has been recognised by the member states as a key component in the creation of the internal European market. Many of the existing European laws impact on procurement decisions either indirectly or directly. However, the Treaty Provisions have been codified into the Public Procurement Regulations 2015. The Regulations apply whenever a contracting authority seeks offers in relation to a proposed public supply, works and/or services contract. Any public body subject to the European Union rules has to invest considerable resource in ensuring compliance with the minefield of regulation in this area.

The European lawmakers are primarily concerned with ensuring that open competition exists across the European Union. The procurement approach is, therefore, of secondary importance here. The regulated area addresses how the winning contractors are selected and whether the bidders for the work had an equal chance of being successful and knew sufficiently well what they were bidding for and the criteria against which they would be judged. These considerations are examined further in Chapter 8 on tendering.

7.10 Conclusion

The procurement strategy decided upon by the employer and its professional team is of vital importance to project success. The initial advice on how to approach a project is a large part of ensuring a rewarding experience in that the client gets the best fit for what it wants and understands the opportunities and limitations in each of the approaches it considers. A well-informed client will examine all options before choosing not only its primary procurement route but also those measures that can enhance the chances of performing well against the brief set and minimising risk for all concerned.

The processes and arrangements adopted in order to procure the services of the construction industry frequently change in response to the increasing demand from clients for projects that meet all their requirements. These requirements are interdependent and sometimes conflicting, and they call for the adoption of strategies that recognise this and strike the optimum balance between the conflicting objectives.

The impact of Building Information Modelling on the field of procurement is providing some interesting recent developments. The focus is shifting from the front end of the building project to the benefits that can be delivered over the whole life of the project. Employers are increasingly able to take decisions based on a complete picture of exactly how projects will be constructed and perform. The ramifications of these changing criteria will encourage the development of new procurement routes to meet the new landscape.

7.11 Further reading

Hughes, W., Champion, R. and Murdoch, J. (2015) *Construction Contracts: Law and Management*, Fifth Edition, Abingdon: Routledge, Chapters 3–7.
Greenhalgh, B. and Squires, G. (2011) *Introduction to Building Procurement*, Abingdon: Spon Press.

Notes

1 The NBS Construction Contracts and Law Survey 2013 indicates that traditional procurement is used most frequently (by 57 per cent of clients). Available at: www.thenbs.com/pdfs/NBS-NationlC&LReport2013-single.pdf.
2 NBS Construction Contracts and Law Survey 2013.

8   Tendering and risk management

8.1 Introduction

The previous chapter dealt with procurement considerations. At this stage, the employer has decided on the procurement strategy in terms of how the project is going to be structured and has appointed his professional team. The employer now needs to appoint a contractor to build the project. The RIBA Plan of Work[1] gives a breakdown of the work stages grouped around the role that the architect, in conjunction with the employer, will need to perform to achieve project completion. The stages are:

- appraisal
- design brief
- concept
- design development
- technical design
- product information
- tender documentation
- tender action
- mobilisation
- construction to practical completion
- post practical completion.

The grouping of these work stages demonstrates that roughly halfway through the process thoughts turn to product information, tender documentation and tender action. This is when appointing the right contractor or consortium to carry out the work at the right price takes centre stage. Consideration is also given to the terms of the contract that will be employed and whether any amendments to the standard form are necessary. The law governing the issue of tendering is a component of construction law which is considered in the first part of this chapter. The second part of the chapter considers the issue of the risk profile the employer wishes to achieve and the theory and practice around risk management.

8.2 Tendering

The purpose of any tendering procedure is to select a suitable contractor, at a time appropriate to the circumstances, and to obtain from him at the proper time an acceptable tender or offer upon which a contract can be let.[2]

In Chapter 2, the rules of contract formation in terms of offer, acceptance and consideration are covered. The construction industry has adapted these rules for its own purposes. The most popular route to procure the services of a contractor is through competitive tender. In this regard, the construction industry is no different to other industries shopping around for the best deal. As every shopper knows, usually through bitter experience, sometimes the cheapest is not necessarily the best in terms of value.

8.3 Introduction to tendering

The UK construction sector comprises over 280,000 businesses and contributes almost £90 billion to the UK economy.[3] The businesses range in size from sole traders to multinational companies employing thousands of people. Varying workloads and the increasing sophistication and complexity of construction projects have led to many of these companies specialising in particular types of project or types of work and offering their specialist skills on a subcontract basis to larger contractors. These, in turn, specialise in management of the process and largely rely on the subcontractors to physically carry out the works.

The fragmented nature of the construction sector makes it very flexible and efficient in its response to fluctuating workloads and also enables it to develop and retain a wide range of specialist expertise within small companies. However, it also makes the selection of the most appropriate contractor (by the employer) and subcontractors (by the main contractor) a challenging and risky process.

Traditionally, contractors and subcontractors have often been awarded contracts based on lowest price. The process of favouring the lowest price causes massive difficulties in the AEC industry. Value is ignored in favour of an all-consuming rush to appoint the cheapest, or the second cheapest, tender. The ineffectiveness of this alone as a basis for selection has long been identified as being at the root of the AEC industry's problems. This was a common finding in both the Latham and Egan reports.[4] This, and the development of new collaborative procurement routes, has led to the development of practices that attempt to identify and select contractors on the basis of best value rather than lowest cost. It is a self-evident false economy to expect contractors and subcontractors to work for tiny profit margins. A client unwilling to pay a fair price for a good job is only storing up problems for later. The notion of open book accounting in modern procurement routes seeks to acknowledge entitlements to decent profit percentages. This recognition that making money is neither underhand nor frowned upon is a massive step towards improving the climate of the construction industry towards open and collaborative behaviour.

In construction, as elsewhere, the maxim of you get what you pay for seems to apply. The interests of both the contractors and the employer need to be dealt with sensitively in the tendering arrangements. The contractors are being asked to put their faith and considerable resource into the tender process. This requires that the tender be run along recognised rules that can be scrutinised. Public procurement, in particular, is subject to strict procedures to ensure that any tenderer can be satisfied that due process had been followed.

The tendering arrangements will be run on behalf of the employer by the architect/contract administrator on a traditional project or by the employer's agent on a design and build project. The architect is probably best placed on a management project to run the competition. Previously, the UK industry could rely on the National Joint Consultative Committee (NJCC) tender rules. This body is no longer active; however, its principles still guide the approaches taken to tendering. For example, the NJCC requires that the tenders submitted are checked for arithmetic errors and that the tenderer be given the opportunity to confirm the price upon having the error brought to their attention. In this manner, the problem of *caveat venditor*, as found in the case of *W Higgins Ltd v Northampton Corporation*,[5] can be avoided. This case was discussed in relation to a unilateral mistake (Section 2.6.2). It is surely not in the employer's interest to ensnare a contractor into a price which is manifestly wrong and on which his profit margin will not operate.

Experience on the part of the overseer is therefore very important. The documents to be disclosed to the tenderers must be uniform and allow the contractors to compete on a level playing field. This was not found to be the case in a House of Commons case where the British government at the time was accused of running a 'buy British' policy. The case of *Harmon CFEM Facades UK Ltd v The Corporate Office of the House of Commons*[6] involved a challenge made by a company employing mainly French resources that the British entrant into the tender competition was unfairly advantaged. The British entrant had been, inadvertently or otherwise, privy to some information not available to the other tenderers. The competition was found to have breached a number of European procurement regulations. Harmon's chances of being awarded the contract were assessed at 70 per cent; the claimant was awarded this percentage of the profit, which would have been earned. This amounted to a few million pounds. This approach to reckoning damages for the loss of an opportunity to win work is clearly extremely subjective.

The importance of a fair competition and due process being followed was writ large in the case of *Blackpool & Fylde Aero Club Ltd v Blackpool Borough Council*.[7] Here, one of the tenderers suspected that his bid had not been considered despite having been submitted in time. In fact, the tender had been hand-delivered to a letter box inside the prescribed time but was not collected until after the deadline. The Court of Appeal of England and Wales decided that tenders and requests for tenders are accompanied by a collateral contract implying that the requestor will inspect the bid. In other words, having bought a ticket for the lottery, the contractor is at least entitled to be in the prize draw. The leading judgement of Lord Justice Bingham is revealing in terms of the essential inequality of the positions of the parties:

> *A tendering procedure of this kind is, in many respects, heavily weighted in favour of the invitor. He can invite tenders from as many or as few parties as he chooses. He need not tell any of them who else, or how many others, he has invited. The invitee may often … be put to considerable labour and expense in preparing a tender, ordinarily without recompense if he is unsuccessful.*

Another case where a contractor felt that all was not as it seemed was that of *William Lacey (Hounslow) Ltd* v *Davis*.[8] The contractor submitted a tender for the rebuilding of war-damaged premises. The tender was not accepted but the contractor was encouraged to go to further expense in preparing additional estimates and schedule for the employer. The contractor was able to recover the costs of the tender as no contract was ever let, and the contractor's efforts were used for the employer's own purposes in negotiation with the War Damage Commissions. The modern legal analysis of this case would be that the contractor's right to payment is in quasi-contract or restitution.

From the contractor's point of view, a good deal of resource and planning is required behind the decision of whether or not to tender. The costs can be extreme – on a design and build contract, as much as 1.5 per cent to 2 per cent of the contract sum can be spent on the tender. No tenderer has a guarantee that these costs will be recovered. The different efforts required in respect of the tender are demonstrated over the three procurement choices as follows.

- In the traditional approach, tenders are relatively quick to process as the contractor is required to react to what has been included in the tender documents. This has historically been in the form of a bill of quantities into which the contractor essentially 'fills the blanks' with his prices or rates. The bill of quantities has seen a decline in recent times. Specifications, schedule of rates and activity schedules are more commonplace, requiring elements of remeasurement in the reckoning of work performed.
- In the management approach, the contractor is being assessed on his knowledge of the buildability of the project and his ability to save time and money in the design process. The innovation and input required to maximise this role will result in higher tender costs for the contractor.
- It is in design and build that the tender costs are likely to be the highest. The contractor will effectively be asked to take over the design from whatever stage the employer has reached – this might be 2 per cent or 95 per cent. The cost of completing the design lies with the contractor. Apart from the cost, another issue that this form of procurement creates is whether the different approaches taken by the contractors allow the employer to fairly and accurately assess one tender against the other. This is likely to be of

academic interest only since the contractor that the employer decides is a good 'fit' is as good a basis for selection as any in the private sector.

The most recent government procurement approach is to encourage the supply side to come together in a consortium of companies (see Section 21.5). One drawback of this approach is that the supply side is subject to potentially even greater costs in terms of formalising arrangements and design decisions between themselves without any guarantee of being awarded the work. The number of consortia that are actually prepared to bid and have the expertise to complete large-scale projects is also limited. This was evident in the competition for the contract to build the Olympic stadia in London for the 2012 Games. This has implications on whether the competition the government would like to see is actually achievable.

In normal tendering competitions (i.e. work not exceeding several million pounds in value), the cost of the tendering exercise for the contractor will reflect the odds of winning the contract that he is likely to be content with. Anecdotally, four to six contractors are a good number to ensure a rigorous completion between the tenderers. The number may be less than this in situations where the design work required is extensive.

Another precaution that the contractor must ensure is correctly observed relates to the tender itself. This will usually comprise a number of different elements including subcontract packages and suppliers. The contractor must take steps to ensure that these supply chain members are correctly aligned behind his tender. Put simply, if the contractor's offer is to be open for acceptance for 21 days then he must ensure the offers made to him are open for a similar length of time. The contractor in the case of *Cook Islands Shipping Co Ltd* v *Colson Builders Ltd*[9] fell foul of this practice. One subcontractor's terms and conditions allowed him to withdraw his price without notice. The contractor was left with the cost of securing an expensive replacement for no extra premium from his client.

The tendering options are considered below. These range from open competition to various forms of restricted competition.

8.3.1 Open competition

This method of contractor selection advertises details of the proposed project in local and trade publications, offering any party interested the opportunity to submit a tender. Contractors who consider they are able to carry out the proposed work request copies of the tender documentation and bid. In theory, this method appears to be the fairest and least restricted approach and to offer the benefit of encouraging new contractors to submit tenders. However, in reality, it suffers from some inherent disadvantages resulting in it being little used today.

The primary drawbacks are the lack of quality control and that, potentially, a large number of contractors will submit valid tenders thereby incurring substantial costs for the assessor. The unknown number also detracts from the attractiveness to the tenderers because they have no idea about how many other firms they will be competing against. In this situation, companies may consider their low chance of winning does not justify the costs they will incur in preparing and submitting a properly costed and considered tender.

The open procedure is supported by the European Union in their desire to protect the open market. The notion that contractors from all over the EU can apply for work in each other's country is their stated aim. This goal is being achieved by the other selective procedures. The top contractors working in the UK are truly multinational in their ownership and approach. The absence of any vetting procedure is seen as too great an issue with this approach together with the need to asses a large number of bids, taking into account financial standing, reputation and capability of completing the works.

8.3.2 Selective tendering

Most employers opt for a selective arrangement to save on the waste inherent for both parties in an 'open' tender competition. This is beneficial in that it limits the number of tenders to those who are preselected, usually on the basis of a pre-qualification questionnaire (PQQ). The PQQ system can be the gateway for a one-off project or a series of projects. Many local authorities run an 'approved list' system, which represents those contractors to whom contracts may be awarded. The approved projects for building are then earmarked for completion by one or more of the contractors on the list. The projects range from 'reactive maintenance work' such as being available for call-out to repair premises and deal with tenant's complaints to major civil engineering and infrastructure projects. These 'framework' contractors are the companies to whom a 'draw down' contract might be awarded.

Recent government initiatives have sought to use one form of PQQ across all government departments. Qualification for a selective tendering procedure is likely to be based on the types of criteria listed below. It is only once the potential tenderers have been checked that they are allowed to proceed to the tender competition itself. A typical range of criteria included in a PQQ is as follows.

- *Quality of work and performance record* – Does the contractor have references from satisfied customers?
- *Overall competence* – Is this the sort of work that the contractor can carry out properly?
- *Health and safety record* – The contractor should disclose its accident rate and good practice areas.
- *Financial stability* – The contractor must submit its annual accounts for scrutiny.
- *Appropriate insurance* – Does the contractor possess sufficient amounts of cover?

- *Size and resources* – Are the turnover and personnel appropriate for the project?
- *Technical ability* – Does the contractor operate the required specialist equipment?
- *Organisational ability* – Can the contractor meet the logistical challenges posed?
- *Ability to innovate* – Is there the possibility to use BIM and other platforms?
- *Design offering* – If the employer requires some or all of the design to be outsourced, can the contractor respond?

The number of firms selected for the tender will depend on the type of procurement adopted and the resources that the tenderers will have to put into the bid. For instance, a design and build contract requiring considerable design input might justify a list of three contractors whereas, for a simpler, fully designed and quantified project, a list of six to eight could be justified.

The inclination to include as many contractors as possible to ensure competition should be resisted because it is likely that if prospective tenderers become aware that they are merely one in a long list, they will devote their resources towards projects that they are more likely to win.

Single-stage selective tendering combines the process of contractor selection with the establishment of a price. This requires the design of the project to be quite well advanced. This form of tendering therefore effectively prevents the involvement of the contractor during the early stages of the project. Where the input of an experienced contractor during the early stages might be particularly desirable (e.g. to accrue the benefits of designing to ensure continuity [and therefore economy] of work, early subletting of specialist items and facilitating the 'buildability' of the design thus ensuring few defects), other methods should be considered.

Once a list of approved or qualified contractors has been established, these potential tenderers can be provided with more detailed information about the scope of works, estimated value, form of contract, etc. to enable them to decide if they wish to be included in the draft tender list. The form of this briefing will vary according to the size or complexity of the project, but it should be exactly the same for all prospective tenderers and also should be designed to ensure that when tender documentation is issued, it will lead to compliant tenders being submitted.

8.3.3 Two-stage selective tendering

An enhancement on selective tendering is to add a second stage to the process. In traditional-type procurement, it is common for the budget to be known at the start of the tender process. In design and build, the finalising of the price may require more detailed input from the contractor. This can be formalised in a two-stage tender procedure. In the first stage, the employer invites draft designs from the contractors with the finer points yet to be decided. The employer selects one 'preferred tenderer' and starts the second round of the process whereby the price and remaining designs are worked up in collaboration with the contractor. Strictly, the contractor has not secured the contract until the second stage is complete and the parties proceed to contract stage. It is rare, although not unheard of, for the contractor who wins the first round of competition not to progress and win the second.

8.3.4 Negotiation

A more radical approach to selecting a contractor and fixing a contract price is offered via negotiation. It has been recognised that many of the adversarial aspects of the construction industry could be attributed, at least in part, to the selection of contractors on the basis of the lowest tender submitted.

Sir Michael Latham promoted an alternative to lowest tender price, stating there is 'scope for awarding contracts in certain circumstances without competition to contractors who had shown particularly good performance on behalf of clients'.[10] This negotiated approach suits employers and contractors who have developed and maintained an ongoing partnering arrangement based on agreed and measurable targets for productivity improvements. The essential philosophy behind partnering is that the mutual benefits resulting from such a relationship tend to encourage both the employer and contractor to focus on the preservation of ongoing business relationships rather than assertion of particular claims or disputes.

Probably the greatest drawback is that a negotiated contract is not subjected to or checked by the open market in the same way as a tender exposed to a competitive tendering process. A client may therefore pay considerably more than the market price for a building than would have otherwise been the case. However, if other aspects of the project brief (e.g. early completion) are met more effectively through the adoption of a negotiated route then this may constitute greater value to the employer than lower cost.

The cost benefits of the competitive tendering process are potentially lost in such negotiated contracts. This disadvantage is more than compensated for by the better quality product and reduced construction periods that result from the improved relationships and mutual understanding produced by the partnering process. It is also possible to introduce an element of competition into the initial selection of potential partners by setting a competitive benchmark.

In addition, negotiated contracts tend to result in fewer pricing errors, can facilitate the contractor's participation in the design process at a much earlier stage, and can result in time saving and cost reduction to both employer and contractor. Incentives to encourage economy can be introduced by basing the contract on a mutually agreed target cost with both contractor and client sharing the 'pain and gain' associated with overspends or savings against that target. The public sector has historically not favoured negotiation as a method of fixing a contract price. This is largely for reasons of public accountability and the need for transparency. These problems are not insurmountable, and government guidance now embraces some aspects of partnering (see Section 21.5).

8.4 Tendering best practice

The selective approaches to tendering should ensure that both parties have a good indication of what they are potentially undertaking and allow the right appointment to be made. There are additional steps that the parties can take to ensure there is a smooth, transparent and fair process for all concerned. The following key principles were suggested by the Chartered Institute of Building.[11]

1 Fair and transparent competition leading to compliant, competitive tenders.
2 Tender lists should be as short as possible.
3 Conditions should be the same for all tenderers.
4 Confidentiality should be respected by all parties.
5 Sufficient time must be given for preparation and evaluation of tenders.
6 Sufficient information should be provided to enable the preparation of tenders.
7 Tenders should be evaluated and accepted on quality as well as price.
8 Practices should discourage collusion.
9 Tender prices should not change on unaltered scopes of works.
10 Standard, unamended forms of contract should be used where possible.

Most of the points made amount to common-sense precautions against poorly prepared procedures. The discouraging of collusion and the use of unamended contracts are two themes which are explored further in later chapters.

8.5 Tender assessment

Following receipt of all the tenders, these should be checked and any non-compliant ones rejected unless they offer an alternative approach to and are received in addition to a compliant tender. Tenders should be assessed on the basis of quality and price and, if necessary, the tenderers may be interviewed to clarify or amplify their tenders. Preferred and next preferred tenderers should be identified. Generally speaking, tender prices should not be altered unless the scope of works has changed or more complete information has become available since the tenders were invited.

Tender assessment in the public sector tends to involve careful scoring of the tenders submitted against predetermined and disclosed criteria. The private sector has a degree of this approach but tends to involve more of a 'gut feeling' for the best match. The overseer may have worked with a contractor before and be prepared to endorse their quote on the basis of their previous experience.

Once a preferred contractor has been identified, they should be notified and the tender formally accepted by the client. Following acknowledgement of this by the successful contractor, all the other tenderers should be provided with a list of tenderers and tenders received. The parties should then proceed to execute the contract and prepare to mobilise as per the RIBA Plan of Works. However, in between awarding the contract to the winning tenderer and signing the contract, there may be delays by reason of unresolved matters holding up progress. Into this category can come issues such as outstanding planning conditions and design details. This interim period can lead to the use of measures such as letters of intent to allow the contractor to make a start on the project pending contract signing. The issues arising around the use of letters of intent are discussed in Section 12.2.

8.6 Public tendering

Public sector tendering uses many of the approaches discussed above with the extra dimension of additional scrutiny being applied by both the tenderers and the authorities. This sector is subject to much more regulated procedure in terms of compliance with the Public Procurement Regulations 2015. The authority must pay meticulous attention to the tender process it runs. The Harmon case referred to above is a salutary tale for what can transpire even where any transgression is entirely unintentional. Society has become increasingly litigious, and this is felt very keenly in public tendering where disgruntled losing tenderers often take out their frustration at losing in procedural challenges.

The principal requirement of the Regulations is that in seeking offers in relation to a public works contract, the following procedures must be used:

- the open procedure;
- the restricted procedure;
- the negotiated procedure;
- competitive dialogue.

The open and restricted procedures are the same as private sector tendering whereby either a procedure open to all is used or one where the tenderers have been preselected by virtue of a pre-qualification questionnaire is adopted. The other two procedures may only be used in a limited number of circumstances.

The negotiated procedure is used where the contracting authority negotiates the terms of the contract with one or more persons selected by it. This procedure may take place in stages. The contracting authority must ensure that the number of parties invited to negotiate at the final stage is sufficient to ensure genuine competition is occurring.

The competitive dialogue procedure operates where the contracting authority engages with selected tenderers in order to define a specification against which the tenderers will tender. This can be useful where the contracting authority is not able to define with sufficient precision the technical specification capable of satisfying its needs. Complex projects like public private partnerships provide an example of where it is not possible at the outset to determine how the project should be structured. A minimum of three candidates must be invited to participate. When a solution that is capable of meeting the authority's needs is identified, the authority invites tenderers to submit final tenders on the basis of the conclusions that have emerged from the dialogue.

The public sector is much more likely than the private sector to use scoring systems to rate performance against criteria. A public body also needs to be transparent on the criteria against which the tenders themselves will be assessed. It is open for the public body to use criteria other than lowest price or to use price in association with other criteria to define what, to them, represents the 'most economically advantageous' tender. In *R* v *Portsmouth City Council ex parte (1) Bonaco Builders Limited & others*,[12] the Court of Appeal confirmed that where no criteria had been specified, a contract must be awarded on the basis of lowest price. Abnormally low tenders may be discounted by a contracting authority provided the appropriate notifications are given.

The criteria for 'economically advantageous' include a non-exhaustive list of quality, price, technical merit, aesthetic and functional characteristics, environmental issues, running costs, delivery date, technical assistance and aftersales service. A contracting authority may state the weighting it gives to each criterion. The definition of 'economically advantageous' also needs careful consideration to avoid discriminating against tenders, whether inadvertently or otherwise. The inclusion in tender information of a requirement that all bids confirm with Irish standards was an issue in what is known as the Dundalk case,[13] which involved the type of pipes to be used in the distribution of drinking water. This specification meant that only Irish manufacturers could comply with the tender requirements.

Inclusion of environmental criteria as the basis of selection was unsuccessfully challenged in the case of *Gebroeders Beentjes BV* v *State of the Netherlands*.[14] This case was a forerunner for the widespread adoption of environmental issues into tender considerations in both the public and private sector.

The biggest difference between the public and private sector – other than the additional clarity required in defining the criteria on which the project will be let – is in the likelihood of challenge in the public sector. Public bodies must use a standstill procedure. This allows a period in which any disgruntled losing contractors can bring a challenge. It is far easier to rerun a tender completion based on the correction of some technical point than to reimburse losing tenderers their tender costs or, potentially, to pay out on profits they would have earned during the contract. The public body must allow a period of at least ten calendar days to elapse between the date of the notification of award and the date on which the contract will be signed. Any unsuccessful tenderer can ask for reasons why it was unsuccessful, and the contracting authority must inform the tenderer of the characteristics and relative advantages of the winning quote. If the unsuccessful tenderer wishes to take matters further then it can commence proceedings under the Public Procurement Regulations in the High Court. The court may:

- suspend the procedure leading to the award of the contract;
- order the setting aside of the decision; or
- award damages to a contractor or supplier which has suffered loss or damage as a consequence of the breach of procedures.

The incidence of illegal behaviour by the supply side in terms of price-fixing and collusion is considered in Chapter 22.

8.7 Tendering and approaches to risk

Tendering considerations involve applying contract law procedures to construction practice. The construction industry has devised some specialist applications for this basic law of offer and acceptance. These include measures aimed at ensuring fairness and transparency and limiting the competition to which the tenderers are subjected. Large sums of money are won and lost when awarding contracts, and the businesses involved are justifiably anxious to ensure that this part of the construction process runs properly.

The tenderers must not overlook the importance of examining the contract documents at tender stage. For this reason, the tenderer will usually take a twin approach to tender competitions. The tenderer will have his estimating team and possibly his design team (if he offers this service) carefully scrutinise the specification and information supplied by the employer to see whether he is interested in this work. At the same time, the prospective tenderer will pass the draft contract over to his legal team (in-house or external) to review the contract terms he is being asked to sign. It is this process of contract assessment and the principles governing it which are considered in this second part of this chapter.

8.8 Risk theory

The tender package includes the information the contractor needs to be able to arrive at a price and to submit the same if he decides upon this course of action. Lurking amongst the enclosures will be the contract he will be called upon to enter into. This is usually stated as being an intention to use a named standard form contract. Reference will also be made to whether the employer intends to issue amendments to the standard form and if so, what these will be.

Amendments to building contracts in this sense are additions and deletions made to the standard wording. These are the 'bespoke' amendments as opposed to the amendments made by the contract providers themselves. The latter type of amendment is usually made to update a contract to take account of changes in the law. The discussion in this section concerns the non-standard bespoke amendments made at the instigation of one or other of the parties.

The prevalence of amended contracts has resulted in a situation where, in the private sector at least, it is extremely rare for the employer to propose that an unamended form of contract is used. The government has sought to mandate that unamended contracts are the norm for public works, and this has had some success. The purpose of the amendments is always to alter the balance of risk between the parties in favour of the employer. The prudent contractor will review these amendments and formulate a strategy for coping with them based on accepting the risk at a price or declining the invitation to sign an amended form.

Risk can be defined as: 'uncertainty and the results of uncertainty…risk refers to a lack of predictability about problem structure, outcomes of consequences in a decision or planning situation'.[15] Risk is the unknown in the sense of issues that might be encountered on a building project. The risks can be identified as 'what if' scenarios such as: What if the ground conditions are not as expected? What if the overseas products do not arrive in time for installation? Most risks can be managed and some eliminated by proper planning. However, the common denominating factor is often that risks cost money to address and the issue becomes which of the parties pays for the risk.

The basics of risk theory dictate that for every transfer of risk, there should be a corresponding transfer of money to compensate the party for the extra risk being taken. Thus the action of the client in seeking to amend the standard form has a reaction that balances the equation. However, this theory breaks down where the employer is reluctant to pay the premium sought and effectively wants the original price and the risk transfer. This places the contractor in an invidious position if it cannot build the premium into its rates. The options for the contractor when responding to a proposal that they take additional risk are to:

- price the additional risk and add it to the price;
- accept the proposal and hope for the best;

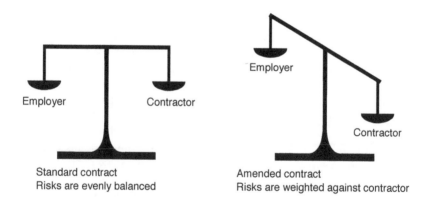

Standard contract
Risks are evenly balanced

Amended contract
Risks are weighted against contractor

Figure 8.1 The balanced and unbalanced scales

- reject the proposal and insist on the standard form;
- seek to pass the risk down the supply chain to subcontractors;
- carefully evaluate each amendment and respond on an item-by-item basis (NEC risk register approach);
- offer to share some of the risks with the employer.

A contractor may choose a single approach or multiple approaches on any given project. All too often the option is taken to pass the risk down the supply chain or to ignore the risk in the hope that it will go away. Anecdotally, what happens next is that the risk is passed down from subcontractor to subcontractor until it lands with someone too busy or too naive to appreciate the nature of the risk they are being asked to take. Insolvency can then ensue as a result of this misallocation of the risk and the fallout from this is then passed back up the supply chain, leaving everyone wishing that they had allocated the risk properly in the first place.

Very occasionally, a far-sighted employer comes along who wants to do things differently. One such was BAA (British Airport Authorities) who, when constructing Terminal 5 at Heathrow, took a contrary approach. This was to reverse the balance of risk; rather than paying a premium to contractors so that they would absorb risk, BAA absorbed the risk itself, thereby lowering the tenders. As risk was removed from contractors, BAA's strategy was to ensure the quality of the build through offering incentives to contractors. The project was a success and the savings were enjoyed by the client. Interestingly, this approach has not caught on. The reasons for this are hard to fathom, but a partial answer may that this approach falls foul of one of Abrahamson's principles[16] (dealt with in Section 8.9) that the party best able to bear the risk should bear it. The risk of such things as non-delivery by subcontractors is, ultimately, probably best borne by the contractor with whom there is a direct contract.

Contractors are thus justifiably wary of amendments the purpose of which is to make them responsible for more than they would normally be accountable for under the contract. The possible responses depend on bargaining position and whether the market is in favour of the buyer of services or the seller. The reality in practice is often not as well rehearsed as this; the contractor will sign, knowing the risks, but backing himself to be able either to broker a deal on the final account or to find a way around the legal situation created. This adds an element of uncertainty into proceedings that is well liked by some in the industry as part of the cut and thrust of business.

8.9 Abrahamson's principles

This section discusses Abrahamson's seminal paper about risk theory and its application to the AEC industry. The basic point made is that risks should be taken by the party best placed to take the risk. In helping decide on this point, consideration should be given to whether the risk:

- is in one party's control;
- can be insured against and, if so, by which party;
- benefits a particular party if it does not happen (the party retains the premium attached to the risk);
- is more efficient for one party to take than another party; and
- if it happens, affects one party more than another party.

Abrahamson acknowledged that the task of balancing these points is extremely difficult. However, he felt strongly enough about the subject to recommend them as a basis for consideration. Essentially, the choice for contract writers is to allocate a risk to either the employer or the contractor or to treat risk as being shared.

Figure 8.2 elaborates on the position of the contractor with regard to the outcome of a contractual risk. It introduces the three categories of risk – employer, contractor and neutral (shared) – and sets out the consequences of each of these eventuating. If the employer risk event causes the problem then this is characterised as 'non-culpable' delay and the contractor is awarded more time (by way of an extension of time) and more money (known as loss and/ or expense). If the contractor risk event occurs then the delay is culpable and the contractor is awarded neither money nor time. The contractor's position is made worse in this situation by the potential for liquidated damages to be levied against him for the period of culpable delay. The third situation is that a neutral event (such as poor weather conditions) causes the issue in which case the contractor is awarded more time but no more money. No liquidated

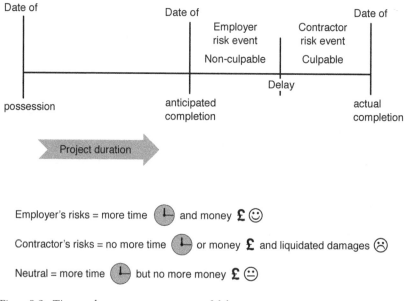

Figure 8.2 Time and money consequences of delay

damages are payable where neutral events are involved. Figure 8.2 reflects situations where only one event impacts on the projects. However, most problems do not occur on their own. There are a number of approaches to what happens where more than one cause occurs concurrently (examined in Chapter 14). The general position in the case of *Henry Boot Construction (UK) Limited* v *Malmaison Hotel (Manchester) Ltd*[17] is that the contractor is given the benefit of the doubt and the employer risk event is given more causal potency when deciding on the action to take.

The arbitrary nature of how some risks are classified under building projects surprised Abrahamson — why should it be that weather conditions are dealt with in one way and labour conditions in another? Another revelation is that one of the most important risks — that of ground conditions — is not expressly allocated to either party in the contract. It is submitted that ground conditions risk ought to be an employer risk although not all commentators or building contracts agree on this point. The land is in the ownership of employers and they should bear some responsibility for the condition of the ground. However, an alternative view is that the contractor impliedly warrants that he can construct the works in accordance with the design provided to him — including having to deal with unexpected ground conditions in doing so. This is very often the first risk that is transferred over to the contractor by employer amendments. The contractor may not like it but often has no choice but to accept it. He can, however, seek to price the risk and take steps to mitigate exposure to it.

The mitigation will usually involve studying carefully the information provided by the employer. The contractor may seek to take a collateral warranty from the producer of a site investigation report. This will allow him the opportunity to seek to recover his costs in the event that the report was negligently produced. He may seek to upgrade the comprehensiveness of the report and pay a premium for a higher standard of report. Or the contractor may simply take the risk, hope that no problems arise and pocket any premium he has managed to put against the risk. Lastly, the contractor can allocate the risk to the ground worker subcontractor. This can be achieved by including a term to this effect in the subcontract. Clearly, the most sensible way to proceed is to carefully scrutinise the information provided and carry out investigations and local searches.

One common amendment to building contracts is, therefore, transferring the risk of ground conditions. Another common risk to transfer on the contractor is responsibility for any design carried out on behalf of the employer before a design and build contract is let. It is also common to reverse the usual priority in the event of a conflict between employer's requirements (ERs) and contractor's proposals (CPs) in design and build contracts. The unamended contract gives the CPs priority over the ERs, the logic being that the CP is responding to the invitation of how best to meet the ERs. By giving the ERs priority over the CPs, the position is arrived at where the fiction is created that the contractor is deemed to know exactly what the employer had in mind when he made his own proposals.

There is virtually no limit to what amendments can be foisted onto the unwary contractor. Responsibility for access to the site and any issues arising out of such things as planning conditions, party wall issues, neighbours' rights and emissions from the site have all been passed to the contractor. However, contractors need to assess these risks carefully and make the appropriate response, whether that is pricing the risk or taking steps to mitigate or renegotiate. It is important for contractors to create strategies to deal with risk, and appreciating the consequences of additional risks starts with understanding the unamended contract forms and the purpose of the amendments being proposed.

8.10 Standard and non-standard contract terms

Familiarity and a working knowledge of contract administration are a necessary prerequisite to any risk response. Standard forms of contract are discussed in Chapter 9. Judges have, on occasion, criticised what they perceive to be poorly drafted contracts when considering cases based on conflicting interpretations of contract terms. The frustrations of the judges are detectable in the pronouncement in *English Industrial Estate Corporation* v *George Wimpey & Co Ltd*[18] where the judge labelled the RIBA form (a forerunner of the JCT) a *farrago of obscurities*. The NEC was not free from criticism either and was described in the case of *Anglian Water Services Ltd* v *Laing O'Rourke Utilities Ltd* as *a triumph of style over substance.*[19]

Criticism of any contract writers, or statute draftsmen for that matter, is harsh in the circumstances. A building contract must provide for all the imaginable scenarios and issues that might arise. It is difficult to come up with a complete solution, and challenges based on conflicting interpretations are likely to be unavoidable. When legal argument is added to the situation, the recipe for disputes is complete. Contractual wording that will cover all the eventualities that can occur and give the parties instruction on how to proceed is a herculean undertaking.

The point of the standard form contracts was set out by the Chairman of the JCT as being to 'give the appropriate risk profile and to give the parties the benefit of precedent'.[20] This statement that case law supports contract interpretation is accurate in that judicial findings have added clarity to some grey areas in contracts. In some instances, this leads to subsequent amendment being made to the contract. This was the case with the two examples that follow – related to inclement weather and the finality of final certificates – which, as well as illustrating the point made, contain some useful elements of construction law knowledge.

8.10.1 Inclement weather

Inclement weather is a good example of a neutral event leading to a sharing of the risk. The contractor is awarded more time but not more money under the JCT form. In JCT terminology, an extension of time is given but not loss and expense. In NEC terminology, a compensation event has occurred for which the contractor may be entitled to money and/or time. The difference between the contracts is negligible. The JCT is arguably clearer for specifying which events lead to just time and which to time and possibly money as well. On the other hand, the NEC acknowledges that money consequences may flow from the delaying event (e.g. standing time for hired plant) and does not rule this out as potential claim.

Inclement weather in JCT contracts is a neutral event – something that gives rise to an extension of time but not to extra money. A neutral event actually shares the risk in that the line has been drawn and the view reached that it would be inequitable to ask either party to bear this risk solely; thus it is shared by the mechanism of neutral events. This is not the only mechanism by which risk can be shared as demonstrated by joint insurance and the NEC risk register. The arrangement here is basically that the employer insists on being named on the contractor's policy and is happy to indirectly pay for the same as this cost should be reflected in the contractor's price.

The cautionary tale in relation to inclement weather comes from the case of *Walter Lawrence* v *Commercial Union Properties*.[21] The contractor considered that he was entitled to an extension of time due to the fact that the weather was unseasonably hot and it was deemed unsafe to work on the exposed roof of the job. Rather than taking an understanding line, the overseer made a pedantic point: the weather was not inclement; clemency is defined as being pleasant and the weather was not unpleasant – it was exceedingly pleasant. The decision not to award an extension of time was upheld by the court, which had no discretion in interpreting the wording in any other way.

The upshot from this unfortunate (from the contractor's viewpoint) set of events was that a position had been arrived at which the contract writers had not anticipated and did not desire. An amendment can be issued to fix the problem in such a situation, and this was promptly done. The JCT definition of bad weather is now expressed as exceptionally adverse weather, which would cover rain and shine. The NEC is quite specific about the records that must be produced to establish that the weather was exceptionally bad. This is known as the '1 in 10' event – the metrological office records must be shown to demonstrate that weather is worse than a 1-in-10 year value.

8.10.2 The finality of final certificates

The second example in relation to case law producing an unforeseen outcome is in relation to final certificates. Chapter 2 describes contracts being created against the background of the common law constituted itself or by a mixture of case law and statute-based law. In the mid 1980s, serious consideration was given to the effect of a final certificate. Was this indeed final and could the employer pursue the contractor for any defective work after the final certificate had been issued? It was never the intention of the JCT contract that the final certificate would displace the background law on limitation and the rights here were not meant to be affected. However, that was not what the judge found in the case of *Crown Estate Commissioners* v *John Mowlem & Co Ltd*[22] where a final certificate was found to be just that – final. The result was an amendment to the contract to the effect that the final certificate was only final in relation to those matters to which the architect had specifically turned his mind. The certificate was final in respect of those matters where an architect had signed off the quality of an item *to the architect's approval*. This phrase and phrases like it disappeared from contract specifications so as not to jeopardise the employer's right to fall back on the common law or to restrict any ability to claim for damages beyond the final certificate.

8.11 Conclusion

This chapter has taken the reader through some of the key concepts shaping the application of construction law. The notion that work is tendered for competitively and that risks are allocated are important considerations in any discussion about construction practice. The reader now has the framework of understanding required to go on and form an appreciation of the detail of the contract provisions. The next chapter is dedicated to examining the role of standard form contracts and introduces the main standard forms currently in use.

8.12 Further reading

Hughes, W., Champion, R. and Murdoch, J. (2015) *Construction Contracts: Law and Management*, Fifth Edition, Abingdon: Routledge, Chapter 9.
Furmston, M. and Tolhurst, G. (2010) *Contract Formation: Law and Practice*, Oxford: Oxford University Press.

Notes

1   Available from the RIBA website at: www.ribaplanofwork.com.
2   Hackett, M., Robinson, I. and Statham, G. (Eds) (2007) *The Aqua Group Guide to Procurement, Tendering and Contract Administration*, Oxford: Blackwell Publishing, p. 27.
3   Department for Business Innovation and Skills (2013) *UK Construction: An Economic Analysis of the Sector*, London: Department for Business Innovation and Skills.
4   Latham, M. (1994) *Constructing the Team: Final Report of the Government/Industry/Review of Procurement and Contractual Arrangements in the UK Construction Industry*, London: Department of the Environment; Egan, J. (1998) *Rethinking Construction*, London: Department of the Environment, Transport and the Regions.
5   [1927] 1 Ch 128.
6   [2000] 67 Con LR 1.

7 [1990] 1 WRL 1195.
8 [1957] 2 All ER 712.
9 [1975] 1 NZLR 422.
10 Latham report, p. 61.
11 Chartered Institute of Building (2009) *Code of Estimating Practice*, Seventh Edition, Chichester: Wiley-Blackwell.
12 [1997] ELR 1/11.
13 *European Commission* v *Ireland* [1987] Case 45/87.
14 Case 31/87.
15 Hertz, D. B. and Thomas, H. (1984) *Practical Risk Analysis: An Approach Through Case Histories*, Chichester: John Wiley and Sons.
16 Abrahamson, M. (1984) Risk management, *International Construction Law Review*, 1(3): 241–64.
17 (1999) 70 Con LR 32.
18 Per LJ Davies in [1972] 7 BLR 122 at 126.
19 [2010] EWHC 1529 (TCC).
20 Hibberd, P. (2004) *The Place of Standard Forms of Building Contract in the 21st Century*, a paper based on a talk given at a Society of Construction Law conference, Wakefield, 11 March: available at: www.scl.org.uk/papers/place-standard-forms-building-contract-21st-century.
21 (1984) 4 Con LR 37.
22 (1994) 70 BLR 1.

9 Standard forms of contract and subcontract

9.1 Introduction

At the end of the procurement and tendering stages, the parties will have a broad understanding about the general apportionment of risk on the construction project. However, it is only after studying the terms of the contract that an appreciation about the detail surrounding the actual allocation of these risks will emerge. The contract is the means by which the employer, contractor and subcontractors agree the detail through the mechanism of the conditions of contract and subcontract applied to the agreements entered into by the parties.

The form of the contract or subcontract used therefore delivers the risk allocation agreed. The form is usually established before contractors are asked to tender. It is important for contractors and subcontractors to fully understand the responsibilities and liabilities allocated to them under the contract in order that estimators and contracts managers can effectively price for and manage the risks they are being required to undertake.

Standard form contracts (SFCs) are prepared and agreed by representatives of all the major participants in the construction process. The Joint Contracts Tribunal (JCT) is constituted of representatives from groups of all the major client, professional and contracting bodies. SFCs recognise the often conflicting interests of these groups and allocate agreed rights, responsibilities and risks in a manner which is acceptable to all the participants in the process.

As negotiated documents, SFCs fall outside the scope of the Unfair Contract Terms Act 1977. It should be noted, however, that a contract based on a SFC but containing substantial amendments might not necessarily be viewed by a court in the same light. SFCs are subject to legal precedent and are regularly revised to reflect precedents set in court decisions or changes in statutes (e.g. all the SFCs have been revised to reflect the requirements of the Housing Grants Construction and Regeneration Act 1996). The fact that aspects of many SFCs have been the subject of court actions removes uncertainty as to how those particular provisions might be interpreted by a court.

Publishers of the various SFCs have often clarified ambiguities in the SFCs by issuing practice guidance notes that provide insight into the application and interpretation of particular aspects of the contracts. These practice notes have no legal standing. However, the courts tend to accept them as authoritative interpretations that effectively reduce uncertainty.

Most SFCs are published in a variety of versions to suit the wide range of procurement routes currently in use. The principles that underpin each 'family' of contracts are usually common across the suite. There is also a wide range of standard documents designed to support the different SFCs, for example, forms of tender, subcontracts, guarantees and warranties. The publication of such suites of contracts, and their regular revision in line with their related SFCs, ensures that any change in risk or liability arising out of a change in precedent or statute is transferred equitably down the supply chain.

9.2 The purpose of standard forms of contract

Standard forms are preferable for the parties to use than bespoke contracts. The benefits of using SFCs can be summarised as follows.

- *Economy* – SFCs avoid the expense of preparation and production of a bespoke contract for each project.
- *Certainty* – The design and construction of buildings is a very complex process often involving the contribution of a range of inputs from a diverse range of sources and the participation of a large number of people. The use of SFCs enables the complex contractual arrangements that are necessary to manage this process to be accomplished with much greater certainty. Certainty is increased and therefore risk is reduced.
- *Familiarity* – The widespread use of some SFCs has made them familiar to many managers and professionals involved in construction. This familiarity is a benefit in that it simplifies the contractual management of projects (the participants are familiar with their roles, responsibilities and rights under the contract before they start).

Against this backdrop of reasons to recommend the use of SFCs, there are certain provisos. As noted, it is unusual to find a SFC being used without some form of amendment to suit the particular requirements of the client. In extreme cases, significant portions of the SFC are struck out and replaced with bespoke clauses. There is a danger here that these replacement conditions might negate the benefits of the SFC and increase the risk to all parties. The imposition of unfair terms or the introduction of contradictions with other parts of the contract are risks that may eventuate.

Despite the publication of many versions of the various SFCs and their potential for amendment, they inevitably reduce the scope for the development of creative solutions to procurement problems. New SFCs are developed to meet the changing demands of clients.

There is also the potential danger that a client or their advisers might choose a particular SFC because they are familiar with it rather than because it best meets the requirements of the project. Most SFCs are complex documents composed of large numbers of clauses. This reflects the need to suit as wide a range of projects and eventualities as possible. Considerable levels of skill and knowledge are therefore required to interpret the contract provisions correctly.

Methods of working, procurement and payment are continually changing and it often takes drafting bodies lengthy periods to respond with the publication of new or amended SFCs to reflect these changes. For example, the response of contract draftsmen to the Building Information Modelling (BIM) working practices has been slow and tentative. Even where the contract provides new updates, they are not universally accepted and may not deliver what the industry wants. For example, the Construction Industry Council BIM Protocol has come in for some criticism and is not widely used. The Protocol has a waiver of liability for the information inputted into the model, which waters down considerably the reliability of the model and procedures used. However, it has to be recognised that protocols and other instruments have a difficult role to fulfil in keeping all interested parties satisfied with the level of commitment required.

9.3 Selection factors

It is important that the contractor and the supply chain are aware of the employer's position with regards to options for SFC selection. The choice of SFC is dependent on a number of considerations which include the following.

- *Identity of the client/employer* – SFCs have been developed to reflect the different requirements and statutory obligations of private clients, local authorities and government bodies. For example, local authorities and government bodies are required to select those contracts promoted by the government because they have potentially different requirements regarding public accountability and they might take responsibility for ensuring particular obligations under the contract are met.
- *Method of procurement* – Various SFCs have been developed to reflect the different administrative processes and obligations of the parties inherent in traditional, design and build, and management contract procurement situations.
- *Source of design* – Design may be done by the contractor, specialist subcontractors, the employer, independent designers or any combination of these, and an appropriate SFC should be selected to reflect the exact design arrangements of each project.
- *Size of project* – Clearly a complex contract is inappropriate for the construction of a house extension and the short FIDIC Green Book would be equally unsuitable for a multimillion pound office development. The complexity of the SFC that is selected should reflect the value of the project.
- *Allocation of risk* – The procurement strategy adopted for a project will have been based on a careful consideration of the likelihood and consequences

of the risk of the cost, time and performance requirements of the project not being met. The allocation of the responsibility for managing those risks (and, by implication, the allocation of liability for them should they occur) is defined by the contract adopted. SFCs are designed to reflect the differing distribution of these risks under different procurement routes.

- *Type of work* – Construction work encompasses a wide range of activities ranging from maintenance work, where there is no design requirement and the amount of work cannot be accurately forecast at the outset, through simple civil engineering and building projects to very large complex projects requiring extensive design input by architects and engineers. The roles and responsibilities of the employers, designers, contractors and subcontractors in each of these will be very different, and the SFC adopted should reflect these differences.

- *Development of design and documentation* – The programme requirements of the project, together with the above factors, will define the extent to which the design had been completed and the detail of documentation available at tender stage and commencement of the contract. Depending on the nature of the project, selection of the contractor may be based on a fully measured bill of quantities, on drawings and specification, or perhaps by clearly defined performance requirements, and the SFC selected should reflect the way in which the project is defined.

Lump sum and shorter lump sum contracts are based on the premise that the design has been substantially completed prior to the contract agreement being entered into. The value of the works is fixed at this point, and appropriate adjustments are made to the contract sum for any subsequent additions or omissions to the works after that point. Deviations from the originally defined scope of works are therefore likely to provide contractors with an opportunity to seek additional time and payment to reflect these changes.

Measurement contracts accept that the precise details of the project cannot be fully determined at the point of contract agreement. They are based on the assumption that the precise nature and quantity of work, and therefore also the payment for the work, will be recalculated on completion. These SFCs are therefore based on pre-estimates of the scope of works, and any deviation from this needs to be significant if adjustment to the contract periods is to be justified.

Cost reimbursement contracts are generally only used where it is impossible to define the type and extent of the work with any certainty.

9.4 Standard forms of contract in use

The different forms of building contract are competing to an extent for the same construction market. Each suite of contracts has its supporters and particular areas of the construction and civil engineering industry where it is preferred. Significantly, only a minuscule proportion (less than 2 per cent) of contracts is based on non-standard forms of contract.

The dominance of the JCT in the UK construction sector stood unchallenged for many years with its market share frequently surveyed as being as high as 80 per cent. In the latest NBS survey available,[1] JCT contracts were found to be the most commonly used contract by 48 per cent of respondents in 2012, though this was a fall from 60 per cent in 2011. The percentage most commonly using NEC had risen to 22 per cent in 2012 while domestic uses of the FIDIC were recorded at 4 per cent and the PPC2000 at 2 per cent. None of the contract writing bodies would necessarily agree with these statistics as being a true reflection of their market share. The NBS surveys are based on increasingly small sample sizes and in recent times have shown greater interest in giving insight into the take-up of new trends like BIM and partnering. Nevertheless, the NBS survey remains the most reliable survey conducted of which forms of contract are the most popular notwithstanding existing doubts around the rigour applied to the survey.

In recent years, there have been significant changes in the popularity of various procurement options with a move away from traditional procurement with the bill of quantity towards contractor design elements and full design and build. These have been accommodated by the use of SFC variants designed to facilitate these procurement options.

The standard forms of contract currently in use can be broadly classified by publisher as described in the following sections. The contract families examined here are the JCT, NEC, FIDIC and PPC2000.

9.5 The Joint Contracts Tribunal

9.5.1 Introduction

This suite of contracts has generally been regarded as the industry standard against which all others are judged. In addition to consolidating all the changes that had been made to the earlier 1963 and 1980 forms, the 1998 version of these contracts also included some fundamental changes suggested by Latham and in part required by the Housing Grants Construction and Regeneration Act of 1996. The contract was further amended in 2005 when the private and local authority versions were merged into one form and then again in 2011 to take into account changes required by the Local Democracy, Economic Development and Construction Act 2009. Its three variants are intended for use in general contracting and 'lump sum' tendering based on bills of quantities, approximate quantities or specifications and drawings.

9.5.2 JCT background and content

The Joint Contracts Tribunal was set up in 1931, consisting of representatives from the construction industry and the professions involved in major building works. The member bodies of the JCT are shown in Table 9.1; they establish the pan-industry appeal and approach of the Tribunal. The best-known form of contract issued by the JCT is the JCT 11 Standard Form of Building Contract. It is intended for use in connection with all types of building works. This form is in three versions, with or without reference to bills of quantities, and also for use with approximate quantities. The latest edition was published in 2011. The 2011 form consists of three sections. These are the articles of agreement, the contract conditions and the schedules.

The articles

The articles are primarily concerned with definitions and set out the names of the employer (the client) and the contractor (the builder). The primary obligation to build and to pay is established as are the contract price and the identity of the architect. The appointment of an architect or 'contract administrator' is a fundamental feature of this traditional method of procurement. The architect is responsible for supervision of the works. The articles also name the quantity surveyor engaged by the employer. The quantity surveyor has responsibility for the valuation of work and reports to the architect/contract administrator.

The conditions

The contract conditions together with the contract bills and drawings form the substance of the contract. The conditions are divided into nine sections where related topics are grouped together. In the JCT form, these sections are:

1 Definitions and Interpretation
2 Carrying out the Works
3 Control of the Works
4 Payment
5 Variations
6 Injury, Damage and Insurance
7 Assignment, Third Party Rights and Collateral Warranties

Table 9.1   Member bodies of the Joint Contracts Tribunal

Professional body	Interests represented
British Property Federation	Employers and lenders
Contractors Legal Group Limited	Main contractors
Local Government Association	Public sector employers
National Specialist Contractors Council	Specialist and subcontractors
Royal Institute of British Architects	Architects/contract administrators
Royal Institute of Chartered Surveyors	Quantity and building surveyors
Scottish Building Contract Committee Limited	Scottish contractors' interests

8   Termination
9   Settlement of Disputes.

The JCT conditions address the basic obligations and the more complex issues alike. The role of the overseer is covered in Section 3 on control of the works. The recognition that the interests of others need to be represented is covered in Section 7. The conditions deal at length with matters such as the contractor's obligations (in Section 2), the architect's powers, particularly in respect of ordering variations (Section 5), provisions for certification and payment (Section 4), and the methods of dealing with any disputes (Section 9). The conditions, when read as a whole, provide a comprehensive package where most eventualities are covered, allowing each party to refer to its rights under the agreement.

The schedules

The schedules to the contract contain additional provisions which the parties can choose to incorporate should they perceive the need to include them. Contractor's design submission procedure, collaborative working, insurance and fluctuations are amongst the schedules included. The fluctuation clause is concerned with any inflationary or deflationary changes which have come about in the amounts paid for labour and in the price of materials since the contract was entered into.

9.5.3 Other JCT forms

In recognition of the varying size and nature of construction contracts and developments in procurement, and in response to demand from clients, a further range of contracts is also published by JCT. These too were fully revised in 2011 and include:

- IC 2011 Intermediate Building Contract – designed for simple medium-sized projects;
- MW 2011 Minor works Building Contract – for simple contracts;
- PCC 2011 Prime Cost Contract – for use where the content and extent of the project cannot be precisely defined prior to commencement on-site;
- DB 2011 Design and Build Contract – the contractor is responsible for both designing and constructing the works;
- MC 2011 Management Contract – also has associated documentation for the tendering and contracting of works contractors;
- MTC 2011 Measured Term Contract – usually employed in maintenance contract situations;
- MC 2011 Client and Construction Management Agreement and Trade Contract – employer employs the trades direct and a construction manager.

In addition to these, JCT also publishes a 'plain English' consumer contract for domestic customers, called the Building Contract for a Home Owner/Occupier.

The Minor Works document is based on the main JCT form but is considered more straightforward. It is not thought suitable for works involving a budget of more than £250,000. The Intermediate form is a popular contract and is reasonably flexible in operation. It is recommended for use where fairly detailed contract provisions are needed but there are no complex building service installations or other specialist work. It can be used in both the public and private sectors and allows subcontract work to be placed with a 'named person'. This will typically involve mechanical and electrical packages of work. This procedure of nominating subcontractors has disappeared from the Standard Form of Building Contract due to reports of the difficulties experienced with operating the nomination provisions in practice.

The other popular form is the JCT With Contractor's Design, which is used in connection with design and build projects. This has a similar layout to and a mostly shared content with the Standard Form of Building Contract notwithstanding the different approaches taken.

9.5.4 JCT strategy

The statements of the principles by which the JCT operates were acknowledged by the Chairman of the JCT in his paper *The Place of Standard Forms of Building Contract in the 21st Century.*[2] This paper represents a robust defence of the position of JCT. This includes the following points with regards to the purpose of the contract.

- The point of a contract is to allocate risk in a recognisable manner, to achieve clarity and give the users the benefit of precedent.
- It is not the contracts that cause adversarial behaviour but the users of the contract.
- The JCT has stood the test of time and has seen other initiatives and innovation come and go.

Central to the JCT approach is that whilst effort should be made to resolve contractual issues as the project progresses, scope remains to revisit issues during the final account and review of extension of time entitlement at the end of the project. The intention to resolve matters as the project goes along is discernible in the valuation of variations where a quotation procedure is applicable as an option and in the inclusion of loss and/or expense claims within the interim application procedure. However, the ability to leave matters for a final reckoning, known as the final account, pervades the contract and differentiates it from the others.

Leaving things until later is not without its merits in terms of 'letting the dust settle' and taking an objective view of what might have appeared, at the time, to be a highly contentious issue. On the other hand, giving the parties scope to argue about time and money at the end of the project where the mutual reliance is less evident might appear questionable.

Similarly, the claim to having the benefit of precedent is also a double-edged sword. Chapter 8 contains instances where case law, and the subsequent amendments made by the JCT, has added further clarity to areas where some grounds for ambiguity have been discovered. However, the scarcity of case law dealing with NEC or PPC2000 wording has the effect of making the JCT approach seem more likely to lend itself to disputes. This is not the whole picture as the JCT has been around much longer than the other forms and existed before adjudication provided a private and confidential means of bringing claims and disputes. The JCT supporters would postulate that the new contracts mean new interpretation problems.

The JCT has not stood still whilst the other contracts have made inroads into its market share. The supplemental provisions to the standard form allow users to benefit from some of the new ideas circulating in the other forms of contract, particularly in relation to collaborative working and acceleration.

The overall impression is that the JCT tries to move with the times and adapt. The bind for the JCT is that it is precisely because of its conservative nature that lending bodies tend to require its use, and it remains a default choice. Lenders and clients not wishing to innovate or seem progressive will stick with the tried and tested JCT on whose terms lenders are content to lend and clients are happy to have their contracts administered for them by architects. This represents a large proportion of private and commercial development and the core market that the other forms would dearly love to penetrate.

The default nature of the JCT has led to some positive discrimination against it in terms of the government wishing to identify itself with newer forms of contract. The government endorsement of best practice has been given to the JCT competitors in what the latter would no doubt see as akin to a superpower intervention on the side of a disputant.

9.6 The New Engineering Contract

9.6.1 Introduction

The Institution of Civil Engineers (ICE) produced this form of contract in 1993. It represented a radical departure from its previous core contracts for use in connection with major civil engineering construction works. The objective of the NEC was to publish a single flexible contract in plain English that would act as a stimulus and guide to good management practice. In practice, its implementation has been slow but, following the sustained support of the Office of Government Commerce, it is beginning to make a significant impact and the 'snowball' effect of adoption continues apace. The form has been so successful that the ICE now promotes this contract as its sole recommendation for use.

The contract works by having main options allowing the employer different ways of pricing the project and different ways of paying the contractor. The main options do not change the rest of the contract, which remains the same. The main options are as follows.

- Options A and B – priced contracts giving a lump sum price for the works described in the activity schedule.
- Options C and D – target contracts whereby the parties agree a target price for the works. If the works cost more than the target price when complete, they share the 'pain' of the cost overrun in pre-agreed portions. Likewise, if the works are completed for less than the target price, the contractor and employer share the 'gain'. This mechanism is known as a 'painshare/ gainshare mechanism'.
- Option E – cost reimbursable contract in which the parties agree the level of the contractor's overheads and profit and the employer then pays the actual costs of the works plus the agreed level of overheads and profit.
- Option F – construction management option. This form is for use by experienced employers only (as per the discussion in Section 7.5).

9.6.2 NEC background and content

The original ICE contract dated from 1945 and the same form was used by private and public employers. Under the ICE contract, the engineer has extensive powers of supervision and control. The NEC is independent of the ICE. NEC was praised in the Latham report as being a means of responding to growing discontent within the industry about long-winded contractual procedures and adversarial attitudes. The NEC has succeeded in being applicable to both building and engineering projects. A further objective for the NEC was to accommodate all varieties of design responsibilities. This is achieved by the flexibility of using core clauses that can be adapted to various types of projects. Secondary option clauses can also be 'bolted on' to the contract to suit the client's requirements. All the clauses are written in plain English, using only common words written in the present tense. The NEC is now in its third iteration, known as NEC3, which first appeared in 2005 and was last amended in 2013.

The overriding aim of the NEC contracts have been to form a stimulus to good management and a collaborative and co-operative method of working. This approach is not without its doubters who prefer the legal certainty approach embodied by the JCT contracts. The NEC3 has been used successfully on some high-profile projects including the Channel Tunnel Rail Link and Terminal 5 at Heathrow Airport; most importantly, the Olympic Delivery Authority procured all work for the 2012 Games using this form of contract.

Notwithstanding the different philosophies of the JCT and NEC, the NEC approach to contractual layout is strikingly similar to that of the JCT and the basic obligations set out above. The core clauses are:

1   General
2   Contractor's Main Responsibilities
3   Time
4   Testing and Defects
5   Payment
6   Compensation Events
7   Title
8   Risks and Insurances
9   Termination.

Under the NEC3, the employer employs the professional consultants (each known as Consultant) and the contractor. Between them, the consultants and the contractor carry out the design and construction of the project. The project manager runs the project on the employer's behalf and takes many of the key decisions under the contract. A supervisor deals with any defects in the works on behalf of the employer.

9.6.3 Other NEC forms

The NEC3 is best described as a suite of contracts and incorporates all the appointments necessary on a project including:

- Engineering and Construction Contract (ECC);
- Engineering and Construction Subcontract (ECS);
- Professional Services Contract (PSC);
- Professional Services Short Contract (PSSC) – newly introduced as part of the April 2013 edition;
- Adjudicator's Contract (AC);
- Term Service Contract (TSC);
- Framework Contract (FC);
- Supply Contract (SC).

It therefore provides for the suppliers, subcontractors, professionals and adjudicator all to be appointed on complementary and sychronised contracts. For smaller projects, there is also a suite of shorter contracts, mirroring the JCT approach of providing a proportionate contract to the project.

9.6.4 NEC strategy

The key points in relation to this newer form of contract are that it can be juxtaposed to the JCT Standard Form of Building Contract position. The essential difference is that the JCT was written by lawyers for architects whereas the NEC is written by managers for managers. This is the underpinning philosophy that good management can improve the situation and lead to greater buy-in from all concerned.

The management ethos appears most prominently in the most popular form of the NEC–Option C. This option uses the target cost mechanism whereby the contractor is encouraged to meet or improve upon the target cost by offering an entitlement to a share of the saving made. This can be a fraction of the saving – say 50 per cent or some other figure. Some of the saving is usually returned to the employer. It is quite common to include other stakeholders in the gainshare arrangement, such as a key specialist contractor. The flip side of the gainshare is the painshare whereby the parties are also expected to contribute to the costs of overruns in the same percentage as the gainshare. This 'pain' arrangement can be instead of or as well as liquidated damages provisions.

The flexibility of the pain/gain arrangement has been much in evidence amongst contract users. The addition of caps and maximum rates of benefit was seen on the Olympic delivery. The cost of the major stadia in the Olympics was subject to such an arrangement and savings were made on the target cost. However, the fact that all of the stadia came in at the point where the contractor maximised his entitlement to a share of the gain left some commentators questioning whether the process had been manipulated by the winning consortium. Whether or not savings to the public purse had been protected by the arrangement or the target costs artificially set continues to be a moot point. It is undeniably true that great care needs to be taken in setting the target price.

The NEC has changed the contractual landscape and many of its ideas have been well received. The risk register is often cited as an innovation that helps with the proactive management of contracts and risks in particular. The central notion here is that the contract's role of allocating risk in a recognisable manner is supplemented by a register on which the specific risks of the project are recorded and discussed. For instance, if the risk of unforeseen ground conditions was particularly acute on the project then measures could be put in place to deal with this risk. This could manifest itself on a remediation project as a requirement for additional soil surveys and contingency allowances.

Other NEC innovations are less obviously novel. It is a self-evident truth that any construction contract must deal with the potential for overruns of time and money and contain effective mechanisms in this regard. In the case of the NEC, this is the early warning and compensation event procedure. In common with that other great engineering-based contract, the FIDIC, the NEC has the requirement for timely notification of claims, either actual or potential.

In law, this element is known as the condition precedent. The meaning of this is that if notice is not given of an event then the contractor is precluded from making a recovery for these sums later. This is the justification of the NEC approach for not having a final account as such – all interim matters are dealt with finally at the time that they arise or in accordance with the deadlines attached to the event. In terms of compensation, events notice must be made within eight weeks of the event occurring. In relation to where an early warning ought to have been given (and the test is whether the reasonably competent contractor would have given the notice), the compensation sought cannot exceed the sums that would have been claimable had the notice been given.

The desirability of the 'sort it out now' approach to construction contracts is self-explanatory. All the stakeholders involved should welcome the chance to deal with matters around the time they arise and not to store problems and disputes for later. The counterpoint to this approach is to recognise that such a regime is resource-hungry with personnel required to be committed to operate the procedure and to keep on top of the many early warning notice and compensation events which may be issued by a contractor aiming to preserve its right to claim for time and money overruns.

Another example of the resource-hungry NEC approach is the prominence given to the programme. The JCT programme is viewed as a contract document which ought to be updated and referred to in terms of requests for extensions of time and keeping the parties informed. In practice, the programme on JCT projects is usually given low importance, and it is only usually referred to when the parties want to argue a point in relation to the same.

The NEC approach is very different with the programme being viewed as the single most important document on the project. Management theory dictates that every effort must be made to have a firm grip on exactly where actual progress is against planned progress. To this end, the requirement on the NEC is to keep the programme as a living, breathing entity that is constantly updated and referred to when any early warning is given and at any point where delay events may occur.

Other threads of good management are apparent in the NEC in terms of its layout and flexibility. The NEC takes a 'mix and match' approach with the different core options and secondary options. Some of these include the partnering Option X12 which allows for the inclusion of key performance indicators (KPIs).[3] A cap on the maximum liability of the contractor is introduced by secondary Option X18. Liquidated (or in this case delay) damages also appear as secondary Option X7 and are not seen as applicable whatever the circumstances. The potential for an overlap between the 'pain' provision and the levy of delay damages could leave the contractor feeling overly put upon in terms of risk for overruns.

The other area in which the NEC makes a change from the norm is in its inclusion of a good faith clause. Clause 10.1 requires the parties to act in a 'spirit of mutual trust and co-operation'. This is a clause which splits opinion as to its purpose and whether it should be included in a contract.

The English law tradition did at one time have good faith doctrine, but this fell into disuse. Good faith provisions are discussed in Section 22.8. Recent cases have struggled to find a meaning for good faith provisions, and a distinction is sometimes drawn between 'aspirational' clauses and the rest of the contract. Traditionalist lawyers would question the benefit of a clause which is aimed at regulating behaviour where the prevailing view is that the parties are free to act how they like provided that the contract terms are observed.

9.7 FIDIC conditions of contract

9.7.1 Introduction

This form is produced by the International Federation of Consulting Engineers (French acronym FIDIC) in association with the European International Federation of Construction. The contract is designed to be used worldwide and is based on the *ICE Conditions of Contract* (4th Edition) with modifications which enable the parties to elect which nation's legal system will be applied to the contract. The contract includes provisions for nomination of subcontractors, settlement of disputes, extensions of time, liquidated damages and all the other complexities usually found in major standard forms of contract. FIDIC forms are used throughout the world and are endorsed by many development banks.

9.7.2 FIDIC background and content

The FIDIC organisation was founded in 1913 by construction professionals coming together in France, Belgium and Switzerland. The UK joined in 1949. The first standard form of contract for works of civil engineering construction was published in 1957 and was known as the Red Book. The most up-to-date FIDIC forms were published in 1999 and are known as the 'rainbow' suite of contracts. Contracts matching different procurement routes are given different colours. The most popular is the Red Book, which is based on traditional procurement whereby the contractor is given the detailed design by the employer.

All of the contract forms include general conditions together with guidance for the preparation of the particular conditions, and a letter of tender, contract agreement and dispute adjudication agreements.

The strucutre and content of the FIDIC contracts are based on ICE conditions. The key role is that of the engineer appointed to act as representative and agent for the client. The contract contains, in 20 clauses, the normal provisions to be found in standard forms relating to building works. The international application of the contract is demonstrated by its flexibility on the choice of applicable laws and the provision for resolution of disputes by arbitration under the International Chamber of Commerce (ICC) rules. The dispute resolution aspects of the standard form is discussed further in Part 4.

9.7.3 Other FIDIC forms

In keeping with the 'rainbow' colour approach, the following contracts are also published by FIDIC (this list is non-exhaustive).

- The Silver Book is for contracts let under design and build contracts and this includes the EPC approach discussed in Section 7.8.1.
- The Yellow Book is also used for design and build, usually for projects with a large element of electrical and mechanical works.
- The Green Book is the short form of contract suitable for fairly simple or repetitive work of short duration.
- The Pink Book, launched in 2005, is an alternative to the Red Book approach that specifically suits projects funded by development banks.
- The Gold Book, published in 2008, is the contract for use with BOOT scenarios as discussed in Section 7.8.2.
- FIDIC Conditions of Subcontract for Construction.

9.7.4 FIDIC strategy

The World Bank and other development agencies have had a long association with this form of contract to the extent that no money is usually lent to organisations and governments unless this form of contract is used.

The argument in favour of the FIDIC is that it is tried and tested, and the World Bank appears content with the level of security and control it can maintain where the contract is used. The most obvious example of this is in the insistence on the use of dispute review boards and international arbitration as the dispute resolution mechanisms of choice.

The FIDIC suite has elements of the other contracts in its make-up. The JCT approach allowing final accounts appears whilst other elements are common with the NEC, such as condition precedent in respect of making claims. In its pedigree, the FIDIC is closer to the NEC as they both have a shared ancestry in the ICE form. This places the contracts more in the realm of the site engineer micromanaging the project rather than the jobbing architect visiting several sites in rotation.

9.8 PPC2000

9.8.1 Introduction

The Project Partnering Contract was published in 2000 by the Association of Consultant Architects (ACA). The latest amendments were issued in 2013. PPC2000 is a form of multi-party contract for procurement capital projects in any jurisdiction. Like the NEC, this contract can be adapted to suit different project sizes and types and different procurement routes. The latest version incorporates all the statutory requirements and includes a BIM amendment, and it can now incorporate public sector payment terms. The contract is published in conjunction with supplementary documentation.

In response to the Egan report,[4] the ACA brought out the first SFC explicitly designed to be used in a partnering context. The principles are fundamentally different from most other SFCs in that the whole project, including the inputs of consultant and key specialist contractor agreements, is governed by a single partnering contract. The contract expressly recognises recommendations on supply chain management and incorporates provisions to encourage the sharing of incentives and non-adversarial dispute resolution processes.

This contract has only enjoyed a modest share of the contracts market in recent times. Nevertheless, it is noteworthy for a number of reasons. This contract is a multi-party contract which moves away from the standard linear method of contracting in the industry. Instead, the PPC2000 involves the novel idea of including all the key players on a construction project in the same agreement. The contract becomes the 'hub' to which the parties are connected as spokes. The linear or supply chain model is shown on the right in Figure 9.1 with the hub-and-spoke model shown on the left.

The market share for the PPC2000 may grow beyond its modest impact to date because the government appears to be heavily promoting this form of contract as part of its drive towards collaborative practice and the adoption of Building Information Modelling as standard. This contract therefore appears set to become more popular on large government projects. The value for the student in studying the contract is to be aware of its embodiment of good practice and to see it as an example of how new ideas can take hold in the industry. These themes are explored in more detail in Part 5.

9.8.2 Other PPC forms

Other forms of contract include:

- the standard form of specialist contract, SPC2000, which underpins the contribution that specialist subcontractors can make by being appointed early and being integrated into the project partnering team;

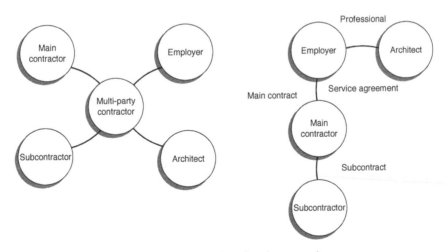

Figure 9.1 Linear approach compared to hub and spoke approach

- PPC International, published in 2007 to provide a form capable of being used overseas;
- TPC2005 Term Partnering Contract for works of a repetitive nature or where multiple projects will be let under one arrangement.

9.8.3 PPC2000 strategy

Good faith clauses are part and parcel of the partnering approach to contracts, which is the underpinning philosophy for the PPC2000. The key points for this section are to reflect on its multi-party nature and its emphasis on collaboration.

In some ways, the PPC2000 can be regarded as having been ahead of its time; it is only with the advent of BIM that its 'hub and spoke' approach can be readily appreciated. Certainly, its market share to date has been disappointingly low with a small but committed minority being regular users, most notably including housing associations.

The lawyers' nervousness around the question 'who is liable to whom for what?' is partly assuaged by the 'net contribution'-type clause, which states that parties should take their share of responsibility as would be assessed on a just and equitable basis. The PPC appears to be a seductively simple contract. However, this would be slightly misleading given the presence of a number of ancillary documents to the main form, including a partnering agreement, a pre-commencement agreement and a joining agreement. Critics of the PPC have pointed out that underneath the partnering guise lie some familiar JCT-type provisions. The remaining uncertainty over multi-party liability has left some doubters questioning the overall benefit of the contract.

9.9 Conclusion

The development of construction contracts over many years has been based on the assumption that the relationships between the parties are essentially adversarial ones. The SFCs that have been developed have tended to reinforce this and have frequently been criticised for encouraging conflict rather than avoiding it. Although the resulting conflict can be creative in some circumstances, it has generally been seen to have a negative effect on projects.

Construction contracts have certain essential features which are present in each of the forms discussed above. The extent to which the different forms put their own approach on these basic ingredients provides the academic interest. In truth, the notion that a client will form its own view on the contract it wants to use is only partially accurate. The government is able to drive good practice at an ambitious rate in the public sector with the risk that a two-tier construction industry might result. Private construction development remains stuck in a business model of amended JCT contracts with an improved risk profile for the employer and its funders. Until the latter are convinced that the means by which they procure their buildings is worth their investment in terms of considering the new approaches then the majority of projects will continue with the JCT. That is not to say that the JCT is, in any event, as reactionary as its detractors make out. Its incorporation of new ideas without abandoning its principles is to its credit. Further moves towards the approach of its competitors remain a real possibility as does the emergence of one jointly approved form of contract.

Another point in defence of the JCT is that it is the one true pan-industry contract. The representatives on the JCT are from all parts of the industry, client side and supply side. This has resulted in a fair contract. The other forms have more of a biased agenda, with the FIDIC and NEC palpably more client friendly. This is discernible by asking the question: on whom does any unallocated risk rest ultimately? In the JCT, it is for the client to take unallocated risk whilst on the other forms, it rests with the contractor.

The partiality of a construction professional to a form of contract can run quite deep. Experience in the formative years of one's career is when the preference is set, and consciously or otherwise, it is the original forms to which one prefers to return. This could explain the curious statistics in the RICS Contracts in Use survey where the JCT minor works forms were used well beyond their prescribed level and two instances of FIDIC contracts were used domestically. It would appear that the clients or advisers involved wanted to follow the old maxim of 'stick with nurse for fear of something worse', which is another aspect of preference based on experience. This unwillingness to embrace change is a telling characteristic at many levels in the construction industry.

9.10 Further reading

Uff, J. (2013) *Construction Law*, Eleventh Edition, London: Sweet & Maxwell, Chapter 11.

Hughes, W., Champion, R. and Murdoch, J. (2015) *Construction Contracts: Law and Management*, Fifth Edition, Abingdon: Routledge, Chapter 8.

Bunni, N. G. (2005) *FIDIC Forms of Contract*, Third Edition, Oxford: Blackwell Publishing.

Thomas, D. (2012) *Keating on NEC3*, London: Sweet & Maxwell.

Clamp, H., Cox, S., Lupton, S. and Udom, K. (2012) *Which Contract?* Fifth Edition, London: RIBA Publishing.

Notes

1   This is the second annual NBS National Construction Contracts and Law Survey, published in 2013 and based on data for 2012: available at: www.thenbs.com/pdfs/NBS-NationlC&LReport2013-single.pdf.

2   Hibberd, P. (2004) *The Place of Standard Forms of Building Contract in the 21st Century*, a paper based on a talk given at a Society of Construction Law conference, Wakefield, 11 March: available at: www.scl.org.uk/papers/place-standard-forms-building-contract-21st-century.

3   This is a system whereby continuous improvement is encouraged by benchmarking performance against preselected criteria, such as fewer defects, fewer accidents, greater

saving and less waste produced. The system is usually seen in connection with repeated work whereby the contractor has the opportunity to demonstrate improvement over the course of a number of projects. Failure to meet KPIs could result in being removed from a framework list of contractors.

4   Egan, J. (1998) *Rethinking Construction*, London: Department of the Environment, Transport and the Regions.

Part 3

Construction contract law

10 Construction contract law

10.1 Introduction

The student has now formed an appreciation of the processes at work in forming construction contracts as well as gaining familiarity with some aspects of the supporting legal concepts. The student should also have established an understanding of why risks are allocated in the ways they are and what the parties can do if they wish to alter this profile. This chapter aims to make some of the key obligations clearer and to give the student insight into the inter-relation and operation of key terms and contractual practices.

10.2 The basic obligations

The basic obligations shown in Table 10.1 are the heart of the commercial bargain struck between the parties. These are the essential obligations undertaken and the rationale as to why a construction contract is required. Construction contracts approach these basic obligations in different ways depending on the importance they place on a diverse range of aspects of project delivery. For example, a contract for the construction of a nuclear power station is likely to be a different undertaking to a contract for a simple house construction. The former needs to provide for maintenance and safety checks and complicated design whereas the latter is required to provide a comfortable home. The emphasis the contracts place on different points can be traced back to the original use and ancestry of the latest contract forms. Whatever the clause being considered states, it is likely to relate back to one of these basic matching obligations.

Table 10.1 establishes what a building contract must provide as a minimum. These 'bare essentials' allow the student to gain an appreciation of the parameters of construction contracts.

Table 10.1 The basic obligations

Pay	Build
Instruct	Obey
Set deadlines	Meet deadlines
Give possession	Take possession
Give design	Follow or complete design

Pay/build

The contract must say when, where and how the contractor is being paid and what for. In return, the services rendered by the contractor must be made clear. Questions can arise around exactly what the contractor is to build and where the design is to be found. This first essential pairing of obligations can be seen as the *consideration* of the contract.[1]

Instruct/obey

The ability to make decisions for and on behalf of the employer needs to be included in the contract. The terminology used varies and the office holder may be known as the architect/contract administrator or the project manager or engineer. The extent to which the contractor is required to obey any given direction needs to be defined. Questions arise around whether the contractor has a right to challenge any decision made and whether this can be before or after taking the compliant behaviour.

The role of the overseer is required, amongst other things, to record when time and money events have happened. Other roles include dealing with discrepancies between contract documents and issuing variations where the employer changes his mind about what he wants. The overseer has other powers such as the ability to exclude unruly subcontractors from site and to order the cessation of work in the event of *force majeure* or the discovery of archaeological remains in the area.

International contracts have a tradition of allowing the resident engineer greater powers in relation to the questioning of his own actions by the contractor. Historically, the first port of call in the event of a challenge to the engineer's decision was to the engineer himself. This was the case despite an obvious conflict of interests arising. This large degree of autonomy brings to mind the imposing (though diminutive) figure of Isambard Kingdom Brunel who had a huge influence on Victorian Britain and has an impressive legacy in the West of England in particular. His role as Emperor of his projects was understood by all concerned, and his treatment of financial backers and workers displayed a disregard for anything that stood in the way of his project.

The difference between engineering and construction contracts is also writ large in the way the standard form contracts envisage operation by the overseer. The JCT depends on the architect's occasional visits to site – an architect could expect to have a number of projects on at the same time and visit them in a fortnightly rotation with the occasional mystery visit. The engineer would be on-site throughout and entirely hands on. This greater emphasis on management explains the approach of the NEC and the desire to be constantly updated and informed of progress and matters that may affect progress.

Set deadlines/meet deadlines

Timing is a crucial feature in building contracts and is often equated to having the same importance as money. The building contract must therefore provide a mechanism for setting the original timescale but, as importantly, must also give a mechanism to move the deadline upon the occurrence of certain events. Questions arise as to who sets the deadlines and the importance of a construction programme.

Give possession/take possession

The importance of estates in land has been covered in Chapter 5. The contractor cannot be expected to carry out and complete the works unless he is given access and can take possession of the site. Directions as to how and when this will happen and the responsibilities that consequently pass to the contractor need to be spelled out in the contract.

Give design/follow or complete design

The contractor must be given some indication of what he is to build. The degree of detail given ranges across the forms of procurements available. Another variable is the extent to which the contractor is required to complete the design from the stage at which it is handed over to him. This procedure is used in design and build procurement. Questions can arise around what happens with ambiguities or errors in design information and the interface between different personnel in the design team.

There are myriad other issues a building contract needs to address, depending in part on the project nature, complexity and value. No study of construction law is complete without a consideration of these wider points, such as insurances, health and safety requirements, subcontracting, determination provisions and liability for the cost of overruns. These core obligations from Table 10.1 are used as the guide for this chapter; each pair of undertakings is unpicked further to describe how construction contract law works.

10.3 Pay/build

This is the central obligation of the contract. The contractor is in business to make money and keep his employees and funders content. The contractor does this by selling his services – which might include design and facility management services as well as construction – to the employer for monetary reward.

10.3.1 Interim payments – general

Building projects are often of a relatively long duration when compared to other commercial transactions. One implication of the length of project is that the contractor is usually entitled to be paid as he proceeds. This contrasts with other industries where payment is made on delivery; for example, buying a car or a consumer good. Progress payments have benefits for both parties. If the contractor was obliged to fund the works up to completion then a good deal of the working capital and/or borrowing of the business would be tied up during the build. Ultimately these borrowing costs incurred by the contractor would be passed on to the employer as part of the tender costs. The employer can avoid these charges if he agrees to pay for work performed during the build. This leads to a discussion about the nature of the intervals used and even whether an advance payment should be made.

Building contracts have many provisions concerning money and a considerable proportion of the contract is given over to this most important of considerations. The contract provides for such monetary considerations as:

- the repayment of advances, usually repaid from earned value during the first few interim payment instalments;
- the retention of a small percentage of earned value to ensure that the contractor returns to remedy any defects coming to light at the end of the project; and
- the right to claim interest on late payments.

Section 6.6 discussed the statutory payment entitlements enshrining the right to interim or stage payments. It has been described how the SFCs were amended to ensure 'adequate mechanisms' were in place to comply with the statutory regime. However, the change introduced by the Housing Grants Construction and Regeneration Act 1996 (HGCRA) in terms of providing for interim payments was already standard practice and had been promoted as long ago as their early versions.

The simple idea is that at the end of the first month of the build, the overseer either values the work himself or calls upon the quantity surveyor appointed to do the same. The valuation process can either start with the overseer's own valuation or be prefaced by the contractor's application. The latter option is preferable for the contactor who can set out what he considers his entitlement to be, plus the overseer has a starting point for his own valuation.

A typical valuation, in its simplest form, contains a claim for labour (whether carried out directly by the employer or subcontracted), materials (whether installed or stored on- or off-site) and an allowance for overhead and profit.

Those unfamiliar with the construction industry are often surprised by the relatively modest amount of profit claimed on a project. Low profit margins can be a product of competitive tendering procedures where the contractors have gone to great lengths to win the tender based on a low price. An unrealistically low tender which then wins the project can be the single element most to blame for subsequent problems on a build.

The valuation process can involve additional processes such as a consideration of earned value and planned versus actual costs. These monitoring provisions of the contract spend provide valuable information for both employer and contractor and are common on larger projects.

The contractor's application and corresponding valuation are the start of the payment cycle. The opportunity then exists for the paying party to deliver its notices in respect of the valuation, comprising the payment notice within seven days of the *due* date for payment and the pay less notice within five days of the *final* date for payment.

The sum valued is payable less any retention of abatement or set-off applied. Retention is a small percentage (usually 3 per cent or 5 per cent) of the earned value retained by the employer to ensure that the contractor returns to fix any defects arising at the end of the project. Abatement is a term used for a deduction of the value claimed by reason of the valuer's assertion that the work

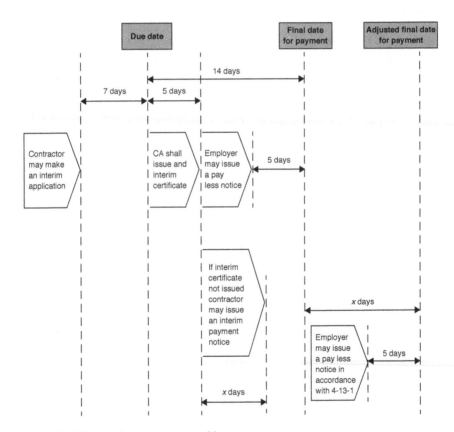

Figure 10.1 The interim payment timetable

is worth less than the amount claimed. It may, for example, include defects against which the deduction is made. Abatement needs to be distinguished from set-off which allows for the deduction of cross-claims from the money otherwise owed. There are two categories of set-off: legal and equitable. Legal set-off sees cross-claims being deducted where they are represented by a clear and incontestable sum, such as liquidated damages. Equitable set-off has a wider scope in that claims can be less well defined but must be very closely connected with the original claim. Building contracts are usually very clear on the exact nature of set-offs, which are deductible.

10.3.2 *Subcontracting and payment*

The position of the contractor with regards to his subcontractors also needs to be considered. Essentially, the role of the main contractor is to oversee and manage the work of the subcontractors and present their accounts with his own to the client for payment. The contractor must ensure that his cash flow and valuation dates accommodate this approach and that he is not liable to pay out money before he has had the chance to claim it.

Managing cash flow was, at one time, much easier for the contractor by virtue of what were known as pay when paid clauses. These clauses were discussed in Section 6.6.2 as now having been outlawed. The approach was, and still is in some parts of the world, that the subcontractor could not seek payment of his account until the contractor had been paid. It would be perfectly valid to say, therefore: 'There is nothing wrong with the work you have performed for me; however, I have not been paid so you will have to wait'. The inequality of this position is self-evident. Unfortunately for the subcontractors, many main contractors were able to bypass the statutory ban and come up with other clauses that have the same effect. New legislation on this point was introduced by the Local Democracy, Economic Development and Construction Act 2009 (LDEDCA) banning all conditional payment arrangements. Whether this led to a final or temporary abolition of pay when paid-type terms is hard to determine.

Subcontractors are sometimes asked to wait for a period of up to three months for their payment, which can, in real time, mean up to four months given that the valuation is usually at the end of the first month's work. This tension on their own supply chain can often result in insolvency pressure as discussed in Section 3.4. The government has for a long time been aware of this inequality in bargaining power and the construction Acts (HGCRA and LDEDCA) were drafted with this in mind. There have also been other initiatives aimed at improving the cash flow, including project bank accounts and removing retention payments. These initiatives are discussed in Chapter 21 as part of the new directions in construction law.

10.3.3 The 'build' obligation

The other aspect of this first matching pair of obligations is the 'build' obligation. Building contracts do not generally particularise this aspect and tend to let the background law fill in any blanks left. The Sale of Goods Act 1979 and the Sale of Goods and Services Act 1982 together with the Building Regulations prescribe the requisite tests, supplemented by the law of negligence if required. The standard required for building can thus be described as being to satisfactory quality and in accordance with Building Regulations and to the standard of a reasonably competent fellow professional. As has been observed, the standard of the build is unlikely to be to the higher 'fitness for purpose' warranty although this is used for some larger projects, power plant engineering contracts in particular, most notably where FIDIC is used. This is one of the differences evident between engineering and construction contracts.

10.4 Instruct/obey

10.4.1 Instruct

The master and servant role has been alluded to earlier in this book.[2] The extent to which the overseer role of the architect/contract administrator or project manager fulfils the master role of 'he or she who must be obeyed' is worth reflecting on. It is the case that the overseer is the employer's representative and they are themselves a servant to that degree. However, as the party to whom all authority is delegated, they can be viewed as the embodiment of the employer's power. It has also been observed that the overseer has a measure of discretion in how they interpret the certifier role and how they ought to resist pressure from an employer who improperly seeks to influence their decision-making in as much as it affects the certification process.

The question arises as to the extent to which the contractor must do the bidding of the employer as directed by the overseer. Most building contracts allow for the contractor to be able to raise any reasonable objection to instructions, but ultimately the contract works on the 'do as I say' basis. The right of the contractor to claim additional time and/or money for acting as directed where a variation is concerned is one of the protections they are given. All instruction should be given in writing in order to create an auditable trail of who told whom to do what and when.

Under the JCT contract, the overseer has the ability to issue the following instructions:

- deductions from the contract sum for errors in setting out and defective work;
- inconsistencies between contract documents;
- expenditure of provisional sums;
- postponement of the work;

- opening up for inspection/testing;
- removal of work;
- exclusion of persons from the work; and
- variations.

The items on the above list are common-sense powers required for effective contract administration. Clearly the overseer is not going to sanction payment for work adjudged to be substandard or defective. The power of the overseer is to give the contractor notice to put the work right and, if the contractor refuses to comply, to then award the work to a third party and contra charge the cost to the contractor. The defective work may have been discovered by virtue of the opening up provisions. This involves the overseer testing part of the works, including those parts now incorporated into the build. This might be, for example, to check if a damp-proof course has been fitted correctly or the correct floor material has been used. Postponement is self-explanatory and could occur where the employer has an issue with funding or planning which requires a temporary cessation of the project. Removing persons from site is an instruction that might be given in a situation where a subcontractor is repeatedly abusive or in breach of warnings on quality of work.

10.4.2 Obey

Variation clauses give a good indication of the limits of the duty to obey and the compromises reached by the building contract. The situation with regard to instruction and obedience is altered slightly where it is the employer who has changed their mind about an aspect of the contract and now wants something else. This unilateral alteration of the contract is not in the same category as the overseer enforcing the contract terms (as dealt with under Section 10.4.1). The contract practice around variation is therefore worth considering in further detail here.

Ideally, the employer should not change either the design or the work required from day one until completion. However, the nature of construction work is such that it is rare for the building owner to know exactly what the final requirement will be. It follows that changes of one kind or another are virtually inevitable as the work progresses.

Contractors often view any change in what they have been asked to do as unwelcome or disruptive to their work. However, it should always be borne in mind that changing one's mind is the prerogative of the payer.

It follows that one of the key elements to successful construction is flexibility. A successful contractor must be flexible in what work they are prepared to do and in their attitude towards being asked to do extra/varied work. Most construction contracts therefore assume that the contractor is willing to do additional or varied work. This assumption is subject to two conditions: that the change made is reasonable and that the contractor is adequately compensated for the change.

There are many different sorts of changes. These range from wholesale changes in design through to minor specification changes, such as the colour of materials ordered. Changes might be needed for instances such as:

- minor design changes and enhancements;
- correction of errors in design documents;
- responding to changes in legislation; or
- other external factors.

Any contractual arrangement between the employer and the contractor needs to cover the whole range of potential changes from the small and seemingly insignificant to much bigger changes in design. If no contractual provision was made for change then, in theory, the contractor could stop work and call upon the employer to deliver a new contract to incorporate the changes. The resulting delay caused by any such renegotiation/retendering would be very disruptive and lead to a huge wastage of time and money. To avoid this, most contracts include a clause enabling the employer to change or vary the contract works. Such provisions are usually called variation clauses. The purpose of a variation clause is to allow change to be made whilst keeping the work within the boundaries of the existing contract. A variation clause also permits any consequential changes to be made to the contract sum.

In most cases, variations mean that the contractor is obliged to perform extra work. The key questions facing a contractor are likely to be:

- Is it extra work?
- Does the contractor have to do it?
- Can the contractor get paid for the extra work carried out?
- What will the contractor be paid for the work?

Is it extra work?

A contractor frequently carries out, or is asked to carry out, work for which he considers he is entitled to payment in excess of the original contract sum. In a widely defined contract, the contractor must complete the whole work for the agreed price. The contractor will have to do whatever is necessary to deliver the building/project required, no matter if the specification is silent or even misleading in parts. The case of *Williams v Fitzmaurice*[3] involved a whole contract to build a house. Flooring was omitted from the specification. The contractor refused to put flooring in the house unless it was paid for as an extra. The contractor was thrown off-site, and his floorboards were seized and laid by a replacement contractor. The court held that the contractor could not recover the outstanding contract sum even though the flooring was not mentioned in the specification. The judge said that it could be *clearly inferred that the contractor was to do the flooring.*

The case of *Sharpe v Sao Paulo Railway*[4] involved a contractor undertaking to construct a railway line in South America from 'terminus to terminus'. The original engineering plan was inadequate. Acting on the engineer's orders, the contractor carried out nearly two million cubic yards of excavation in excess of quantities set out. The contractor claimed entitlement to additional payment for the extra excavation work. The judge in this case held that these were *not in any sense* extra works and refused the contractor's claim.

The situation is different with exactly defined or bills of quantities contracts. The bill of quantities is the long-established practice whereby the quantity surveyor surveys the quantities and 'takes off' the amounts of labour and materials needed to construct what is shown on the architect's plans. The resulting bill of quantities is given to contractors to price and return as a tender for the works. The bill of quantities with the contractor's agreed prices then forms one of the contractual documents, and any divergence from it can form the basis for a variation claim/calculation of extra work involved.

Where the work is exactly defined, the contractor is still bound to carry out work in excess of that stated in the bill of quantities if required so to do. The difference is that such extra work should result in additional payment to the contractor. The JCT 2011 family of contracts are based on a bill of quantity approach and therefore this principle is applicable to their use. It is less important to distinguish what is extra work in 'measure and value'-type contracts. Here, the work performed by the contractor is measured at the end of the contract and payment made at the rates set out in the contract. Any extra work which was not originally contemplated can be swept up in the measurement at the end of the project.

Does the contractor have to do it?

Under the traditional route of procurement, the role of the architect is (amongst other things) to instruct the contractor what work they should do. However, the architect has no general power to do this and they must be enabled by a contract clause to issue variation/extra work orders.

Such a power is included in most construction contracts. If it is not then the contractor needs to tread very carefully. Acting on an architect's instruction where they do not have power to instruct you will not saddle the employer with liability for the extra work. Interestingly, the architect may be liable to the contractor for breach of warranty of authority.

What if the extra work required involves large-scale and significant changes to the nature of the work? Is the contractor still obliged to do it? The basic principle which has emerged from case law and practice here is that variations that go to the root of the contract are not permissible without the contractor's consent. A contractor does not necessarily have to perform a major change to what was originally contracted for. Clearly the contractor may decide to perform the varied work anyway – particularly if the contractor considers the work to be lucrative.

A good example of changes going to the root of the contract and beyond is supplied by the case of *Blue Circle Industries Plc* v *Holland Dredging Co.*[5] An instruction was given by the architect under a dredging contract to dump material dredged from the bottom of a Scottish loch. The instruction was to use the material to form an island bird sanctuary in the middle of the loch. The contractor objected and the employer sought a court order to compel the contractor to obey the instruction. The court held that the instruction was outside the scope of the original contract and therefore the work had to be carried out under a separate contract. The change ordered here had gone beyond the root of the contract, which was to dredge, not to form bird islands.

The rule has emerged therefore that the contractor cannot refuse to do work which is reasonably ancillary to the contract works. The JCT contracts provide a right to object whereupon the matter can be adjudicated upon in default of agreement.

Can the contractor get paid for the extra work carried out?

It is only right and proper that the contractor will require payment having gone to the trouble of performing additional work at the architect's request. The principles which have emerged support this view. In order to receive payment for additional work, the contractor must carefully follow the contractual procedures and keep meticulous records of expenditure and justification for the steps taken.

The general rule is that the absence of written order or other condition precedent means that the contractor cannot recover money even though the employer has had the benefit of the extra work. The exception to the rule is where there is an implied promise to pay or a waiver of formalities – a very common occurrence on a building site. This requires the contractor to show that the employer was abusing his position and seeking to take advantage of the contractor's willingness to do extra work without being given a written instruction.

What will the contractor be paid for the work?

Extra work of the kind contemplated by the contract will be paid for in the manner provided by the terms of the contract. Payment will usually be at or with reference to the contract rates. If there are no relevant rates then a reasonable amount will be payable. What constitutes a reasonable amount depends on the circumstances.

Most construction contracts have a mechanism for valuing variations. The JCT contracts have rules based on how similar the work and conditions are to the contractual rates. Work in more difficult conditions and of a more demanding nature can attract an enhanced rate. Valuing variations is one aspect of how construction contracts have sought to improve on the general position at common law to add clarity and certainty to what would otherwise be grey areas – what, after all, does reasonable actually mean?

10.5 Set deadlines/meet deadlines

10.5.1 Set deadlines

Aspects of commencement and completion are usually described in the early sections of SFCs given the importance of these terms. The dates for possession and completion are entered into the contract particulars in order to ensure that both parties focus their attention. The duty on the contractor is to proceed regularly and diligently to progress the matter to completion on or before the completion date itself.

An interesting question stems from the extent to which the employer has to assist the contractor to finish the contract in advance of the completion date. The case of *Glenlion Construction Ltd* v *The Guinness Trust Ltd*[6] established that where the parties had agreed a process of design release, the employer was not under a duty to release the detail earlier simply because the contractor had reached a state of readiness for that detail ahead of schedule. In most instances, the employer would have been pleased at the prospect of early completion; but this is to miss the point that time certainty does not necessarily mean 'as soon as possible' as taking early possession of a building has implications for the employer who has to insure, secure and take back premises that have reached practical completion. This can incur costs for the employer for which they have budgeted to run from a certain time and not before. In rare instances, the employer does not want the building back for the foreseeable future if their plans for the building have fallen through. This might be an instance where an unscrupulous employer may seek to use their influence to delay the giving of practical completion, which should be resisted by the certifier as unwarranted meddling in his professional duty.

Unforeseen events leading to delay are common not only before commencement but during the works themselves. The operation of these clauses usually involved notice of delay to be given in writing by the contractor with supporting information estimating the likely effect on completion. The architect then considers the matter and, if satisfied that the delaying event is not something the contractor should be responsible for, will notify the contractor of a new completion date.

The grounds for allowing an extension of time are known as the 'relevant events' and concern acts of impediment, prevention or default by the employer. The occurrence of a relevant event does not automatically result in an extension of time and there is a requirement to show a link between the cause of delay and the effect. This can be straightforward but at times becomes extremely complicated, particularly where there is more than one cause of delay and these occur concurrently. Delay analysis is dealt with in Chapter 14.

The relevant events that will trigger an extension of time include:

- *force majeure* (otherwise known as acts of God but frequently including acts of man such as war or riot);
- exceptionally adverse weather conditions;
- loss or damage arising from employer insured events;
- compliance with architect's instructions;
- delay in receiving instructions or of design release; and
- delay by nominated suppliers and employer's employees.

These occurrences are employer risk events as discussed in Section 8.8, and they lead to additional time and potentially more money for the contractor (with the exception of a neutral event such as exceptionally adverse weather conditions).

10.5.2 Meet deadlines

If the contractor fails to meet the deadlines and is in culpable delay[7] then the fact that the contractor has not completed on time must be duly recorded. Under JCT, this requires a certificate of non-completion to be issued as the precursor to liquidated damages potentially being deducted or otherwise recovered by the employer. The right to claim liquidated damages arises out of being denied access to the finished project. A reduction or elimination of the contractor's potential liability for damages arising out of non-completion of the works can be achieved by obtaining an extension of time.

When the contractor has progressed with the works to the stage that the employer can take back possession, this requires a certificate of practical completion to be issued. Practical completion marks the start of the defect liability period. At the end of this period, the architect issues a certificate of making good defects, which is usually a precursor to the final certificate itself. Once this last certificate is issued, the architect's role has finished and he is *functus officio* as referred to in the case of *H. Fairweather Limited* v *Asden Securities Limited*.[8] This case involved an attempt by the architect to compel the unwilling contractor to obey instructions after the final certificate had been given. The contractor's right to refuse was upheld by the court.

Some projects never have the final certificate issued as to achieve this stage requires the final account and all outstanding issues, such as the completion of the health and safety file and operation manuals for all the mechanical and electrical installations, to be handed over.

The importance of programmes

Giving notice, whether under JCT or NEC, involves the contractor giving an estimate of the knock-on effects of the delay. This involves a consideration of the programme.

Many construction professionals will be familiar with the bar chart programme where each construction activity appears as a bar of time. The

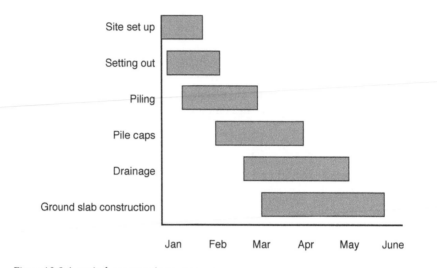

Figure 10.2 A typical construction programme

programme allows the parties to gain an appreciation of which trades require access at what time and the extent to which they can be overlapped. A major benefit of the BIM approach is that rather than being supplied with that 2D bar chart, the programme can be run to see the construction in a time-lapse representation of what will happen in real time. The models allow the project team to assess what the position will be on-site at any time in the build and assess the state of completion projected at that stage.

The importance of the programme is writ large in the NEC approach to contract administration. The JCT treats the programme as a contract document which is provided at the start of the project and potentially updated through the process although there is no primary responsibility on the contractor to do this. The JCT would use the programme as the starting point for the assessment of whether an extension of time claim was reasonable, but it is usually seen as a planning tool.

With NEC, the programme is a living document that is kept up to date and planned actual comparisons are carried out throughout the build to ensure that, at a glance, the knock-on effects of any delay can be considered and mitigated. Again, the downside of this is the resource needed, but if the project is large enough then the resource is an investment well made. Whichever approach is taken, it is certainly the case that the forecast of a delay is likely to be more accurate and easier to arrive at where the programme has been kept up to date with actual progress.

The early warning system of the NEC is often lauded as a breakthrough in terms of contract administration. The FIDIC contract has a similar system, and both approaches treat notice of delay as a condition precedent to being able to claim. This is in stark contrast to the JCT where the ability to finally resolve the time and money aspects of the contract is left to the final account, post practical completion window. This departure in how the contracts deal with the issue marks again the distinction between the 'hands on' management approach desired by the engineering forms and the more *laissez faire* approach of the JCT.

Both approaches have something to commend them. The NEC advantage of not storing up problems for later and addressing them at the time they occur is helpful in so far as it goes. However, the additional administration and resource-hungry reprogramming can be a headache for the contractor. The propensity for a good deal of communications to be generated through this system is also something which doubters have been quick to seize upon. Furthermore, when neither party follows the procedure, it is not uncommon for the NEC process to revert to a JCT approach but without any contractual basis on which to value the time and money aspects. That said, a competent NEC contractor and employer well versed in the heightened sense of project management can benefit from the security built in to the approach of addressing problems as you go along. The benefit of this additional emphasis on project management is also felt in such mechanisms as the acceleration quotation procedure.

Acceleration

Acceleration has historically been treated with suspicion by contract writers as being more trouble than it was worth. However, if the contractor is treating the programme as the management tool, it can be that an acceleration quote should be relatively straightforward to generate. The employer is not bound to accept the quotation and can seek to negotiate. The flexibility in this approach is commendable. The JCT added the acceleration procedure option to its latest versions.[9]

Whilst comparing the two main forms of contract, mention should also be made of PPC2000. This has a similar approach to the JCT with the addition of a partnering timetable for pre-construction activities. There is also a project timetable. The contract uses terms such as 'commencement' and 'completion date' and 'extensions of time' for matters beyond the contractor's control.

FIDIC contracts join the other SFCs in using another device which involves the contractor giving notice of when he thinks he has finished and inviting the engineer to agree this fact. Early acceptance of completion is without prejudice to the employer's right not to necessarily facilitate early completion.

The importance of pre-construction activity is likely to increase with the industry movement towards BIM and related technology. The timetable for developing the model with its attendant data and supply issues must be added to the overall project duration. The risk is that inadequate time will be given for proper development, resulting in difficulties down the line for the project.

10.6 Give design/follow design

10.6.1 Give design

Under the traditional procurement route, the tasks of design and construction are kept separate from each other. The requirement is on the designers to give the design to the contractor in sufficient time for the latter to be able to programme, resource and order the materials and specialists necessary to deliver the project. The traditional route describes how design must be complete before the contract is let. The need to have the design completed is lessened by the impact of such practices as the Information Release Schedule (IRS). This applies in a situation where not all the information required to construct the works has been prepared or issued when the tender documentation is issued. The IRS gives final dates for the release of information from the consultant team. Schedules should be kept up to date and may be changed to reflect the construction programme. The IRS may need to include specific release protocols where BIM is used. Failure to keep to the dates set out in the IRS may then be a matter for which the contractor attempts to claim an extension of time and loss and expense.

The ownership of the design is of importance to the designers. Chapter 4 discussed the protection of intangible property rights through the law of tort. This usually involves a consideration of the infringement of the design copyright.

Copyright is vested in the designer and is governed by the Copyright, Designs and Patent Act 1988 as amended. The designer retains copyright in the majority of cases and the client has a licence to reproduce the design in the form of a building. This is usually set out in the form of appointment. The licence is likely to be dependent on the designer's fees having been paid. The impact of BIM on the copyright of the constituent designs is the subject of some interest. The ease with which a model can be used to replicate a building elsewhere is of concern in some design quarters.

10.6.2 Follow design

Provided that the contractor constructs exactly as drawn or specified, he is likely to have little liability for the result unless any defect is so obvious that the contractor should have warned the employer of the potential danger.[10] The position changes if the contractor is not provided with sufficient information. In this situation, the contractor should ask the architect for the missing information. If the contractor, thinking he knows what is required, presses on with the work then he is placed in a difficult situation and may have assumed responsibility for the design. This was the case in *CGA Brown* v *Carr & Another*,[11] which involved the contractor's solution to a leaking roof problem for which the architect had given only sketchy instructions. The contractor was held liable for the whole of the re-felting work as he had created a roof vulnerable to leakage because of the inadequacy of the joint the claimant decided to make.

A different issue arises around design and build procurement and in particular around the requirement for the contractor 'to complete' the employer's design. Design and build contracts involve the taking over of responsibility for design at some point in the project. The taking over of responsibility raises some difficult questions such as who will now be responsible for that part of the design done before responsibility changed hands. The position was thought to be that the employer's designers remained liable for the work done before the appointment of the design and build contractor. However, the case of *Co-operative Insurance Society Ltd* v *Henry Boot Scotland Limited*[12] changed this perception. The design in question involved piled walls relating to the demolition and reconstruction of a property in Scotland. The contractor's obligation was to complete the design, which the judge interpreted as including an obligation to check the design that had already been prepared to make sure that it worked. The contractor cannot therefore assume that the existing design is correct.

The JCT took account of this judgement when drafting the 2005 and 2011 versions of its design and build contract. A new clause (2.11) was included which provides that the contractor is not responsible for the employer's requirements, nor for verifying the accuracy of any design contained in them. The standard form amendment reinstates what the popular perception of the position was previously. It is not uncommon for a bespoke amendment to be introduced to a design and build contract to achieve the outcome decided in the *Henry Boot* case.

10.7 Give possession/take possession

10.7.1 Give possession

The failure on the part of the employer to give possession of the site on the date specified amounts to a breach of a major term of the contract. Without possession, the contractor cannot execute the works and may be entitled to accept a repudiatory breach, rescind the contract and commence an action for damages. Several contractual mechanisms apply to possession of the site, underlining the importance of the issue.

Deferment of possession may be used where last-minute issues arise resulting in the employer not being able or willing to give possession of the site to the contractor at the appointed time. Examples of delaying events include satisfying planning conditions and funding being secured. Rather than allowing the contract to be frustrated by the failure to start and/or the deadlines to be compromised before the start, the system of deferment permits the employer to effectively have the contractor on standby and ready to commence when instructed. The consequences for the employer are that the contractor may claim for an extension of time and the employer may have to pay for this privilege as this is usually a recoverable head of damage for loss and expense. The contract envisages only a maximum of six weeks for deferment, but longer periods may be negotiated depending on project specifics.

Postponement is a separate contractual mechanism that may be used once the works have commenced. The difference here is that the contractor is in possession of the site throughout a period of postponement. Again, the period of any postponement will result in costs being incurred by the contractor and borne by the employer.

10.7.2 Take possession

One issue arising out of possession is the extent of the site which must be given over to the contractor to effectively start the project timetable. The contractor will request as much of the site as possible. The contractor is in control of how the works are to be programmed and is entitled to a free hand in their planning. The case of *Whittal Builders Co Ltd v Chester-le-Street DC*[13] underlines the point that piecemeal possession of a site is not sufficient. In this case, the contractor was entitled to additional time in which to perform the works. In another case, *London Borough of Hounslow v Twickenham Garden Developments Ltd*,[14] squatters (a man, a woman and their dog) had occupied part of the site and this caused delay and disruption to the contractor. Although only part of the site was occupied by the squatters, this was enough to restrict unhindered progress and the contractor was successful in claiming additional time and money.

The longer forms of the JCT allow for partial possession and completion of the work in stages. A piecemeal handover and completion of the site is complicated in practice, involving such things as insurances and the problems of moving people in and out of different parts of the site. Most contractors prefer to insist on possession of the whole site where possible.

The contractor's right to possession lasts only for the duration of the contract. If there is culpable delay for returning the site to the employer then liquidated damages may be levied against the contractor. The measure of liquidated damages is concerned with the occupation of the site and the income the employer would have generated had the contractor finished on time. Once the project is finished and free from known defects then practical completion can be certified. This is not the same as requiring that the completion must be carried out to the last detail, however trivial and unimportant. The judge in the *Westminster* case defined practical completion as *almost but not entirely finished*.[15]

Liquidated damages

Liquidated damages are a major area of difference between normal contracts and construction contracts. This is not to say that this type of damage clause is unique to construction as other industries specify the consequences in terms of delay for not abiding by the terms of the contract. The purpose of liquidated damages is to focus the minds of the parties on what will happen if the contract is delivered late. Lateness can often be seen as being as unwelcome for the employer as an overspend. Plans are made around delivery schedules – there are, for example, tenants to find, stock to sell and a time frame for delivery – often backed with commitments to financial institutions such as banks and funders.

The purpose of liquidated damages is that the effects of the delay are quantified into a measurable amount and translated into money sums. Liquidated damages are measured weekly or even hourly. In the case of an airport re-tarmacking, the knock-on effects of diverting planes to other airports might include demands for landing fees from those airports on top of the cost of cancelled flights and compensation to passengers. This could well make the cost to the airport in excess of £100,000 per hour. The conventional way to price this is by way of weekly fees with an instruction that the employer may charge £x per week or part thereof. The status of liquidated damages as a 'may' rather than 'must' clause gives the employer the discretion not to charge them in the event that the damage does not occur despite having been envisaged at the time the contract was entered into. The concept appears bizarre in that the employer may deduct liquidated damages even if the loss envisaged does not occur. This is, however, in keeping with the philosophy of contract law, which is to provide in the contract for the consequences of certain actions. Enforcing these promises is the role of the contract.

Liquidated damages are an emotive issue in contract negotiations and are frequently challenged. The role of liquidated damages is to assist the parties and lead them away from a position where the damages are unliquidated. In this sense, construction law is no different to the normal contract law discussed in Chapter 2. The purpose of damages is to put the claimant back in the position they would have been in had the breach of contract not happened, as enshrined in the *Hadley* v *Baxendale* two-headed test.[16]

Liquidated damages save the employer the time, uncertainty and expense of having to prove what is or is not a loss flowing from the breach and what was or was not in the parties' reasonable contemplation. By pre-agreeing the amount of liquidated damages, the parties are expressly contemplating the issue and recording their consensus on the point in the contract.

Liquidated damages must be a genuine pre-estimate of loss and must not be *in terrorem* as in the *Dunlop* case.[17] The pre-estimate can be rough and ready, but the exercise must be carried out. There must be some method behind the calculation and some thought given to what will happen if the building is late.

Two cases are worth reflecting on in terms of good contract administration and checking. In *Temloc Ltd* v *Errill Properties*,[18] the rate of liquidated damages was stated to be '£nil'. The court held that the liquidated damages clause constituted an exhaustive remedy, entitling the employer to nil damages in respect of delay. The employer was not entitled to claim general damages as an alternative. In *Jarvis Brent Ltd* v *Rowlinson Construction Ltd*,[19] the architect overlooked the requirement to issue the correct certificates (the equivalent of the certificate of non-completion) as the precursor to being able to deduct liquidated damages. The judge found that the oversight was fatal to the ability to deduct liquidated damages.

Some recent forms of contract have sought to offer alternatives to liquidated damage clauses, which they regard as being too blunt an instrument to incentivise the contractor's performance. One alternative is to share out the costs of the overrun based on the actual expenses encountered. However, the main forms considered in Chapter 9 all provide for something similar to liquidated damage arrangements.

The most straightforward way that the contractor can avoid having to pay liquidated damages is by seeking an extension of time. This 'picking up and moving of the finishing post' has been discussed in Section 10.4.1. If the contractor can establish that a relevant event has occurred then he can legitimately expect the architect to give him an extension of time as is fair and reasonable in the circumstances. The architect may insist that the contractor puts his case with contemporaneous evidence in support and goes some way to establishing 'cause and effect'. These two requirements are of paramount importance in law generally and nowhere more so than in construction law. Proving entitlement is the theme for the forthcoming chapters in this part of the book once the extraneous matters of construction contract law have been considered.

10.8 Conclusion

Contract administration is a complicated and involved area of study. Professionals administering contracts for the first time can find themselves overwhelmed by the terminology and formalities required to protect their client's or employer's interests. The simplification of these procedures into their key obligations as set out in this chapter should assist the first-time administrator. Ultimately, if all obligations are traced back to their origins then the student can maintain the overview necessary to appreciate the context of the matter in hand.

Not all construction contract law is capable of being attributed to the basic obligations set out in Table 10.1. The next chapter considers some of the outstanding issues that must be addressed to be able to operate in this field of practice.

10.9 Further reading

Uff, J. (2013) *Construction Law*, Eleventh Edition, London: Sweet & Maxwell, Chapter 9.

Hughes, W., Champion, R. and Murdoch, J. (2015) *Construction Contracts: Law and Management*, Fifth Edition, Abingdon: Routledge, Chapters 14 and 15.

Ashworth, A. (2012) *Contractual Procedures in the Construction Industry*, Sixth Edition, Abingdon: Routledge.

Notes

1   See Section 2.4.2.
2   Section 6.2.
3   (1858) 3 H & N 844.
4   (1873) 8 Ch App 597.
5   (1987) 37 BLR 40.
6   [1987] 39 BLR 89.
7   See Section 8.9.
8   (1979) 12 BLR 40.
9   Clause 2 of Schedule 2 of the JCT 2011 Standard Form of Building Contract.
10   *Lindenberg* v *Canning and Others* [1992] 62 BLR 147.
11   [2006] EWCA Civ 785.
12   (2002) 84 Con LR 164.
13   (1987) 40 BLR 82.
14   (1970) 7 BLR 81.
15   *Westminster Corporation* v *J Jarvis & Sons* (1970) 7 BLR 64.
16   See Section 2.10.
17   *Dunlop Pneumatic Tyre Co Ltd* v *New Garage Ltd* [1915] AC 79.
18   (1987) 39 BLR 30.
19   (1990) 6 Const. LJ 292.

11 Other aspects of construction contract law

11.1 Introduction

The previous chapter introduced the reader to the basic contract machinery working towards ensuring a successful build and a dispute-free project. This chapter adds some further detail around the key points covered to ensure that the student has an appreciation of some of the finer and more complicated nuances of construction contract law.

11.2 Priority of contract documents

The construction contract usually refers to other documents which contain relevant information and other terms agreed by the parties. The contract should identify which of the contract documents has priority in case of conflict or contradiction. The general rule is that any written words specifically agreed by the parties will prevail over printed or standard words. This is a rebuttable rule and can be overridden by an express term of the contract. Clause 1.3 of the Standard JCT Design and Build Conditions states that: 'The Agreement and these Conditions are to be read as a whole but nothing contained in the employer's requirements, the contractor's proposals or the contract sum analysis shall override or modify the Agreement or these Conditions'.

In the case of *Fenice Investments Inc v Jerram Falkus Construction Ltd*,[1] this was interpreted as giving priority to the amended conditions. The case involved two different interim payment mechanisms, and the parties could not agree which one should have effect. The judge settled upon the one that was given precedence as per the priority of documents clause.

The NEC contract also views the conditions and completed contract data as the most important document in resolving any ambiguity or inconsistency. The FIDIC approach is that the documents forming the contract are to be taken as mutually explanatory of one another. Where this results in inconsistencies, the sequence of priority is the contract agreement; the letter of acceptance; the letter of tender; the particular conditions; the general conditions; the specification; the drawings; the schedule; and any other documents forming part of the contract.

A non-exhaustive list of documents which might be referred to as 'contract documents' includes:

- pre-let meeting minutes;
- bill of quantities;
- plans;
- contractor's proposal;
- employer's requirements;
- health and safety plan;
- preliminary breakdown;
- tender;
- invitation to tender;
- schedule of valuation dates.

The sheer number of documents and the different stages at which they are prepared can be seen as a recipe for ambiguities and inconsistencies. An integrated approach to contract administration can remove the risk of these problems. Recent technological advancement in this area is provided by software such as CEMAR which can greatly improve the prospects of running a problem-free project. One of the benefits of an integrated approach to contract administration – such as the approach used in the CEMAR programme that is designed to support NEC3 contract management[2] – is to remove the risk of these problems.

11.3 Determination provisions

Construction contracts will typically provide a mechanism for identifying the circumstances in which the defaults of either party will entitle the innocent party to determine the contract. Parties need to take extra care when deciding to use these provisions. Serious consequences flow from their use. For instance, if a contractor has wrongfully determined the contract then the employer can seek to retain any money that is owed to the contractor in order to pay a replacement contractor to do the work.

Discharge of a contract was examined in Section 2.9. Here, more detail is given with respect to the general position at common law. 'Termination' is the term used to describe a contract coming or being brought to an end. The terms determination and termination are used interchangeably in the AEC industry. Termination does not create a situation as if the contract had never existed. Instead, it excuses both parties from further performance of their primary obligations. However, where a party has defaulted in performing its primary obligations prior to the termination then a secondary obligation to pay damages to the non-defaulting party arises.

Either party might wish to terminate a contract for the following reasons:

- one party's performance is unsatisfactory in certain critical respects;

- one of the parties is refusing to perform the contract; or
- one of the parties has gone into or is about to go into insolvency.

Termination of a contract is a drastic step, and the parties might attempt instead to agree a variation of the contract or to renegotiate the relationship. Alternative forms of dispute resolution considered in Part 4 can also avert a crisis that might have devastating effects on the businesses moving forwards.

Positive action is generally required to terminate a contract and usually involves the giving of notice. The notice of termination must be unequivocal in order to be effective. The notice must also be given within the time frame stipulated by the contract. Any error in calculating time periods leading to a premature termination may backfire on the party seeking to terminate. This can put the notice giver in the position of being viewed as the party who has repudiated the contract by giving the premature notice. This amounts to a reversal of the roles of the parties with the party originally in default now in the position of being able to accept the repudiatory breach of the notice giver. This may allow the original defaulter to elect to treat the contract as having been repudiated and to claim damages from the notice giver.

The courts may sometimes imply a term of reasonable notice where the contract contains no express provision for it to come to an end. What the courts consider reasonable depends on the individual circumstances of the case. Case law[3] suggests that the degree of formality in the relationship, its length and the amount of dependency on the other party for revenue are factors to be considered here.

Under the JCT provisions, the contractual consequences of termination based on contractor default, insolvency or corruption are as follows.[4]

- The employer may appoint others to carry out and complete the works.
- The contractor can be required to remove any temporary buildings, plant and equipment.
- Further sums falling due to the contractor from the employer are no longer due (subject to the Housing Grants Construction and Regeneration Act 1996 as amended).
- There is a reckoning up of any sums due between the contractor and employer after completing the works and making good the defects.

The employer does not have things all his own way in respect of termination. The contractor's common law rights remain intact in a situation where the contractor has not been paid. Thus, in the case of *CJ Elvin Building Services Ltd v Peter and Alexa Noble*,[5] it was held that the employer was in breach of contract not only by refusing to pay sums as they became due but also by threatening to make no further payment until the job was completed. The contractor was held to be justified in suspending the works and could have elected to accept the repudiatory breach in leaving site. Issuing the correct notices would have added some clarity to an uncertain position.

The best advice for any contractor considering leaving site in the face of a deteriorating relationship with the employer or their advisors is to stay put. Seeing the project through to the end preserves entitlements to payment which largely disappear once the contractor leaves.

11.4 Defects

Defective work is a frequent occurrence in both residential and commercial construction projects. Defects range from the trivial 'snagging items' at practical completion to undetected major issues such as problems with foundations comprising the structural integrity of buildings. Defect is defined in the case of *Yarmouth* v *France*[6] as *anything which renders the [building] unfit for the use for which it is intended, when used in a reasonable way and with reasonable care.*

Practical completion will be certified by the overseer when the building appears to be free from patent defects. It does not mean that the building is free from latent defects. This in itself is a grey area and the question emerges as to when a defect is patent and when it is non-patent. The case of *Sanderson* v *National Coal Board*[7] introduces an objective test to the effect that if the defect was observable then it is patent.

Latent (and unseen patent) defects coming to light during the defects liability period must be put right by the contractor. The contractor is obliged to return to site and remedy any defects that arise within a certain period after practical completion (normally 6–12 months). Engineering contracts such as those involving power plants have longer periods (typically 24 months in duration).

The question about whether or not a defect is patent or latent is relevant not only to the issuing of practical completion but also to the contractor's liability for defects moving forwards in terms of the reckoning of time. The right to rectify defects is both a benefit and a burden for the contractor. The contractor usually prefers to put right its own defects rather than facing deduction from sums due to him or money paid to a replacement contractor to do the same. The replacement contractor is likely to charge a good deal more than the cost to the original contractor who may insist on the chance to fix the snags provided the contract is still in place. The right of the contractor to insist on his right to fix his defects was established in the case of *Pearce and High Limited* v *Baxter and Baxter.*[8]

Once the defects notified have been rectified, the overseer issues a certificate of making good defects and the remaining half of the retention is released to the contractor. If a latent defect appears after the end of the contract then it is adjudged to be a breach of contract for which damages are payable. The measure of damages is the 'cost of cure' in putting the work right. An alternative measure of damages is the diminution in value of the property. This is appropriate where the defect is irreparable or where the cost of reinstatement is out of all proportion to the benefit gained.

Some employers have sought to insure against the potential risk of latent defects occurring and the non-performance of the contractor in this connection. Latent damage and defect insurance are two of the products considered in the next section.

11.5 Insurance

Abrahamson's risk principles[9] established that the availability of insurance was a key consideration in allocating risk between the parties. Insurance is necessary to protect both the employer and contractor on the occurrence of certain events. Insurance is a contract in which the insured pays a premium to the insurer in return for financial compensation if a specified event occurs. The party with the insurance is 'covered' to the extent that if the specified event occurs then the insurance will meet the claim provided that the conditions of insurance are met. The following types of insurance are common on construction and engineering projects.

- *Contractor's all risk insurance* – This insures physical damage to the work and site materials and is maintained by the contractor until practical completion of the project. There is usually a requirement that the policy be held in the joint names of the contractor and employer to ensure straightforward access to the benefits of the policy for the employer in the event that the contractor cannot or will not make a claim.
- *Public liability insurance* – This insures liability arising from death or personal injury and property damage to third parties such as visitors, passers-by and neighbouring property owners.
- *Employer's liability insurance* – This insures a party against liability for injury or disease arising out of employment and is required by the contractor and any subcontractors with employees on the project.
- *Product liability insurance* – This insures against liability to third parties or damage to their property arising out of products supplied by a business. In the context of construction, this might be relevant for such items as lifts or escalators, which will carry their own separate insurance arrangements.
- *Professional indemnity insurance* – This insures against liability arising from professional negligence. In construction projects, professional negligence can arise from negligent design, surveying, project management or contract administration. Contractors and specialist contractors providing design services are required to carry this cover. The cover is usually limited to the lower standard of reasonable skill and care rather than a fitness for purpose warranty (see Section 6.2).

Commercial insurance contracts are based on the legal principle of utmost good faith. This principle imposes the burden of risk on the proposer to make sure that the insurer has all the information they require before entering into the insurance contract. This includes an obligation to make disclosure of material facts even if not specifically addressed on the point. Attempts have been made to introduce a similar high standard of good faith in construction contracts, and these are considered in Part 5. The key consideration for the parties when negotiating insurance is to ensure that the requirements are 'back to back' with their obligations under the contract. Hence, if the contractor is carrying out design works through its subcontractor then the level of professional indemnity insurance cover needs to be at a consistent level between the two.

11.6 Subcontracting

11.6.1 Introduction

Generally, a subcontract is a separate contract that does not affect the primary responsibilities of the parties to the main contract. The privity of contract rule prevents any link arising between the subcontractor and the employer. Nevertheless, a subcontractor could be liable in tort; for example, where the subcontractor is found to owe a duty of care to the employer and is negligent in carrying out the work and causes physical damage to the party's property or person.

The ability to subcontract depends on the nature of the main contract itself. If the main contract involves the provision of services personal to the supplier then subcontracting is not permissible. This occurred in the case of *Davies* v *Collins*[10] where a contract for the cleaning of a soldier's tunic before the Victory in Europe celebration parade was held to be personal to the dry cleaner and should not have been subcontracted. The defendant's argument based on transferring liability onto the subcontract therefore failed. In construction contracts, the parties should check whether the main contract permits subcontracting. Employers and their agents recognise the importance and prevalence of subcontracting and usually allow it subject to the approval of the list of subcontractors proposed.

11.6.2 Reasons for subcontracting

Government reports as far back as the Simon report in 1944[11] have observed that it had become impossible for any single architect or builder to have specialised knowledge and experience to deal effectively with all aspects of major projects. As a result, the need arose to appoint specialist firms. This requirement was recognised as long ago as 1944 when the Simon report noted that the increasing complexity and specialisation of construction work required a higher proportion of the work to be carried out by highly specialised firms.

According to the Simon report's estimates, two-thirds of work was carried out by specialist firms. In the post-war period, the growth and extent of specialist work have been accelerated further by the development of new materials and technology and the increasing complexity of construction works. The figure of subcontracted works is now well over 90 per cent of construction work.[12]

The relationships between contractors and subcontractors are usually defined by written contracts and in many instances these take the form of standard forms of subcontract, which have the same strengths and weaknesses as the SFCs previously noted. Parts of a construction contract are usually subcontracted because, for a variety of reasons, either the contractor or the client/architect decides that this strategy will assist them in achieving their project or wider business objectives. Where the selection of a subcontractor is at the discretion of the main contractor, these are usually described as 'domestic' subcontracts.

The advantages of subcontracting include the following.

- *Specialism of subcontractor* – The subcontractor is a specialist in a particular type of work which the main contractor may be unable to undertake economically or because they do not have the necessary technical skills or resources. Examples range from the subcontracting of mechanical and electrical installations, which often require relatively scarce high-level technical skills, to more straightforward trade subcontracting, such as bricklaying and plastering. In the latter case, this type of work can often be achieved much more economically by subcontractors with lower overheads and greater flexibility of operation. Subcontracting can thus offer the main contractor a competitive advantage over his competitors.
- *Risk* – The contractor may wish to transfer part of his risk exposure onto a third party through a subcontract; for example, the subcontracting of groundworks where a contractor can transfer risk – such as that surrounding the disposal of waste or the cost implication of delays as a result of inclement weather – away from both themselves and the employer and onto the subcontractor. If this is the motive for subcontracting, the contractor must take great care to ensure that the subcontract agreement entered into does effectively transfer these risks and that the subcontractor is realistically able to manage and carry them.
- *Flexibility* – The use of subcontractors reduces the need to directly employ labour and enables main contractors to quickly adjust the workforce in response to peaks and troughs in workload. It also eliminates or reduces the need for contractors to ensure an even flow of work is available to keep their own workforce fully employed. Again the risks associated with direct employment of a labour force (i.e. finding, training and retaining staff and keeping them economically employed) are effectively transferred onto the subcontractor.

The architect, sometimes at the specific request of the client, often requires the contractor to subcontract part of a project to a particular 'named' or 'nominated' subcontractor. This may be for reasons of control where the architect (or the client through the architect) may wish to exert some control over who carries out certain parts of the works. This may be in order to ensure specialists of proven reliability and repute are used or where there is a desire to specify a particular subcontractor for technical or commercial reasons.

Commercial reasons might apply where a subcontractor is a part of the same parent group as the client.

Another reason for naming subcontractors is in relation to the programme. Where there is a design element to work of a specialist nature or a long lead-in to manufacture, it may be desirable to commence this in advance of the main contract in order to ensure that programme requirements are met. The early selection of a subcontractor, prior to the appointment of the main contractor, allows for the design and pre-ordering of work. This pre-ordered work can subsequently be integrated into the project through the mechanism of a subcontract.

The standard forms of subcontract available generally fall into the three categories of domestic, and 'named' or 'nominated' (the last two being known as specified) subcontracts. The intention of these subcontracts in their unamended form is, where appropriate and equitable, to transfer the risk the main contractor carries for cost, time and performance on to the subcontractor (and in the case of nominated subcontracts, on to the client).

11.6.3 Domestic subcontracting

The form of subcontract applying in this situation is often left up to the parties to agree and is frequently based on a contractor's own terms and conditions. Although these are often designed to favour the contractor, they may also run the risk of being subject to the Unfair Contract Terms Act 1977. There are, however, some standard forms which have gained widespread acceptance.

- *JCT Standard form of Subcontract SBCSub/a and SBCSub/c* – This subcontract, comprising recitals, articles and particulars (/a) and the conditions (/c), reflects the terms and conditions embodied in the main contract forms and is designed to impose them 'back to back' on the subcontractor.
- *JCT ICSub/a and ICSub/c* – This subcontract is designed to be used with IFC 2011.
- *NEC3 Engineering and Construction Subcontract ECS* – This is designed to be used back to back with the NEC3 Engineering and Construction Contract.

Domestic subcontracts rarely involve the overseer appointed under the main contract. Agreement of payments, extensions of time and claims for loss and expense are usually dealt with by the contractor.

The JCT 2011 contracts stipulate particular JCT subcontract forms for domestic subcontracts where considered appropriate. They require the inclusion of certain terms in respect of ensuring access for the architect and the management and ownership of unfixed material on-site. The likelihood of these not being included and the main contractor becoming liable for the consequences of this breach is substantially increased where a non-standard form is used.

11.6.4 Named and nominated subcontractors

The nomination or naming of a subcontractor by the contract administrator (architect or engineer) generally reduces the potential risk exposure of the contractor further than is the case in a domestic subcontract situation. In return for the reduction in contractor choice that results from nomination or naming, the SFCs require the employer to retain some of the risks that they would normally transfer to the main contractor. The overseer also has a much more significant role to play in these subcontracts.

Naming subcontractors allows the employer to influence the contractor's selection of subcontractors, whilst leaving responsibility for their performance with the main contractor. This is an alternative to nominating subcontracts as the contractual relationship is less complicated.

The client identifies a list of potential subcontractors for a particular package of work. The client may invite these potential subcontractors to submit tenders for the package. The client then makes a shortlist of acceptable subcontractors and appends this to the tender documents as a provisional sum. Once appointed, the successful contractor seeks tenders for the package from the named subcontractors. Once the main contractor has selected and appointed a subcontractor, the provisional sum is replaced with the actual price agreed. Under this arrangement, the main contractor assumes responsibility for the subcontractor's performance. In effect, the named subcontractor becomes a domestic subcontractor.

A nominated subcontractor is selected by the employer to carry out an element of the works. Nominated subcontractors are imposed upon the main contractor after the main contractor has been appointed. The contractor is not held responsible for the failure of the nominated subcontractor to perform given that the contractor had no choice in the selection. The employer should ensure that there is a direct warranty from the nominated subcontractor to guarantee performance. The contractual arrangements allowing nomination are very complicated as they attempt to cover all possible eventualities both between the employer and contractor and also between the contractor and nominated subcontractor. Some forms of contract (such as JCT) no longer include provision for the nomination of subcontractors. The use of named subcontractors is generally considered to be a simpler alternative.

11.6.5 The position of subcontractors

Subcontracting in the construction industry is to experience life at the sharp end. The risks and rewards are greater at this tier of the supply chain. A subcontractor can expect to earn on average a higher profit margin than a main contractor whose role is essentially to co-ordinate the subcontractor activity. Conversely, the risk of insolvency and disruption to the business for non-payment is greater here.

The provisions of the Housing Grants, Construction and Regeneration

Act 1996 have improved the subcontractors' position greatly in terms of them being able to take steps to ensure access to their money. Advances by progressive main contractors in risk management/allocation and collaborative working have also been made. However, several areas of deep concern remain for specialist contractors, foremost amongst them the continued imposition of onerous non-standard terms of subcontract. The continued use of these terms undermines the advances made elsewhere.

For the average specialist, issues of survival and continuity of work still dominate their decision-making process and unless these firms can be convinced that partnering can improve their chances, it is unlikely to have a significant impact. Whilst attitudes towards subcontractors have improved over the past 20 years, they have not improved nearly enough.

11.7 Conclusion

This chapter has considered some of the particular features of construction contracts required to achieve contractual certainty and process. Defects, termination and subcontracting are important areas to focus on in this regard.

The role of a construction contract is to attempt to cover all the possible eventualities that arise on a project in sufficient detail to achieve the required level of certainty. The contract alone does not contain the whole cornucopia of legal arrangements and is supplemented by the background law. According to one commentator, the advice to read the contract is only partly correct. The better advice is to learn the law and then read the contract.[13]

The length of construction contracts can be daunting to the reader. The length of the contract is largely unavoidable if the documents are going to adequately cover all the extraneous matters required. The advent of virtual design and construction may alleviate the need to represent the contract and the design in 2D format and lead to a more interactive and user-friendly medium for defining roles and responsibilities.

11.8 Further reading

Uff, J. (2013) *Construction Law*, Eleventh Edition, London: Sweet & Maxwell, Chapter 8.
Hughes, W., Champion, R. and Murdoch, J. (2015) *Construction Contracts: Law and Management*, Fifth Edition, Abingdon: Routledge, Chapters 22 and 23.
Ramus, J., Birchall, S. and Griffiths, P. (2006) *Contract Practice for Surveyors*, Fourth Edition, London: Butterworth-Heinemann.

Notes

1 [2009] EWHC 3272 (TCC).
2 www.cemar.co.uk.
3 *Jackson Distribution Ltd* v *Tum Yeto Inc* [2009] EWHC 982.
4 JCT SBC11 and JCT DB11 Clause 8.7.
5 [2003] EWHC 837 (TCC).

6   19 QBD 647.
7   [1961] 2 QB 244.
8   [1999] BLR 101.
9   See Section 8.9.
10 [1945] 1 All ER 247.
11 Simon, E. (1944) *The Placing and Management of Building Contracts: Report of the Central Council for Works and Buildings*, London: HMSO.
12 Gray, C. and Flanagan, R. (1989) *The Changing Role of Specialist and Trade Contractors*, Ascot: Chartered Institute of Building Papers.
13 Starzyk, G. F. (2014) Alliance Contracting: Enforceability of the ConsensusDocs 300 Mutual Waiver of Liability in US Courts, in A. Raiden and E. Aboagye Nimo (Eds), *Proceedings of the 30th Annual ARCOM Conference*, 1–3 September, Portsmouth, UK, pp. 547–56.

12 Construction law in the wider sense

12.1 Introduction

The previous chapter introduced aspects of construction contract law necessary to give the student an overview of the procedures involved in building contracts beyond the basic obligations. Discussions around termination, insurance and supply chain considerations were all covered in a demonstration of the wide nature of issues potentially arising. This chapter pans out further on the view of construction contract features to take in some of the ancillary contract matters that arise. The chapter discusses letters of intent, collateral warranties, novation and assignment. All of these features have an important role to play and understanding their purpose provides some important linkages between the topics already covered.

12.2 Letters of intent

12.2.1 Introduction

Construction projects involve a host of technical and practical elements that need to be put in place before physical work can start. These include:

- development of design and the co-ordination of the various design team members;
- supply chain issues, the identification of the right people/products, and the ordering of the same;
- access to site and planning arrangements to ensure the build can go smoothly once commenced.

The employer usually spends pre-commencement time focused on agreeing these issues rather than formalising the contractual arrangements needed. Suddenly, the momentum gathers to complete the deal. An end date appears to which the client is contractually committed and on which a good deal of money rests. Issues outstanding at the front end of the contract, such as planning conditions and approvals, have not been resolved but the contractor must start if the end date is not to be jeopardised. Negotiations are still ongoing with the contractor at this stage. Under such circumstances, the parties may consider using a letter of intent.

The idea of a letter of intent is straightforward. It acknowledges that the employer is not in a position to enter into a contract for the work but that the project must begin. The work is to be carried out in accordance with the drawings and specification issued to date. If the employer has to stop the work then the contractor will be paid for what has been carried out, subject to any limits agreed. It is not uncommon for the limits introduced to be incrementally raised if the obstacles to signing the full contract remain in place yet the contractor has reached the limit set. The benefits for the parties of using a letter of intent are as follows.

- For a contractor, a letter of intent provides some comfort relating to payment for work to be carried out before the contract documents are signed. The contractor appointed under a letter of intent is overwhelmingly likely to be awarded the whole project.
- For the employer, a letter of intent allows the contract works to start; it is possible to gain some breathing space as to the delivery of the project on the date by which the works are to be completed and some income can be produced.

The issue arising from the use of a letter of intent is that the contractual position created can be uncertain. The point of having a standard form contract in place is for the benefits it brings in terms of risk profile and contractual certainty for the parties. This can be missing with a letter of intent. The courts have dealt with numerous cases where parties have undertaken a project on a letter of intent and subsequently discovered that their contractual position was different to what they thought. Letters of intent carry a health warning: proceed with caution. In the words of Lord Clarke, the parties should be wary of the *perils of beginning work without agreeing the precise basis upon which it is to be done. The moral of the story is to agree first and to start work later.*[1]

Several contractors and architect practices have a blanket ban on the issuing of letters of intent. How rigorously these are applied is open to question.

12.2.2 Categories of letter of intent

The categories of letters of intent range from non-binding arrangements to fully functioning complete contracts subject to a cap or time limit. In between these extremes is a third category, described here as a quasi-contract. The key distinction for a letter of intent is whether or not it is binding on the parties. A non-binding arrangement is of limited value to both parties.

The potential for how a contract can be viewed is shown in Figure 12.1.

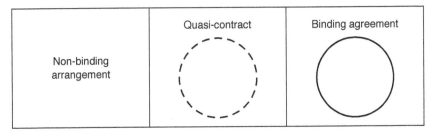

Figure 12.1 Categories of letters of intent

Non-binding arrangement

A contractor starting work on a letter of intent which does not contain any right to remuneration would be foolhardy indeed. A letter of comfort, which simply expresses the employer's current intention to enter into a contract at some point in the future, would fall into the category of a non-binding arrangement. These types of letters are often headed 'subject to contract', which means that unless and until they enter into a formal contract, their negotiations (even when concluded) are not legally binding. These words can be sufficient to prevent a contract from being entered into notwithstanding the presence of the ingredients for a contract being present.[2] If there is no promise of payment then a judge will be hard pressed to imply contractual terms into a non-binding arrangement. In the absence of a contract, the law of restitution may apply. This is an equitable remedy that can be brought in the courts and is based on the law of unjust enrichment – why should the employer have the benefit of the contractor's services without paying for them? This usually involves a consideration of the law of *quantum meruit*. The pricing of extra work may therefore be by analogy to the contractual rates or some other starting point based on industry practice.

Chapter 2 identified the situation where contracts can be augmented by the introduction of implied terms into an incomplete contract. The Sale of Goods Act 1979 and the Sale of Goods and Services Act 1982 will operate to introduce minimum standards and, crucially for the contractor, the right to reasonable reimbursement for the job performed. These rates and prices might be more favourable to the contractor than anything discussed at the pre-contract stage. The major benefit for the contractor is that liquidated damages will not apply in the absence of a contract term stating that they will.

Quasi-contract

The perforated or quasi-contract applies where there is a sort of contract in existence, but one that is unable to cover the eventualities that may materialise. Hence if the quasi-contract did not provide for a means of recompensing loss and expense to the contractor then a term will be implied into how the contractor is to obtain his compensation. Once again, the law of *quantum meruit* is likely to apply, leaving the contractor without certainty as to the sum he should be paid. The parties or the courts will still have to determine the rates and prices to be used and the effect of any delay and defects in the works on the amount to be paid.

The lack of certainty for the employer (and to some extent the contractor) leads to the sage advice that it is always better to have the contract agreed before work commences on-site or at the very worst very soon thereafter.

Binding arrangement

For a letter of intent to be binding, it should be clear in its terms and ensure that:

- the letter records the agreement of both parties and is signed and dated;
- the letter provides for consideration, which is payment for the works to be carried out;
- the intention to create legal relations is expressed in the terms of the letter.

It is standard practice for the letter of intent to identify the SFC that will govern the project. The contractual mechanisms in the SFC may therefore be used in seeking to avoid the need for *quantum meruit*-type arguments.

The advantages of a binding arrangement are that it provides increased certainty for both parties in relation to factors such as cost and time and can be tailored to the employer's needs in terms of limiting the sum to be paid and the scope of works to be carried out. The problem is that once a letter of intent is issued, both parties tend to forget what it says and simply get on with the project as though a contract has been signed. This state of affairs continues until a dispute arises which sends both parties back to the letter and their different interpretations about what it might mean. This was the case in *Mowlem plc* v *Stena Line Ports Ltd*.[3] The financial limit in an original letter of intent on the project for the construction of a hoverport terminal in Wales had been incrementally raised 14 times during the works. The last letter sent by the defendant stipulated a maximum of £10 million and a date for completion. The claimant's work far exceeded this amount but the defendant refused to pay. The argument was brought that the defendant had allowed the claimant to continue even though it was clear that the limit was exceeded. The court was unimpressed with this argument and upheld the terms of the letter – the cap of £10 million applied regardless of whether or not the claimant continued to work beyond the limit set and the due date set.

The practice of issuing multiple letters of intent is not in anyone's interest and can result in an action in negligence against the overseer if the employer is exposed to additional costs. This was the case in *Trustees of Ampleforth Abbey Trust* v *Turner & Townsend Project Management Ltd*.[4] Eight letters of intent were issued, none of which contained the right for the employer to claim liquidated damages in the event of a late finish by the contractor. This absence of an important mechanism to keep the contractor in check was sufficient for the overseer to be adjudged negligent for not having pointed out the risk to the employer. The damages applicable were reduced on the basis that it would have been open to the contractor to argue that the liquidated damages figure was a penalty.

The judge found that the advice to issue the initial letter of intent was not negligent; it was the advice to continue to issue further letters that was questionable. This recognises that there are occasions when a letter of intent is justified, though the practice must be used sparingly. Any letter of intent should be as clear as possible as to its application and any limits imposed.

12.2.3 The contents of a letter of intent

The scope of the letter must be established at the outset and may include such tasks as starting site preparation or enabling works prior to entering into a formal contract for the main contract works. Where used, the letter should specify the terms and conditions applying to the work and the requirement for quality and timing of the work required. Other issues to be covered include the insurances to be maintained, a copyright licence in respect of any contractor design, a dispute resolution clause, termination arrangements and payment terms compliant with the construction Acts. Boilerplate clauses such as governing law and jurisdiction should also be included. The model form produced by the City of London Law Society is in effect extremely close to a complete contract.

12.3 Collateral warranties

12.3.1 Introduction

Collateral warranties are another unusual feature particular to construction law. Construction projects involve many different parties often brought together for a one-off project. The contractual nexus that needs to be created on even a straightforward project means that numerous collateral warranties are likely to be required. A collateral warranty creates a direct contract link between parties who appear remote or at least 'one step removed' from each other in the contractual nexus.

A collateral warranty solves a recurrent problem on construction projects. The problem arises where defective design or workmanship by a professional consultant or contractor causes third party losses. Section 6.3 describes the linear arrangement featuring extended links in terms of those performing the work and those on whom the consequences of poor performance of the work would fall. Thus, the employer's tenant might be the party suffering the consequences of the contractor's negligent design. A contractual remedy in this situation is more satisfactory than relying on the vagaries of the law of tort.

Discussions around whether the warrantor (the person giving the warranty) owes a duty of care to the person suffering the loss are avoided. For example, whilst the employer (landlord) should be proximate enough in terms of the neighbour test to be identifiable as a person to whom a duty is owed, will the same thing be said about the tenant or subtenant of the building? What about the funder who has exercised step-in rights to complete the building upon the insolvency of the original employer? These people may not be said to be proximate enough in law to be owed a duty of care. The collateral warranty overcomes this issue by providing the direct link.

The other shortcoming of the law of tort is the limitation around the recovery of pure economic loss. Pure money claims are not recoverable under the law of tort as demonstrated in the case of *Spartan Steel & Alloys Ltd* v *Martin & Co Contractors Ltd*.[5] Drawing a distinction between economic loss and other categories of loss can seem illogical given that ultimately all losses come down to money as the means of recompense. This distinction is avoided by adopting collateral warranties that transform the damages payable into contractual damages.

12.3.2 Privity of contract

The law of tort does not give adequate protection to those potentially suffering the consequences of negligent design or workmanship. The economic loss exception and the uncertainty of whether duties of care are owed mean that contract law is a much safer bet in terms of protecting interests. The obvious choice for contractual enforcement would be through the building contract itself. However, the third party cannot claim the benefit of a contract to which they are not a party. Section 6.6.4 sets out the position in relation to privity of contract, which can be summarised as: only a party to a contract can bring a claim under it. This means that a contract involving parties A and B does not provide a remedy for party C (who is not party to the contract) even if he is affected by what happens between parties A and B. The only way around this (until recently) was to insist on a separate contract between A and C whereby A is given the same rights to pursue C as if C had been a signatory to the original contract. This collateral warranty or side agreement is almost exclusively for the benefit of A (the employer) and gives comfort that in the event of default by C then A has a contractual remedy.

This was not the position in Europe where the legal traditions are more aligned to the notion of third party rights under a contract. The Contracts (Rights of Third Parties) Act 1999 (discussed in Section 6.6.4) was seen as a way to introduce the European tradition on this point and do away with the multitude of collateral warranties on projects with a consequent saving of everyone's time and money. SFC writers were slow to embrace this change but now do so with third party rights schedules, which can be used instead of collateral warranties. Many lenders, despite the additional resources involved, still prefer the security of collateral warranties, in particular the step-in and intellectual property rights licences they offer. Chasing missing or incomplete collateral warranties is something that construction lawyers still spend a good deal of time doing.

12.3.3 Typical contents of a collateral warranty

The key clause in any collateral warranty is that the warrantor confirms that reasonable skill and care has been used in the fulfilment of the work it has performed. This is, put simply, the main point of the warranty. The warrantor is saying: 'I did my job properly and you can rely on this'. This is nothing more than the warrantor has already confirmed in their original contract. Most professionals in the construction industry are content to give collateral warranties subject to their professional indemnity insurer's approval. Some negotiating may occur around some of the terms (set out below), in particular the amount of insurance carried by the firm and the acceptability of the limitations on liability sought.

In addition to the core reasonable skill and care clause, warranties also typically include:

- licence on the warrantees or their assignees to use the designer's intellectual property rights;
- confirmation that professional indemnity insurance is maintained (and will be maintained) at the limit required;
- confirmation that banned or hazardous products have not been used in the build; and
- step-in rights in the event of the insolvency of the employer.

Limitations on liability is an interesting area and illustrates some of the recurring themes in construction law. These are effectively the protections that the warrantor insists upon as a *quid pro quo* in return for giving the warranty to those with whom he would not otherwise be in contract. In practice, the warrantor may be able to gain the benefit of *some* of these limitations but rarely *all* of them unless they are in a strong bargaining position.

- *No greater liability* – This states that the professional consultant or contractor cannot owe the purchaser or tenant greater duties under the collateral warranty than they would owe the employer under the professional appointment or building contract.
- *Equivalent rights of defence* – The warrantor may use any defence they had under their original contract to defend a claim from the beneficiary under the collateral warranty.
- *Net contribution* – This clause limits the recovery made against the warrantor to those losses which it is 'fair and reasonable' or 'just and equitable' to levy against the warrantor given the involvement of the other consultants and

contractors on the project. In effect, the 'contribution' of the warrantor to the problem experienced is considered.

• A limit on the types of losses that the beneficiary may recover to the 'reasonable costs of repair, renewal and/or reinstatement'.

Typically, collateral warranties are actionable for a period of 12 years (if signed as a deed) from the date of practical completion of the building contract.

12.4 Assignment and novation

12.4.1 Introduction

This chapter has discussed several instances where the contractual network established on a construction project extends beyond the main building contract. The position of third parties such as funders, purchasers and tenants has been discussed with particular reference to collateral warranties. The purpose of a collateral warranty has been described as giving a third party a side agreement with the warrantor, usually a designer, allowing the third party a contractual remedy in the event of a breach of their original agreement with the employer.

Assignment and novation cover a similar yet distinct area. If the third party is not content with, or provided with, a side agreement then they might insist on having the benefit of the whole original contract transferred to them. This is where assignment and novation provide the solution sought, which may occur during or after completion of the construction works. This area involves a discussion around 'benefits' and 'burdens' of contracts, illustrated here using a situation where an architect provided initial design for a developer who then sells the project to a purchaser who wishes to retain the architect to work up the detailed design. In this situation, the benefit to the developer is the warranty from the architect that it has designed the project with reasonable skill and care. The benefit to the architect is the right to be paid by the developer. The burden for the architect is the obligation to design with reasonable skill and care. The burden on the developer is the obligation to pay the architect.

12.4.2 Assignment and novation distinguished

Assignment is the transfer of an interest from one party to another. In a construction context, this usually implies that the benefit of the contract is transferred. Hence, a collateral warranty may be assignable two times, permitting assignment from the first to the second tenant. Only the benefit of the warranty is being transferred, which in this case could mean the right to pursue a designer for a faulty design. A key feature of assignment is that it can occur without the express consent of the party whose interest is being transferred under the assignment. The assignment is usually bipartite in that any contract recording the assignment does not need to be entered into by the party who is effectively being passed from one employer to another.

Any limits on assignment in the contract will prevent any purported transfer having effect. The assignment or novation should be dated at the same date as the sale of the property to ensure there are no gaps in the contractual network.

This can be compared to novation where the benefit *and* the burden of a contract are transferred and the consent of *all* the parties involved is required. This involves a tripartite novation agreement. The difference is that a completely new contract is created rather than the situation where the new party replaces the old party. A novation comes with the fiction that the original contract never was and that all obligations under the contract are viewed *de novo* (i.e. from new). This means that the original party is released and discharged from any potential liability dating from a time before the novation happened. This is preferable for the party stepping out of the contract as they cannot be pursued for liabilities said to have accrued before novation.

In the above example involving the designer, purchaser and architect, the following positions might exist:

- on assignment between the developer and purchaser, the developer must still pay the architect as the burden has not transferred;
- on assignment, the purchaser may sue the architect;
- on novation, the purchaser now takes the burden and must pay the architect; and
- on novation, the purchaser may sue the architect in contract as the benefit has been transferred.

12.4.3 Novated design and build

Chapter 7 discussed procurement routes and introduced the three classic choices of traditional, design and build, and management. A popular variant on design and build procurement is known as novated design and build. This can be described as a hybrid procurement route whereby the employer starts the pre-construction phase using traditional procurement by hiring a design team to develop its brief and produce tender drawings. At some stage through the design process, the design team is novated across to the contractor. This can be viewed as giving the employer the best of both worlds as its designer designs the building the employer wants rather than the building the contractor wants to build. The employer, from the point of novation, enjoys single-point responsibility in respect of the construction project as it continues. The position of the contractor is usually protected in respect of pre-novation design errors by a collateral warranty from the architect under which the contractor can pursue the architect for pre-novation errors. Novated design and build is not without its critics. One issue surrounds the difficulty for the architect in fulfilling his post-novation role where he becomes subservient to the contractor and sees design ideas potentially compromised.

Blyth & Blyth Ltd v *Carillion Construction Ltd*[6] involved the situation (as described above) where the employer worked up the early design with his team of professionals under the traditional route and then sought to novate. The facts concerned the novation to the contractor of the terms of appointment between an employer and a consulting engineer for the construction of a leisure complex near Edinburgh. After the novation, disputes arose between the engineer and contractor for unpaid fees, and the contractor was counter-claiming for additional construction costs arising out of alleged inaccuracies in the engineer's initial design given to the employer and provided at tender stage to the contractor. Confusingly, the document transferring the engineer to the contractor was headed 'consultant switch' which pointed at it being an *assignment*.

The Scottish judge found against the contractor in relation to the arrange-ments. The transfer was a *novation* as it was signed by the three parties, and its effect was to extinguish the original contract, meaning that the contractor could not bring a claim against the engineer for pre-novation losses. As a consequence, where contractors are taking over responsibility for the design process, they should protect themselves with appropriately worded collateral warranties.

In a further twist, this case gives some insight into the 'black hole' argument, which is raised here only briefly. Assume that the judge had found that an assignment was in place and that the contractor could pursue its benefit by suing the engineer for breach of its duty to conduct the design with reasonable skill and care. Assume, which is relatively common, that the 'assignment' contained an equivalent right of defence clause to the effect that the consultant may use against the contractor any defence it could have used against the employer. The engineer could then have used the no loss or black hole argument. This argument would go: 'I can use against you (contractor) any defence I could have used against the employer. In this situation, I would use the defence that the employer has suffered no loss because any loss that might have arisen has been suffered by you, the contractor. I therefore do not owe you anything and your case must fail'.

There is judicial support to rule in favour of this interpretation, and this provides an example of the involved and esoteric debates construction law can generate. One way around this problem is to draft the equivalent rights of defence clause in such a way as to expressly exclude the no loss argument.

12.5 Complex structure theory

12.5.1 Introduction

This consideration is an offshoot of the rules around the exclusion for the recovery of economic loss in tort cases. The illustration of the problem was given by Judge Akenhead[7] by analogy with the facts of the famous case of *Donoghue* v *Stephenson*:[8] *the purchaser of a ginger beer bottle which contains a snail may recover for personal injuries caused if she drinks the ginger ale but not for the cost of the bottle.* The latter amounts to pure economic loss. The claimant is restricted from recovering in tort for damages to 'the thing itself' unless it can be established that the damage was to 'other property'.

Complex structure theory suggests that a claimant should be able to recover for damage where a structure is so complex that individual elements of it cannot be distinguished *so that damage to one part of the structure caused by a hidden defect in another part may qualify to be treated as damage to 'other property'.*[9]

12.5.2 Complex structure theory

When a contractor is responsible for building the whole building then the project can be looked upon as one integral unit. There is therefore no question of damage to a part of the building being classed as 'other property'. However, if a subcontractor installs a component (for example, plumbing supplies) that is defective and the defective component results in, say, a flood, there is a question as to whether that loss is recoverable on the basis that the damage was to 'other property'. The subcontractor could therefore, in theory, be pursued for the losses and would not be able to rely on the economic loss exemption.

This approach has been largely discredited and criticised by the judges, who have found it to be unhelpful when arriving at their judgements. The availability of other solutions to the relief sought such as product liability insurance and contractor's all risk insurance ought to provide sufficient cover for the parties concerned about component failure.

12.6 Conclusion

This chapter has delved into some of the more contentious and complicated areas of construction law. The student should form an appreciation that a discussion of construction law involves issues which are not limited to the obligations of the main contract. Third parties have divergent interests that the law has sought to protect in different ways, which can result in competing interests. The skill for the practitioner in this area is to work through these arrangements in a logical and clear-headed way to ensure that the contractual network protects those interests adequately.

The next chapter and those following in this remainder of Part 3 focus on the claims brought in construction law relating to overruns of time and money. Issues of causation and entitlement are the first topics covered.

12.7 Further reading

Uff, J. (2013) *Construction Law*, Eleventh Edition, London: Sweet & Maxwell, Chapter 7.

Hughes, W., Champion, R. and Murdoch, J. (2015) *Construction Contracts: Law and Management*, Fifth Edition, Abingdon: Routledge, Chapter 17.

Chappell, D. (2015) *Construction Contracts, Questions and Answers*, Third Edition, Abingdon: Routledge.

Notes

1   *RTS Flexible Systems Ltd* v *Molkerei Alois Muller GmbH & Co KG* [2010] UKSC 14.
2   *Whittle Movers Ltd* v *Hollywood Express Ltd* [2009] EWCA Civ 1189.
3   [2004] EWHC 2206 (TCC).
4   (2012) 144 Con LR 115.
5   [1973] 1 QB 27.
6   (2001) Scottish Court of Session.
7   *Linklaters Business Services* v *Sir Robert McAlpine Ltd* [2010] EWHC 2931(TCC).
8   [1932] AC 562.
9   *D & F Estates Ltd* v *Church Commissioners for England* [1989] AC 177.

13 Causation, the recoverability of damages and global claims

13.1 Introduction

Returning to the origins of the rule of law covered in Chapter 1, the role of construction law can be perceived as being twofold:

- to protect the rights of the parties and to allow them to allocate risk and obligations and to provide certainty, as far as possible, in the dealing between the various stakeholders involved including, but not limited to, the main contractor, employer, subcontractors, design team, purchasers, tenants and funders; and
- to provide rules and procedures by which any claims and disputes between the parties can be resolved and their entitlements with regards to each other properly calculated.

The previous chapters have examined the arrangements relating to the first role of construction law. This chapter signposts a change in emphasis as this work now targets the second function of construction law. This is achieved by examining claims for overruns of time and cost and in Part 4 on dispute resolution techniques. Together, these chapters provide the detail necessary for the student to form an appreciation of the issues involved on the contentious side of construction law.

Later chapters in Part 3 discuss the specific issues of delay and disruption and the heads of claim that result from them. This first chapter examines some more fundamental issues in terms of proving causation as a concept at law and the limits on the recoverability of damages. The chapter concludes with a consideration of whether global claims are the answer to establishing entitlement to money under a construction contract. These general issues provide a framework for a wider appreciation of some of the complexities encountered before looking in more detail at the issues raised.

13.2 Causation

Construction is a risky business and unforeseen events occur with unerring frequency. These events can have a delaying or disrupting effect on the construction process. This delay or disruption will often result in the original completion and handover of the project being missed. As a consequence, additional costs are likely to be incurred by one or both parties to the contract.

These overrun costs may be directly associated with the project; for example, labour and overhead costs over an extended period and overtime payments, plus additional wage costs associated with increasing the workforce in order to accelerate the works. Alternatively, they may be more remote; for example, loss of rental income to the employer resulting from a delayed handover.

The mere fact that unexpected difficulties have been encountered or that the project is proving much more expensive than predicted does not automatically entitle the contractor to be compensated by the employer. Similarly, completion of the project after the agreed completion date does not automatically entitle the employer to compensation from the contractor. In both cases, a discussion around causation is important in establishing the facts and the resultant responsibility.

The issue of causation is often bypassed by those practising in the construction industry in a headlong rush to discuss the various methods of delay analysis and their particular merits. However, examining the subject of causation from the legal perspective highlights its importance and provides insight into how case law has influenced thinking in this area of construction law.

'Cause' is one half of the mantra repeated by those responsible for preparing construction claims. The other half is 'effect'. Causation is the relationship between cause and effect. Every effort must be made when preparing a claim to demonstrate what caused the incident being complained of and how that cause had the effect claimed. This is another way of viewing the three-point test of negligence, which involves a consideration of duty, breach and loss. The duty here is already established by virtue of the construction contract being clear on responsibility for the various eventualities. The remaining components required to establish liability are breach (or cause) and loss. It is self-evident most of the time what the cause was. The situation becomes more complicated when competing causes are involved. Competing causes, real or merely alleged, can occur frequently in complex construction projects. The matter on which the contractor seeks to rely must be linked, without interruption, to the loss suffered.

One of the leading cases on the subject of competing causes is *Leyland Shipping* v *Norwich Union Fire Insurance Society*.[1] A ship had been torpedoed but had managed to make port, where she was regularly grounded between tides before sinking. The court had to decide which event had caused a ship to sink: was it the consequence of an insured risk (action of the sea) or an uninsured risk (being torpedoed)? Lord Shaw of Dunfermline set out the meaning of establishing causation as follows:

> *We have had a large citation of authority in this case, and much discussion on what is the true meaning of [proximate cause]. Yet I think the case turns on a pure question of fact to be determined by common sense principles. What was the cause of the loss of the ship? I do not think that the ordinary man would have any difficulty in answering she was lost because she was torpedoed.*

Another competing cause situation was addressed in the case of *Quinn v Burch Bros Ltd*.[2] The defendants failed, in breach of contract, to supply the claimant plasterer with a stepladder. As an alternative, he used a trestle table which slipped, causing injury to the claimant. The Court of Appeal took the view that the injury was the claimant's own fault and that the defendant's breach of contract did not cause the injury. The breach of contract had merely given the claimant the opportunity to injure himself – it was impossible to say the defendant had caused the injury in law.

As a consequence of these decisions, the courts view the resolution of such issues as one of common sense. This approach has stood the test of time; in the more recent case of *Knightley v Johns*,[3] this common-sense approach was once again relied upon when Lord Justice Stephenson stated: *Questions of remoteness or of causation have to be answered not by the logic of philosophers but by the common sense of ordinary men.* Although, in this particular instance, it is more likely to be men familiar with the construction industry for as Judge Bowsher stated in *P&O Developments Ltd v The Guy's and St Thomas NHS Trust*:[4] *The test is what an informed person in the building industry (not the man in the street) would take to be the cause without too much microscopic analysis but on a broad view.*

From these authorities, it can be deduced that the successful party is the one who establishes their case on the basis of the balance of probabilities. In other words, where there is more than one plausible cause of a delay, the one which is more likely than not to be the root cause should be preferred. The implications of this approach are considered in context in Chapter 14.

13.3 The relevance of causation to construction claims

Extensions of time have particular importance for a contractor. An extension is a precursor to recovering time-related expenditure and relieves the contractor from the liability to pay liquidated damages. In order to establish entitlement, the contractor must demonstrate that the delay affecting progress to the completion date was caused by an event for which the employer was at risk under the contract.

There are three steps to establishing an extension of time and thereby proving causation.

1 An employer's time risk event must have occurred.
2 A delay to progress must have been caused by the employer's time risk event.
3 A delay to completion must have been likely to be caused by the delay to progress resulting from the employer's time risk event.

These three statements are often referred to as the 'three-part chain of causation', in which the three parts are not independent causes, but successive links in a single 'chain'. For a claim to be successful, it is necessary to prove each one.

The first hurdle for a contractor to overcome in establishing entitlement to additional time and possibly money is to convince the overseer that the claim is legitimate. The overseer is under a contractual duty to ascertain the amount of additional time and money properly payable. One of the questions the overseer will address is whether or not the events alleged have been sufficiently well established and whether they caused the lost time or financial consequences claimed. Where there are several possible causes, the burden of proof is on the contractor to show that one cause is more likely than the others.

The duty on the overseer to go through a proper process in relation to causation was established in the case of *John Barker v London Portland Hotel*.[5] The case focused on the issue of justifying extensions of time on a hotel refurbishment project. In this case, the claimant's approach was endorsed while the architect's methodology was rejected in the following terms:

> the architect's assessment of the extensions of time due to the Claimant was fundamentally flawed because he did not carry out a logical analysis in a methodical way of the impact which the relevant matters had or were likely to have on the Claimant's planned programme; he made an impressionistic rather than calculated assessment of the extensions ... where he allowed time for relevant events the allowance made in important instances bore no logical or reasonable relations to the delay caused. Therefore, although there was no bad faith ... on the part of the architect, his determination of the extension of time due to the Claimant was not a fair determination nor was it based on a proper appreciation of the provisions of the contract. Therefore, it was invalid.

If the overseer decides, rightly or wrongly, that causation has not been made out sufficiently well then the claim will be rejected. The contractor then has a choice about whether to escalate matters and formalise the claim in a dispute resolution forum. This is likely to require additional effort and expense in preparing the claim properly, most likely with the assistance of a claims consultant.

The contractor's burden of proof will still require discharging. Proving causation will require the employment of alternative approaches as discussed in the next chapter. In complicated situations, one of the first steps is to 'filter' irrelevant causes. One technique that can be employed to do this is the 'but for' test which is defined by Clerk and Lindsell[6] as follows.

> The first step in establishing causation is to eliminate irrelevant causes, and this is the purpose of the 'but for' test. The 'but for' test asks: would the damage of which the claimant complains have occurred 'but for' the negligence (or other wrongdoing). Or to put it more accurately, can the claimant adduce evidence to show that it is

more likely than not, more than 50 per cent probable, that 'but for' the defendant's wrongdoing the relevant damage would not have occurred. In other words, if the damage would have occurred in any event the defendant's conduct is not a 'but for' cause.

The 'but for' test can be helpful in removing some causes but it does not assist in situations where both (or more) of the alleged causes have equal causal potency. Neither is it enough to merely eliminate some causes and leave two or more likely causes. This was demonstrated in the case of *Fosse Motor Engineers Limited and others* v *Conde Nast and another.*[7] The case concerned how a fire at a warehouse owned by the first claimant started. The options were:

- by a cigarette discarded by four agency workers employed by the second defendant, who were the only people working when the fire was discovered;
- by someone working earlier;
- by an intruder; or
- by some electrical equipment.

Expert evidence was not able to identify the cause of the fire, so the case depended on the evidence of witnesses of fact. The judge accepted the evidence of the agency workers and concluded that the fire was not attributable to them. He also discounted the electrical cause as improbable on the evidence. However, he could not determine on the balance of probabilities whether the fire was caused by either of the two remaining causes. Fosse had therefore failed to prove its case.

In reaching his decision, the judge emphasised that it is inappropriate to rank possible causes in terms of probability and select the most probable. He remarked that to do so is a *dangerous and generally a fruitless occupation.*

Some other approach is therefore required if causation as a matter of fact is not readily established. More sophisticated methods of establishing entitlement are required in this situation. This topic is covered in Chapter 14 in the discussion around concurrent delay. Approaches such as taking the dominant cause, viewing causation as a network of events rather than a chain, and support for proportionality are all examined.

13.4 The extent of recoverable damages

Another area concerning claims involves the measure of damages properly recoverable under a construction contract. Chapter 2 established that the normal remedy for breach of contract in English law is to pay damages. The purpose of damages is to put the claimant back in the position they would have been had the breach not occurred.

However, the ability to recover damages for all losses is likely to discourage commercial transactions. Such considerations become increasingly relevant where there is a series of contracts, as is often the case in the construction industry. With the widespread employment of subcontractors and sub-subcontractors, the exposure to significant risk for unlimited damage claims can potentially increase.

Therefore it is common when negotiating contracts that the party providing goods or services will seek to limit their liability arising from a breach. There are various ways in which liability can be limited. It may be by way of providing a financial cap on liability or by defining the types of losses that the party will be liable for as a result of a breach.

In the absence of any limit in the contract, the scope of damages is interpreted by applying the *Hadley* v *Baxendale* tests.[8] This section examines in more detail the implications of the test in relation to construction claims. Before the landmark *Hadley* v *Baxendale*[9] decision, the case of *Robinson* v *Harman*[10] had sought to extend the scope of damages. A wider liability for damages seemed to be suggested in the statement of Judge Alderson: *where a person makes a contract and breaks it, he must pay the whole damage sustained.*

This is clearly a quite far-reaching obligation and would, in the words of Lord Justice Asquith, in effect provide the claimant with *a complete indemnity* de facto *resulting from a particular breach, however improbable, however unpredictable.*[11] It therefore raises questions in the commercial world of business as to whether it is realistic for one party to expect to recover all of its loss from the other party to the contract and whether it is equitable to do so.

A modern take on where the limits should occur was discussed in the case of *Balfour Beatty Construction (Scotland) Ltd* v *Scottish Power PLC*.[12] The claimant was employed to construct a section of the Edinburgh City Bypass, and in order to undertake this, it entered into contract with Scottish Power for the provision of a temporary electricity supply to a concrete batching plant. Part of the works required the construction of a new aqueduct to carry the Union Canal over the road, and during the construction of this aqueduct, the power supply failed, causing the concrete batching plant to shut down. It was a requirement of the specification for the construction of the aqueduct that it was done with a continuous concrete pour, and the shutdown of the concrete batching plant prevented this from happening. It was therefore necessary to demolish the partly constructed aqueduct and rebuild it, so Balfour Beatty sought to recover from Scottish Power damages incurred through the demolition and rebuilding works.

The House of Lords overturned the earlier decision and found that Scottish Power were not liable for the costs of demolishing the aqueduct and rebuilding it. They should have been aware that in the ordinary course of events, an interruption to the power supply to the batching plant would have resulted in abortive costs for items such as lost concrete, remedial works, standing time, and the like. However, the requirements for a continuous pour in constructing the aqueduct were special circumstances, and if they were to be liable for damages arising from these special circumstances then they should been made aware of them at the time of entering into contract. Scottish Power had not been made aware of these special circumstances so were not to be held liable for the losses arising from them.

The 'special circumstances' in the above case appear to equate to the 'reasonable contemplation' test in the second head of *Hadley* v *Baxendale*. *Hadley* v *Baxendale* has received widespread and enduring acceptance over the years and has been described as 'a fixed star in the jurisdictional firmament'.[13] Despite the judgement being handed down 160 years ago, the two-limbed test of *Hadley* v *Baxendale* remains the overriding principle for establishing liability for damages as a result of a breach of contract. This provides that a party to a contract shall be liable for those losses arising from a breach of that contract which first arise naturally, in accordance with the usual course of things, and second are in the contemplation of the parties at the time of entering into the contract as the probable result of the breach. Those losses that occur in the usual course of things are determined by the imputed knowledge, which a reasonable person is taken to know, and also the actual knowledge of special circumstances that will be in the contemplation of the parties at the time of entering into the contract.[14]

The measure of damages payable upon the establishment of an entitlement to claim is therefore limited to the *Hadley* v *Baxendale* tests. Chapter 15 discusses loss and expense and explains how these rules have been interpreted in deciding the quantum of claims.

13.5 Global claims

The third issue examined in this chapter concerns global claims. It is by taking notice of the approaches to this subject that more can be learnt about the accepted ways to present claims and those that are frowned upon.

It will often be very difficult for a contractor to attribute particular losses to particular periods of delay or disruption events. In such circumstances, the delay or disruption claim may be presented on a 'global basis'. In this type of claim, the detail is missing and the attempts to link cause and effect through the records available are extremely limited. This type of unparticularised claim was in the sights of one commentator when he stated:

> some of the elaborate literary efforts put forward as disruption claims are works of fiction rather than fact ... provided lawyers create the maximum confusion at the arbitration about what actually happened on the works (and the arbitration procedure seems to help) the Claimant can hope for a substantial award.[15]

'Global' claims are frequently also referred to as 'total loss' claims or 'rolled up' claims.

There is a significant amount of case law that discusses the 'global' claims issue. The first case to be considered is *Wharf Properties* v *Eric Cumine Associates (No. 2)*[16] where the client's actions against their architect for negligent design and contract administration were struck out as incomplete and therefore disclosing no reasonable course of action. Per Lord Oliver:

> the pleading is hopelessly embarrassing as it stands ... in cases where the full extent of extra costs incurred through delay depend upon a complex interaction between the consequences of various events, so that it may be difficult to make an accurate apportionment of the total extra costs, it may be proper for an arbitrator to make individual financial awards in respect of claims which can conveniently be dealt with in isolation and a supplementary award in respect of the financial consequences of the remainder as a composite whole. This has, however, no bearing upon the obligation of a claimant to plead his case with such particularity as is sufficient to alert the opposite party to the case which is going to be made against him at the trial.

The courts confirmed in *John Holland Construction and Engineering Pty Ltd* v *Kvaener RJ Brown Pty Ltd*[17] that pursuing a 'global claim' will only be permissible in cases where it is impractical to disentangle specific losses that are attributable to specific causative events.

In *Laing Management (Scotland) Ltd* v *John Doyle Construction Ltd (Scotland)*,[18] John Doyle was engaged by Laing to carry out certain work packages for the construction of new headquarters for Scottish Widows, pursuant to an amended form Scottish Works Contract. In the litigation, John Doyle sought an extension of time of some 22 weeks and made a substantial claim for loss and expense. The loss and expense claim was calculated on a classic 'global' basis by comparing pre-contract estimates and actual costs.

The following guidance points can be summarised from the judgements.

- A contractor must normally plead and prove individual causal links between each alleged breach or claim event and each particular delay for a delay claim under a construction contract to succeed.
- If the consequences of each of the alleged delay events cannot be separated, and if the contractor is able to demonstrate that the fault for all of the events on which he relies is the employer's, it is not necessary for him to demonstrate causal links between individual events and particular heads of loss.
- However, if a significant cause of the (cumulative) delay alleged was a matter for which the employer is not responsible, the contractor's global claim will fail.
- There may be a delay event or events for which the employer is not responsible. Nonetheless, provided that these are insignificant, an apportionment exercise may be carried out by the tribunal.
- Pursuing a 'global claim' will not be permitted if it is advanced in lieu of proper pleading. The fundamental requirements of any pleading must be satisfied, namely:
  - Fair notice must be given to the other party of the facts relied upon

together with the legal consequences that are said to flow from such facts.

- So far as causal links are concerned, in a global claim situation, there will usually be no need to do more than set out the general proposition that such links exist.
- Heads of loss should be set out comprehensively.

In the more recent case of *Walter Lilly & Company Ltd* v *Mackay*,[19] the defendant argued that Walter Lilly & Company's (WLC) loss and expense claim was a global claim and therefore not enforceable. Here the court considered the previous case law on global claims. The following principles can be identified from the judgement.

- Claims by contractors for loss and/or expense must always be proved as a matter of fact. The contractor must show, on a balance of probabilities, (i) that events occurred which entitle it to claim loss and/or expense, (ii) that those events caused delay and/or disruption to the works and (iii) that the delay/disruption caused it to incur loss and/or expense. There is no set way of proving these three elements – it is for the contractor to present evidence that it considers is sufficient to prove its claim.
- Although there is nothing wrong in principle with a global claim, such claims do create added evidential difficulties for the contractor to overcome.
- In order to bring a global claim, the contractor does not have to establish that it is impossible for it to identify cause and effect, but the contractor must nevertheless prove that it is entitled to recover loss and/or expense.
- If a global claim is to succeed, the contractor will have to demonstrate that the loss it has incurred (i.e. the difference between what the contractor has been paid and the costs it has incurred) would not have occurred in any event. This will require the contractor to establish that its tender was sufficiently well priced to have resulted in a net return.

The court found that WLC's claim was not global because WLC had sought to identify the specific additional costs it incurred and to link those to the causes of delay/disruption relied upon.

The emerging picture then is that, *in extremis*, a global claim may be permitted by the judge. Understanding the circumstances in which this approach is possible and the steps that need to be exhausted before it is deemed permissible needs an appreciation of the standard approaches. These are covered in Chapter 15.

13.6 Conclusion

This chapter has introduced the reader to some of the difficult questions which exist around the consideration of construction claims. Although answers are available and the law has been rehearsed, there is scope for much argument.

What does common sense mean in relation to proving causation? What level of damages can be said to be in the parties' contemplation? When is a global claim acceptable? The reader will know that to expect definitive answers is to remain ignorant as to the nature of legal debate and discussion. Whilst it is possible to recite conventions and favoured approaches, the chance remains that any given judge on any given day may decide the issue in an unpredictable manner or side completely with one's opponents. The importance of seeking specialist advice is therefore writ large in this area.

13.7 Further reading

Hughes, W., Champion, R. and Murdoch, J. (2015) *Construction Contracts: Law and Management*, Fifth Edition, Abingdon: Routledge, Chapter 21.

Carnell, N. (2005) *Causation and Delay in Construction Disputes*, Second Edition, Oxford: Blackwell Publishing.

Notes

1 [1918] AC 350 (HL).
2 [1966] 2 QB 370.
3 [1982] 1 All ER 851.
4 [1999] BLR 3.
5 (1996) 83 BLR 31.
6 See Jones, M., Dugdale, A. and Simpson M. QC (Eds) (2014) *Clerk & Lindsell on Torts*, 21st Edition, London: Sweet & Maxwell.
7 [2008] EWHC 2037 (TCC).
8 See Section 2.10.
9 [1854] EWHC J70.
10 (1848) 1 Ex 850, 855.
11 *Victoria Laundry (Windsor) Ltd* v *Newman Industries* [1949] 2 KB 528, 539.
12 (1994) 71 BLR 20.
13 Gilmore, G. (1974) *The Death of Contract*, Columbus, OH: Ohio State University Press.
14 *Victoria Laundry (Windsor) Ltd* v *Newman Industries* [1949] 2 K.B. 528.
15 Abrahamson, M. (1979) *Engineering Law and the ICE Contracts*, Fourth Edition, London: Spon Press, p. 515.
16 [1991] 52 BLR 8.
17 [1996] 82 BLR 83.
18 [2004] BLR 393.
19 [2012] EWHC 1773 (TCC).

14 Delay and disruption

14.1 Introduction

This chapter considers in more detail the link between delays, the granting of extensions of time, and the money claimed as a result of these events. The approach taken by JCT, FIDIC and PPC2000 has been to treat the concepts of time and money as separate. This allows some events to be treated as neutral – entitling the contractor to more time, and thus avoiding the responsibility for liquidated damages, but not more money. Exceptionally adverse weather is the standard example of this approach.[1] The NEC approach has always been that time and money are run together in the compensation event clause. The rationale here is that, where appropriate, an event should be compensated with what the situation requires, be it time, money or both. The NEC core clause 60 sets out a long list of compensation events that may give rise to an adjustment in the contract sum or to the contract programme. Clause 61.3 deals with the notification of the compensation event, which may come from the project manager or the supervisor or from the contractor. The contractor must give notification of the compensation event within eight weeks of becoming aware of the event. This is therefore a condition precedent to the right to claim.

Logically, it is appropriate to consider the principles governing the grant of an extension of time before considering loss and expense, which are covered in Chapter 15. Reimbursement of direct loss and/or expense and the grant of an extension of time are separate and distinct matters. Both, however, involve a discussion about causation and the standard of proof required. The purpose of the provisions for money claims and those for extension of time are different in their outcomes. The grant of an extension of time merely entitles the contractor to relief from paying liquidated damages from the date of completion stated in the contract. There is no automatic entitlement to compensation because the supervising officer or architect has determined an extension of time.

In order to provide a framework for discussion, the principles of extension of time need to be examined first.

14.2 Extensions of time

Most building contracts contain express provisions allowing the period agreed in the contract for the completion of the project to be extended where the delay is the consequence of events or factors that are outside the control and responsibility of the contractor. The purpose of these provisions is twofold.

1 *To relieve the contractor from liability for damages resulting from a delay wholly outside the contractor's control*
   Where the delay is outside the control of either the contractor or the employer, the facility to extend the contract period benefits the contractor by relieving them from any liability to pay damages for the delay in the event that the contract is completed late.

   Common examples of this are provisions to extend the contract where it has been subject to exceptionally adverse weather conditions, where there has been delay on the part of statutory authorities or where the works have been damaged by an insurable 'peril' (e.g. lightning, flood, fire, etc.).

   If no such provision is included in the contract, the contractor will be required to take on the risk and will pass this on to the employer, in part at least, by means of the inclusion of a risk premium in his price.

2 *To protect the employer's rights to deduct damages for late completion*
   The second purpose of the inclusion of such provisions is to protect the position of the employer where their actions or those of their agents result in a delay to the progress of the contractor.

   In common law, the contractor's responsibility for completing the contract by a specified date is removed where the employer (or their agents) causes the contractor to be delayed. If the employer was to cause such a delay and no provision for extending the contract period was included in the contract, the contractor would only be required to complete 'within a reasonable time'. In this situation, the completion date and any consequential liquidated damages for late completion would become unenforceable, i.e. time would be 'at large'.

   The inclusion of an extension of time clause allows the employer to set a revised completion date, taking account of the delay the employer (or any agent of theirs) has caused, and still preserves the right to deduct liquidated damages if the revised date is not met.

'Reasonable time' for completion of a project is notoriously difficult to define. However, a reasonable time could be legitimately established on the basis of the original programme suitably adjusted or by reference to other similar projects.

An extension of time can only be given where the contract expressly permits it. Where the contract has been delayed by an event which is not covered by the contract, the contractor cannot claim an extension of time, nor can an employer grant an extension in order to preserve their rights to deduct liquidated damages. For example, the power to extend time for delays arising out of orders for additional work can only be applied where the work has been 'properly' ordered.

In *Murdoch* v *Luckie*,[2] it was held that the relevant instruction was given orally instead of in writing as required by the contract and was not therefore properly ordered. The employer was not able to extend the contract. In this situation, it is likely that time would be 'at large' with all the consequences noted above.

The contract must clearly state the circumstances under which an extension can be granted, and general terms like 'other unavoidable circumstances' or 'circumstances wholly beyond the control of the contractor' will not generally be enforced by the courts.

Although some standard forms of contract allow extensions of time to be granted for factors which to some extent are within the control of the contractor (e.g. inability to secure labour or materials in JCT 2011), in practice it is common for these to be deleted. In addition to having the same rights to extensions of time as main contractors, subcontractors usually have additional rights of extension if delays have been caused by the main contractor.

In general, an extension of time can only be granted where procedures laid down in the contract have been strictly adhered to. Standard forms of contract usually set out detailed procedures relating to the application for and granting of an extension of time.

Awards for extensions of time is an area which underlines the importance of contractual notices. The contracts set out requirements for the submission of notices and the submission of details (sometimes within stipulated time limits) to enable the extent of the delay and extension due to be calculated. Failure to comply with these will not necessarily lose a contractor his right to an extension in most contracts. However, if the contract expressly states that service of a notice is a 'condition precedent' to the granting of an extension of time and the contractor fails to submit notice on time then he will lose his rights. NEC3 is an example of such a contract. However, if the contractor submits an early warning notice and specifies the knock-on effects of the delaying event then the project manager will be unable to challenge this if he does not respond in the time limits with his own reckoning. This represents a double-edged sword for the project manager.

Standard forms of contract usually place the responsibility for deciding the validity of a claim for an extension of time and the period to be granted on the certifier. The employer thus has no right to extend the contract period except through the delegated power of the certifier. Contractors are required to use their best endeavours to mitigate the effects of delays, and the decisions of the overseer will take this into account when assessing the impact of delaying events on overall completion.

There is some variation in the list of delaying events for which the employer is responsible in construction contracts. In the UK, the most complete list is illustrated by the list of 'relevant events' found in clause 2.29 of JCT 2011.

These relevant events can be classified as those which are outside of the control of either party or those which are caused by the employer and/or the employer's agents.

14.3 Types of delay

It is useful here to distinguish between 'delay' and 'disruption'. Delay is usually understood to mean anything that puts back the progress of the works. Delays may be concurrent, affecting several different activities simultaneously, or serial where the delays have a cumulative effect on the programme although not necessarily on the completion date. The effect of delays on the completion date will depend on whether they are critical or non-critical. A delay to any activity need not necessarily have an effect on the overall programme and final completion date unless it impacts on a critical path activity.

A delay claim is essentially a claim for prolongation, either of the project or of particular activities. Certain resources (such as an office or management) may have been required for a longer period, or particular activities may have taken longer so that resources (such as plant or labour) were required for that activity for a longer period, thus incurring additional cost.

Disruption is more generally understood to mean costs or loss and expense incurred as a result of uneconomic working or loss of productivity due to work not being carried out in a planned or a logical sequence. Disruption therefore may not affect the completion date but may alter the sequence of activities or the method of work adopted. In proving a disruption claim, however, the contractor will have to show that he was obliged to carry out works in a less efficient manner as a result of acts or defaults of the employer. There may be periods or areas of work where the contractor's progress was interrupted or where employer 'events' necessitated the use of additional resources. Disruption claims are more difficult to prove as not all disruption will trigger the payment of compensation. A contractor will only be entitled to recover such cost if it can prove that the employer prevented or caused hindrance to the execution of the works.

Indeed, proving disruption is notoriously difficult. The process is made easier if the 'measured mile' approach is available. This involves identifying a period and rate of progress in the project where the contractor has free rein to proceed in accordance with his planning. This is established as a measured mile against which slower progress, caused by the impact of the employer's disruptive events, can be compared. It is still open to the employer to suggest that other factors, not down to their actions, impacted on the project and made the measured mile an unachievable outcome.

Either delay or disruption can result in a contractor incurring additional costs which, depending on the cause of the delay or disruption and the contract being used, the contractor may be entitled to recover from the employer and/ or their subcontractors.

14.4 Causes of delay

The completion of a construction contract can be delayed as a consequence of a variety of occurrences or factors. Although dependent on the detail surrounding the causes and the express provisions in the contract, the responsibility for an outcome falls into three categories.

- *The actions (or inactions) of the contractor* – for example, failure to manage the works competently or an inability to secure adequate labour and repair of damage caused by subcontractors – the responsibility for the management of subcontractors lies with the contractor and therefore this delay would fall within this category.
- *Events that are neither the fault nor the responsibility of the contractor or the employer* – for example, exceptionally adverse weather conditions or storm damage to works in progress – unless the works were inadequately protected in which case this would fall into this category.
- *Delays caused by the employer or the employer's agents* – for example, the issue of a late instruction by the architect – assuming the architect was aware of the requirement for the information at the given time then this would contractually be the liability of the employer.

The rights of a contractor to have the contract period extended or to recover any additional costs incurred will depend on the source of the delay as categorised above. Generally, the standard forms of contract and subcontract clearly define the allocation of risk of delays and the additional costs that naturally flow from them. However, in the absence of express provisions within a contract and unless the delay or disruption is due to the omissions or actions of the employer or the employer's agents, either the contractor or the employer must bear the risk of the project being delayed. The liability for the consequential costs arising from these unallocated risks will normally not qualify as a claim that can be brought against the other party.

14.4.1 Delay factors outside the control of either party

Force majeure

This term comes from French law and translates as 'superior force'. In French law, the term has a definite meaning which is missing in English law. The term is used in construction contracts to describe exceptional unforeseen events beyond the reasonable control of both parties. Such an event makes execution of the contract wholly impossible and, in context of the major standard forms of contract, it is also likely to be an event which would enable one or both the parties to determine the contract. In JCT 2011, the term has a restricted meaning as most of the events that would be likely to have this result (e.g. war, fire, exceptionally adverse weather) are dealt with elsewhere in the contract.

Exceptionally adverse weather conditions

In addition to delays that might result from exceptionally cold or wet weather, this relevant event can be extended to include those arising from exceptionally hot and dry weather conditions. The NEC 3 requires there to be a weather measurement recorded within a calendar month before the completion date for the whole of the works and at a contractually defined place, the value of which, when compared to the contractual weather data, is shown to occur on average less than one in every ten years. Exceptionally adverse weather conditions are best established, where possible, by reference to Meteorological Office records for that locality supported by detailed site records. Clearly the weather conditions must have had a demonstrable effect on the progress of the work so that, for example, a claim for an extension of time under this heading when a building is weathertight and the bulk of the work is indoors is likely to fail.

Loss or damage occasioned by specified perils

The 'specified perils' in JCT 2011 are essentially those also covered by the insurance provisions of the contract, including fire, flood, lightning, explosion and storm. Where the contractor's actions have partially contributed to the event, an extension of time would not necessarily be automatic. For example, 'bursting of pipes' was a specified peril in the contract in the case of *Computer & Systems Engineering plc* v *John Lelliot (Ilford) Ltd.*[3] However, the burst pipe was a direct result of a subcontractor dropping a purlin and rupturing a high-pressure water main. The subcontractor was held responsible for the delay and no extension of time was granted to the contractor.

Civil commotion, strikes, lockout, etc.

This applies to delays arising from industrial action taken by any group of people employed on the works or those supplying or transporting goods and materials to be incorporated into the works.

The exercise by the government of any power which directly affects the works.

An example of this was the government restriction of electricity provision to three days a week in 1974. In practice, this is a little-used entitlement.

Contractor's inability to secure labour or materials

This entitlement is often struck out of contracts with the risk for consequent delay being carried solely by the contractor. In any event, to be successful, a contractor may have to demonstrate that this difficulty could not have been reasonably foreseen.

Carrying out of work by a statutory power

This ground for an extension of time is restricted to the carrying out of work by a local authority or statutory authority in pursuance of its statutory powers. It does not therefore cover the situation when the statutory undertaking is acting as a subcontractor.

14.4.2 Delays caused by the employer and/or the employer's agents

Compliance with contract administrator's instructions

In JCT 2011, the contract administrator is given wide-ranging powers to issue instructions to the contractor during the progress of the works. Any resulting delay would entitle the contractor to an extension of time provided that the instructions do not relate to the non-compliance of the contractor with other provisions of the contract.

Opening up, testing and inspection of completed work

Assuming no defective work is identified, any delay arising out of an instruction to open up or test completed work would entitle the contractor to an extension of time.

Delay in the supply of information

There is an implied term that the contract administrator will keep the contractor supplied with adequate information to progress the works although this is subject to certain provisos. First, the contractor must request the information in writing and, second, the request must not be unreasonably distant from or unreasonably close to the time when it is required. Although it has been held that the contractor's programme can constitute a proper request, the preparation of an agreed schedule of due dates for information is, in practice, a much more certain basis for a subsequent request for an extension of time.

The execution of work not forming part of the contract

The contractor must allow the employer (or any other contractor in their employ) access to undertake parts of the works that are outside that contract. If the contractor is delayed as a consequence of this additional work, the contractor has an entitlement to an extension of time for any impediment, prevention or default, whether by act or omission, by the employer, the architect/contract administrator, the quantity surveyor or any of the employer's persons.

The supply of materials by the employer

Where the employer has undertaken to provide some of the goods and materials for the works and is late in doing so and this results in delay to completion, the contractor has an entitlement to an extension to the contract period. However, to ensure this entitlement, the contractor must ensure that the employer is fully aware of the dates when these goods and materials will be required.

Failure by the employer to give access over employer's land

This ground only applies where the employer has undertaken to give the contractor access to land or buildings adjacent to the site. Failure to give access to the site itself is covered elsewhere in the contract.

14.4.3 Further examples of delay

The analysis of who is to be held responsible for a delaying event is sometimes far from straightforward. The approach appears simple enough but complicating factors exist. Until 2005, the JCT had a procedure known as nominated subcontractors whereby the employer was able to select a subcontractor for the contractor to appoint. This nomination limited the contractor's ability to have a free hand in the appointment; the *quid pro quo* was that the employer should be, in some ways, responsible for the default of the nominated subcontractor. Nomination continues in a watered-down format in the intermediate form where it is known as named subcontractor procedure. Nomination is still common under FIDIC conditions and in the Middle East.

Nomination is an example of how the standard approach can become more complicated. The position arrived at is that although responsibility of the supply chain lies with the contractor, there is an exception if a nominated subcontractor is involved. Deciding the extent to which the nominated subcontractor caused the delay and the impact of other delaying factors for which the contractor might be liable demonstrates the complexity that can arise. Other examples of situations giving rise to complexity include delay in giving the contractor full access to the site and fire damage to completed works. In the latter case, unless the fire is a consequence of the negligence of the contractor, and in the absence of specific conditions in the contract, responsibility for the costs and delays arising from a fire would lie with the employer. For this reason, most standard forms of contract include provisions to ensure that the works are adequately insured to cover such an eventuality.

14.5 Critical and non-critical delays

Delays will often occur on a project, but not all project delays will cause delay to the completion date. Therefore, it is important to distinguish between critical and non-critical delays. The construction programme identifies all the activities that are critical to the works progressing as envisaged. A 'critical activity' is an activity necessary to achieve the completion date. The 'critical path' is the combination of the critical activities in a construction programme, which determines the overall project duration. Critical delays are those delays which extend the overall project duration and the completion date because they cause delay to subsequent critical activities. A non-critical delay is an activity that is not necessary to achieve the completion date and that does not impact on the critical path. Non-critical delays may become critical if they are excessively delayed.

The parties may not agree on the criticality of the delays caused. Critical path analysis refers to the practice, usually retrospective, used to decide the criticality of the delays occurring. Delay analysis experts may be employed to forensically investigate delays. The position can be further complicated by the existence of concurrent critical delays.

14.6 Delay analysis

The starting point for analysing delays is to decide on whether the approach taken is based on the planned programme or the actual impact of the delays experienced. The decision about which approach to take usually depends on the quality of the records available to the contractor to attempt to prove causation and to win his case on the balance of probabilities. The better the records, the more likely it is that the contractor will pursue an actual approach. All methods of delay analysis have their limitations as they rely on subjective and theoretical assessment.

As-planned versus as-built

This form of delay analysis involves a simple comparison between the contractor's original plan and what actually occurred. This can be presented as two bar charts side by side and can be substantiated by witness evidence. This method is useful in identifying the periods of delay but does not identify the causes or the consequential impact on the completion date.

As-planned impacted

This method of delay analysis adds the delay to the as-planned programme entries in order to update the programme as it progresses. This can be done prospectively during the programme or retrospectively afterwards. The prospective approach is used in NEC project management to pre-agree the compensation event consequences. This approach tends to give a contractor-friendly outcome as no account is taken of the contractor's own delays.

Collapsed as-built

The focus in this method of delay analysis is the actual situation encountered. The as-built programme is the starting point. Delay events are then removed from the as-built programme in reverse chronological order. By stripping out the delays for which the employer is responsible, it is possible to identify the dates when the works would have been undertaken 'but for' the employer delays.

Window analysis

This approach breaks the project down into manageable time slices (windows) to enable the process to be reviewed and investigated. The extent of the delay at the start and end of each window is determined by comparing the as-planned and as-built dates for the activities along the identified critical path.

Time impact analysis

This method of delay analysis involves updating the as-planned programme and then 'impacting' it with the effect of a delay event. The aim is to analyse the position of the project at the time of the delay event. This approach has the advantage of tracking the as-built status through the project based on delays as they occur. This is the approach favoured by the Society of Construction Law's Delay and Disruption Protocol.

14.7 Concurrent delay

Concurrent delay refers to the situation where both an employer risk event and a contractor risk event cause delay to the construction project at the same time. Many contract draftsmen (including the authors of JCT, NEC, FIDIC and PPC2000 contracts) prefer to leave the problem of concurrent delay unaddressed. There is little guidance available as to how the overseer should make an assessment. The potential for a dispute to arise and be referred to adjudication or arbitration is therefore present.

Judicial pronouncements on concurrent delay have traditionally promoted the 'dominant cause' approach. This approach essentially gives the benefit of the doubt to the contractor. Where there is an employer risk event causing delay then this automatically becomes the dominant cause, provided that the delays are concurrent.[4] This can be justified by applying the prevention principle. The prevention principle provides that a party may not require the other party to comply with a contractual obligation in circumstances where that party has itself prevented compliance. If the employer has caused a delay then it cannot penalise the contractor for not meeting the completion date.

The law in this area is not settled and another approach, known as the apportionment approach, is also supported in case law. This does allow the overseer to have regard to the concurrent delay caused by the contractor and to carry out an apportionment of the delay accordingly. This was the approach in the case of *H Fairweather* v *Borough of Wandsworth*[5] where it was held that there cannot just be identification of the overall dominant cause of a delay. Rather, each cause has to be assessed on its overall effect (i.e. the 'but for' test) on the critical path and if there has been another significant delay caused by the other party. These factors will then determine whether the employer has a right to claim damages and whether there is a valid extension of time for the contractor.

The apportionment approach was applied in Scotland in the case of *City Inn* v *Shepherd Construction*.[6] However, the approach was rejected in the English case of *Walter Lilly & Company Ltd* v *Mackay*[7] and should, therefore, be treated with caution. The dominant cause approach was preferred, and the contractor was able to recover substantial sums in a situation analogous to concurrent delay.

The common-sense approach in deciding the appropriateness of an extension of time is recommended. However, the onus remains on the party to establish its case properly, and this cannot be done satisfactorily without decent record-keeping.

14.8 Record-keeping

Contractors are usually well aware of the importance of keeping records, from updating the programme with actual progress through to keeping minutes of all meetings at which decisions are taken with regard to the project. The contractor usually has an advantage over the employer unless the overseer is as scrupulous as the contractor in maintaining records. The category of documents that can support a claim for an extension of time include progress meeting records, marked-up drawings, correspondence, labour allocation sheets, daily work area records, daily site diaries, handover records, daily weather records and progress photographs.

The judge is bound to favour the version of events supported by contemporaneous records in a situation where competing accounts are put forward. The judge may prefer one witness of fact to another but the witness able to refer to the documentary records of what happened is much more likely to be believed. Many claims have failed due to poor records in circumstances where the requirement that a case be proved on the balance of probabilities has been left unsatisfied. Equally, a seemingly poor claim can achieve a substantial recovery where it is supported by good records.

14.9 Float

A final issue to consider before leaving the subject of delay concerns float. Float is generally taken to mean the difference between the total time available to undertake an activity and the total estimated time of the activity duration.

In effect, it is the spare time or 'wriggle room' in the programme and is sometimes used to provide a buffer between activities to prevent the critical path being adversely impacted.

The question arises as to who owns the float on a construction project. The relevance of the question is apparent when a contractor seeks an extension of time – should they be required to show the float in their programme and exhaust this first before establishing their entitlement to more time? The contractor may argue that the float was added for his own protection and is for his benefit.

The approach taken by the available case law suggests that the float is owned by whoever requires it first. In the case of *Ascon Contracting Limited* v *Alfred McAlpine Construction Isle of Man Ltd*,[8] the float was seen as being for the benefit of all parties on the construction project. This can seem harsh on the prudent contractor who has given himself breathing space at the end of the job. It is for this reason that contractors do not typically show the float on a programme.

14.10 Conclusion

The presentation and assessment of extension of time claims is a complicated area based on some simple rules. A 'claim' for extension of time is the usual precursor to a prolongation claim, to the extent that many contractors and architects believe that unless the contractor is given an extension of time first, he is not eligible for a prolongation payment. It is often convenient for the contractor to get his extension of time first because the evidence in support of entitlement for extension of time will often be the same as the evidence required to establish an entitlement to loss and/or expense, although it will not establish the quantum.

14.11 Further reading

Thomas, R. and Wright, M. (2011) *Construction Contract Claims*, Third Edition, Houndmills: Palgrave Macmillan.
Pickavance, K. (2010) *Delay Disruption in Construction Contracts*, Fourth Edition, London: Sweet & Maxwell.

Notes

1 JCT SBC clause 2.29.7 lists exceptionally adverse weather as a relevant event.
2 (1897) 15 NZLR 296.
3 (1950) 54 BLR 1.
4 *Henry Boot Construction (UK) Ltd* v *Malmaison Hotel (Manchester) Ltd* (1999) 70 Con LR 32.
5 (1987) 39 BLR 106.
6 [2010] CSIH 68.
7 [2012] EWHC 1773 (TCC).
8 [1999] 66 Con LR 119.

15 Loss and expense

15.1 Introduction

'Loss and expense' is a term frequently misunderstood in the context of construction law. The simplest approach to this subject is to think of loss and expense as the financial consequences to the contractor of 'mini breaches' by the employer under the contract. The term 'mini breach' is used because the eventuality of the contractor being able to make a claim is expressly provided for in the contract, which prevents the events experienced becoming actionable as breaches of contract. Put another way, the contract acknowledges that overspends are likely to happen and provides a mechanism for them to be claimed without needing to have recourse to a damages claim. Loss and expense is simply terminology for what would be the position under the background law – the damages to which the claimant is entitled for what would amount to breaches on the part of the employer. The term 'loss and/or expense' emanates from the JCT contract. For the purposes of this book, this is simplified to 'loss and expense'. However, the ability to claim is universal on most construction contracts hence the interchangeability of the term with compensation events under the NEC and additional payment claims under FIDIC.

The previous chapter discusses claims for extensions of time. This is often seen as the precursor to making a financial claim for loss and expense. In this sense, the money claim can be seen to accompany the time-related claim. This presupposes that the money claim is composed of time-related events, which is usually the case. A claim for loss and expense arises out of the express terms of the contract allowing the contractor to claim in certain circumstances. In addition to the express terms of the contract, there are also implied terms that the employer will not hinder or prevent the contractor from carrying out his obligations under the contract. In the absence of an express term, it would be possible for the contractor to pursue his claim under an implied term.

15.2 The nature of claims

The term 'claim' in the context of construction contracts can be defined as the assertion of a right to payment arising under the express or implied terms of a building contract other than under the ordinary contract provisions for payment of the value of the work.[1]

The contract landscape only permits certain categories of claim to be brought.

Contractual claims

These are the claims that arise out of the express provisions of the particular contract (e.g. 'direct loss and/or expense' in JCT 2011). The entitlements under the contract may extend beyond damages for breach of contract (e.g. under JCT 2011, a contractor can claim additional costs incurred as a result of complying with an architect's instruction which is not a breach of contract).

Common law claims

These are sometimes referred to as ex-contractual or extra-contractual claims. These are claims for damages for breach of contract and/or legally enforceable claims for breach of some other aspect of the law (e.g. for breach of copyright). The contractor's entitlement to claim in common law is expressly preserved in the major standard forms and in some cases, this remedy might provide a fallback where, for example, a contractual right has been lost through failure to give required notices.

Quantum meruit

These provide a remedy where work has been carried out but a contract has not been formed and no price has been agreed. The legal principle in these situations is that the party carrying out the work is entitled to a reasonable payment for work carried out in good faith. In the construction industry, it is common practice for employers to issue letters of intent to enable works to be commenced while the final details of the contract are resolved. A *quantum meruit* entitlement might be created by a letter of intent situation. Care must be taken in the wording of such letters of intent to avoid the inadvertent formation of a contract with all its potential liabilities.

Ex gratia claims

An *ex gratia* (out of kindness) claim is one that the employer is under no obligation to meet. In the context of construction contracts, its application is unusual, and an *ex gratia* claim from a contractor is unlikely to be met by an employer unless that employer is likely to accrue some benefit in return for the payment. An example might be where an employer might agree to an *ex gratia* payment to prevent a contractor going into liquidation in order to avoid the potentially greater cost of employing another contractor to complete the works.

In a claim situation, an overseer can only certify payment of sums for which the contract gives them express power to do so. Under the major standard forms of contract, the overseer is not authorised to certify amounts in respect of common law, *quantum meruit* or *ex gratia* claims.

What is recoverable under the loss and expense claim should equate exactly to the damages recoverable at common law under the *Hadley* v *Baxendale*[2] principles, which were stated in Chapter 2.[3] Although a contractor may incur additional costs as a direct result of the action of the employer or the employer's agents, any contractual claim they may make under the contract will generally be limited to the express (i.e. stated) rights conferred on them by that contract together with any terms that might be implied by a court. The contractor may pursue an employer for compensation for other breaches of contract but this will be done through the courts rather than under the contract provisions. This chapter restricts itself to consideration of the basis on which contractual claims may be made.

The contractor must present his contractual claim based on the occurrence of a loss and expense event, together with substantiation of such losses. The overseer will then assess and ascertain if any sums are payable and issue a certificate if so required.

15.3 Heads of claim

Once the entitlement has been established in principle, the claim needs to be calculated. It is clear from established contract law that in order for a claim to succeed, the courts usually require any loss and expense to be adequately demonstrated and supported. This may occur once the overseer has been satisfied about the duration of the prolongation period. The next step is for the contractor to establish the loss and expense suffered as a result. It should be remembered that entitlement to an extension of the completion date does not automatically entitle the contractor, or subcontractor, to additional payment for the extended period of the works.

Actual costs incurred by the contractor are preferred to a substantiation based on planned cost. This is discernible in relation to preliminaries (on-site overheads). It is not usually sufficient to merely recite a figure appearing in the bill of quantities or tender documentation. The actual cost to the contractor of being on-site for the extra period of time must be established.

15.3.1 Site overheads

This head of claim is sometimes called preliminaries. This head includes such items as hutting, electricity, standing plant, small tools and site supervision. When the contractor does not own the hutting and plant, his entitlement will normally be based on presentation of hire invoices, including any charges from a sister company. A difficulty arises when the contractor owns his own huts and plant. Overseers faced with a claim for contractor-owned items will usually allow a reasonable hire rate.

The principles are the same for supervisory staff and non-productive labour on-site as for hutting and other plant. Hired staff can be claimed against invoices and employed staff claimed on the basis of cost. Cost in this instance can be taken from the contractor's accounts and wage information.

In some instances, parts of general overheads can be identified directly with a particular project. Typical examples would include special premiums for a professional indemnity insurance where taken out for one contract, or for a bond, or the hire of off-site storage facilities. The general principles described for other heads in this section apply equally here. Any increase in value affecting premiums would normally be deemed to be covered by the overhead and profit element in the valuation of variations. However, a prolonged construction period may increase the premiums paid. In those circumstances, the contractor would be entitled to reimbursement as a prolongation cost, again based on actual additional cost incurred.

15.3.2 Head office overheads

Head office overheads relate either to the cost associated with running the contractor's business or to the contribution required by the contractor from each of his contracts towards such cost. The former should be considered as expense and the latter as loss. On JCT contracts, the difference is immaterial since the entitlement is to loss and/or expense. Items falling within this head include offices, support overheads such as buying and accounts departments, rent, rates, heating and telephone bills, and indeed anything going towards the cost of maintaining the business operation as a whole as opposed to an individual project.

The basis of claim is that because of delay on the project, the company workforce was deprived of the chance of earning contribution or recovering its overhead cost from elsewhere. Small companies with limited staff resources can sometimes demonstrate by reference to correspondence that they have had to turn away new work, but it is difficult for a large national company to show that a delay on a single project had a significant effect on the whole company's ability to accept new work.

The position is succinctly put in *Keating on Building Contracts*:[4] 'But for the delay, the workforce would have had the opportunity of being employed on another contract which would have had the effect of contributing to the overheads during the overrun period'.

It is important to distinguish between these two elements of overhead costs, however calculated. One set of overhead costs is costs that are expanded in any event: rates, electricity and the like. The other is managerial time that is directly allocable to the project in question and to no other.

Once the claim is established in principle, the contractor is then obliged to provide information to enable the loss or expense to be calculated. A popular means of calculation among contractors is the use of a formula. Formulae in common use are the 'Hudson formula', the 'Emden formula', and the 'Eichleay formula'.

The Hudson formula is the best-known formula. It calculates loss as an average overhead and profit allowance per week – based on the contract period and percentage mark-up included in the bills – multiplied by the length of delay. The formula can be criticised on the grounds that it relates to tender allowances (i.e. value) rather than actual costs.

(Head office overheads + profit) ÷ 100 × contract sum ÷ contract period in weeks × delay in weeks

The Emden formula differs from the Hudson formula only in that as a means of calculation, once entitlement in principle is established, the overheads are taken as an average percentage from the contractor's accounts.

(Total overheads and profit/turnover %) ÷ 100 × contract sum ÷ contract period in weeks × delay in weeks

The use of a formula for calculating overheads was approved in the case of *JF Finnegan Ltd* v *Sheffield City Council*.[5] The court held that the contractor's loss in respect of lost overheads was to be calculated by looking at the fair annual average of the contractor's overheads and profit as a percentage figure, multiplied by the contract sum and the period of weeks of delay, divided by the contract period. This approach is consistent with the Emden formula. However, the judge also concluded that a recovery of such costs is only possible if the contractor can demonstrate that he could have been employed on another contract that would have the effect of funding the overheads during the period of delay.

The Eichleay formula approaches the problem from a different direction, represented by a shortfall in contribution. The average weekly contribution necessary to run the company is calculated from the company accounts and then multiplied by the period of delay. The product, which represents the total contribution required from all income in order to run the business during the delay period, is reduced pro rata to reflect the share of the total contribution required from the project in delay to pay its way. The pro rata deduction is made by comparing the turnover during the period of delay with total turnover (value) on all projects during that period.

15.3.3 Loss of profit

Loss of profit claims are under the same principles as head office overheads. As with overheads, it is for the contractor to establish that he was deprived of the opportunity to earn profit elsewhere. It is not sufficient to add a percentage to the net value of the other heads of claim.

The loss of profit, which the contractor would otherwise have earned but for the delay or disruption, is an allowable head of claim. It is recoverable under the first part of the rule in *Hadley* v *Baxendale*. It is only the profit normally to be expected which can be claimed and if, for example, the contractor was prevented from earning an exceptionally high profit on another contract, this special profit would not be recoverable unless the employer was aware of the exceptional profit at the time the delayed or disrupted contract was entered into. This would bring the claim under the second head of the *Hadley* v *Baxendale* tests.

In the case of *Walter Lilly & Company Ltd* v *Mackay*,[6] the claimant was engaged by Mr Mackay to build a five-storey luxury home for Mr Mackay and his family in London. Substantial delays to the project were caused by ongoing design decisions. Walter Lilly & Company (WLC) claimed over £276,000 for loss of profit and overheads. This claim was based on the fact that as a result of the delay to this project, WLC was unable to take on other projects and lost profit as a result. In addition, WLC lost the opportunity to spread the cost of its head office overheads on to those other projects. At trial, the number of tendering opportunities that WLC had declined were *precisely detailed on a comprehensive schedule*. The court was duly impressed and upheld WLC's claim for lost profit/overheads in its entirety.

15.3.4 Inefficient or increased use of labour and plant

This type of cost may arise from the need to employ additional plant and labour to carry out the same work or from plant and labour standing idle or being underemployed. Again, in practice, it is often impossible to establish the exact amount of additional expenditure or to relate it directly to particular events and so a global approach, which identifies the overall cost effects attributable to a combination of various causes, may be acceptable. The limitations of global claims are discussed in Chapter 13.

This head of claim may include the claim for 'winter working'. This argument runs that where delay to a project pushes it into a period of less favourable working, the overspend is claimable. As an example, where delay pushes the programme for internal finishings from the summer (when ambient temperatures and natural ventilation are sufficient to dry out the 'wet' trades) into a winter period requiring heating and dehumidifiers, the cost of taking these additional steps would normally be payable. The vagaries of the British weather are one of the factors counting against this argument, which rarely succeeds.

15.3.5 Financing charges

Finance charges by way of interest expended are allowable as a head of claim. In *FG Minter Ltd* v *Welsh Health Technical Services Organisation*,[7] the Court of

Appeal recognised the realities of the financing situation in the construction for those seeking credit and reflected this in the judgement:

> *[In] the building and construction industry the 'cash flow' is vital to the Contractor and delay in paying him for the work he does naturally results in the ordinary course of things in his being short of working capital, having to borrow capital to pay wages and hire charges and locking up in the plant, labour and material capital which he would have invested elsewhere. The loss of the interest which he has to pay on the capital he is forced to borrow and on the capital which he is not free to invest would be recoverable for the Employer's breach of contract within the first rule in* Hadley v Baxendale *without resorting to the second.*

The financing charges recoverable are the actual costs incurred by the contractor whether or not these charges were thought unreasonable when viewed objectively. The point here is that if the employer obliges the contractor to seek credit then he is prevented from criticising the terms on which the contractor was able to secure the lending.

15.3.6 Loss of bonus

Some contracts provide for the payment of a 'bonus' should timely or early completion be achieved in order to incentivise swift, efficient progress on-site. There may be circumstances where the contractor complains that his recovery of the bonus payment was scuppered as a result of employer events causing delay.

Bywaters & Son v *Curnick & Co*[8] concerned refurbishment works carried out to the defendant's restaurant in 1903. The contract, agreed in early 1903, provided for a bonus payment of £360 if the works were completed within nine weeks. The progress of the works was held up by a party wall dispute with the adjoining property with the result that possession was not granted until May 1903, and the defendants refused to pay the bonus. The Court of Appeal held that the party wall dispute constituted an act of prevention and that the bonus payment was recoverable.

The NEC Option C contains the target cost arrangement, which rewards the contractor for making savings on the amount of the contract sum by passing on a proportion of the savings made. It would be interesting to run the argument in a situation where the employer's default was the only issue preventing savings being made to see whether the contractor is entitled to claim the would-be savings.

15.3.7 Third party settlements

Whenever a project is delayed or disrupted, it is likely that the contractor's subcontractors will be affected. A difficulty arises for contractors when a subcontractor presses his claim long before the contractor's entitlement is finally ascertained. The contractor will probably be faced with the argument that the settlement is irrelevant to the contractor's entitlement, or he might be put to the task of proving the subcontractor's claim.

In the case of *Oxford University Press* v *John Stedman Design Group*,[9] the principles emerged that the law encouraged reasonable settlement and the avoidance of the costs of litigation. A party could therefore be reimbursed the sum(s) paid to settle third party claim(s) provided such claims had a reasonable prospect of success against them. In determining whether or not it was reasonable to settle a third party claim, it would be relevant to disclose that legal advice had been obtained recommending settlement. The legal advice itself need not be disclosed.

15.3.8 Cost of claim collation

It is generally accepted that the contractor is not entitled to reimbursement for any cost he has incurred in preparing the claim since he is not required to prepare a claim as such, but merely to make a written application to the architect that is backed up by supporting information. Similarly, fees paid to claims specialists or to independent quantity surveyors or other professional advisers are not in principle allowable as a head of claim at law. Where a claim proceeds to arbitration or litigation, the contractor is entitled to claim his costs and the arbitrator's award or the judgement of the court can condemn the employer in cost.

The approach was approved in the case of *Bolton* v *Mahadeva*.[10] Here the claimant installed a hot water system at the defendant's property. The defendant alleged that the same was faulty and obtained an expert report which backed up his allegations. The Court of Appeal held that the costs of procuring an expert report would not be recoverable as damages:

> So far as the defendant's claim in respect of fees for the report which he obtained from his expert is concerned, it seems to me quite clear that that report was obtained in view of a dispute which has arisen and with a view to being used in evidence if proceedings did become necessary, and in the hope that it would assist in the settlement of the dispute without proceedings being started. In those circumstances, I think that the judge was right in reaching the conclusion that that report was something the fees for which, if recoverable at all, would be recoverable only under an order for costs.

Appointing a lawyer effectively starts the clock on the recoverability of costs as part of a claim. Using the lawyer to appoint the claims consultant and expert witness has the added feature of extending privilege to the documents produced. In other words, the correspondence and draft copies surrounding any report produced need not be disclosed to the other side given that it is made in the contemplation of legal proceedings.

15.4 Claims management

Good contract administration dictates that claims should be dealt with on almost a monthly basis and there should not be any need for the single consolidated claim at the end of the project, which is more common in practice. Handling of claims under the NEC should not allow issues to proceed unresolved and will see them dealt with in accordance with the strict timetable.

However, although interim extensions of time may be granted and payments made on account during the progress of the works, it is often difficult to establish the total extent of delay and/or cost at the time the causal events occur. Also, the final agreement of extensions of time and loss and expense is often combined with the agreement of the final account.

Whenever the claim is submitted during or at the end of the contract, the need for a claim that is well presented and properly justified and formulated remains. The following offers some suggestions regarding format and content to assist in achieving a positive outcome.

Particular points to remember are as follows.

- *Quality of submission* – The claim should be prepared on the basis that the reader knows nothing about the project. Indeed, if agreement cannot be reached, it is likely that the claim document may form the central part of any adjudication or other formal dispute resolution process and as such should set out the contractor's case as clearly and objectively as possible.
- *Proof* – The burden of proof is usually on the claimant; claims must therefore be well documented and, where appropriate, include cost records, programmes and progress schedules, correspondence, records of meetings and site diaries to support assertions made in the claim.

A well-presented claim should comprise the following.

1 *An introduction* – This should identify the parties, the basis of the contract including the contract conditions that apply, the tender date, the contract sum, dates of possession and completion and details of cost and time applications previously made and the awards, if any, that followed.
2 *Contractual basis of claim* – This should identify the specific provisions of the contract under which any claim for additional time or reimbursement of additional loss and expense is being made. This may be linked to the detailed description (below).
3 *Detailed description of the factors which disrupted and delayed the project, fully documented where appropriate* – In the first instance, the issues of extension of time and the causes and effects of factors leading to delayed completion of the project should be separated as far as possible from any claim for costs associated with delay and disruption to the project.
4 *Evaluation of claim* – As with the detailed justification of the principles of entitlement (outlined above), evaluation of the impact of delaying factors

on the completion date and evaluation of the costs should be clearly separated.

15.5 Conclusion

Construction projects, by their very nature, are subject to delay and disruption from a wide range of sources. This delay and disruption invariably carries with it costs that have to be distributed amongst the participants in the process.

Construction contracts are designed to identify and allocate these risks; they set out in advance liability for cost and time should certain events occur or should one or other of the parties fail to meet their obligations under the contract.

Some of these risks are unpredictable. Others could be avoided if sufficient time is made available to consider and manage the risks in advance of a contract being entered into.

This chapter has looked at how construction contracts deal with changes in the way in which the contract is carried out in practice and how that affects the liabilities of the parties to the contract with regard to additional time and cost. Finally, the chapter has offered some guidance as to how a contractual claim arising out of the above may be put together and presented.

15.6 Further reading

Hughes, W., Champion, R. and Murdoch, J. (2015) *Construction Contracts: Law and Management*, Fifth Edition, Abingdon: Routledge, Chapter 21.

Powell-Smith, V. (1990) *Problems in Construction Claims*, Oxford: BSP Professional Books.

Notes

1 Chappell, D. (2007) *Understanding JCT Standard Building Contracts*, Eighth Edition, Abingdon: Taylor & Francis.
2 [1854] EWHC J70.
3 See Section 2.10.
4 Furst, S. and Ramsey, V. (2001) *Keating on Buildling Contracts*, Seventh Edition, London: Sweet & Maxwell, pp. 332–3.
5 (1988) 43 BLR 124.
6 [2012] EWHC 1773 (TCC).
7 (1980) 13 BLR 7 CA.
8 (1906) HBC 4th Edition, Vol 2, p. 393.
9 (1990) 34 Con LR 1.
10 [1972] 2 All ER 1322.

Part 4

Dispute resolution

16 Choices for dispute resolution and features of claim preparation

16.1 Introduction

There are many words used to describe the situation where two or more parties enter into a dispute. The title of this section refers to *dispute* resolution, but references to conflict, disagreement or difference would be equally valid. Disputes usually involve money whether in the guise of underpayment or overpayment or rectification costs based on defects in the build. In its simplest form, this involves non-payment of an application or invoice for services rendered or goods supplied. Recovery of money through the more formal dispute resolution techniques is called an award of damages. The damages are claimed as a result of alleged breach of contract or breach of tortious duty. The intention of damages is to put a party to a contract into the position it would have been had the dispute not arisen.

The construction industry often refers to disputes as claims. Claims were addressed in Part 3 and were defined as any sum of money claimed by one party from the other which was not included in the original contract price. A claim is merely one party's assertion of its rights whereas a dispute is a rejection of any claim made and possibly the assertion of a counter-claim. The key point with a dispute is that a situation has been arrived at in which the parties are unlikely (for the time being at least) to be able to resolve their disagreement without third party assistance.

The word resolution means pursuing something to the point where a result is achieved. That result may be by way of a settlement or payment of a judgement following a trial/arbitration. In both cases, the dispute will have been resolved. Resolving a dispute is not the same as ensuring that the resolution will be actually carried out or enforced. If the dispute resolver ordered a payment to be made from one party to another then enforcement action might be necessary to actually ensure the outcome sought occurs. This may involve court proceedings and may extend as far as insolvency proceedings.[1]

Part 4 is concerned with the choice of dispute resolution options available; these are introduced in the chapters following this general discussion about the characteristics and general nature of disputes. Dispute resolution can be seen as the culmination of the preceding chapters. A successful party in a formal dispute resolution procedure will have to work carefully to establish their case in terms of:

- *the law pleaded* – Which rules of contract law (for example) will need to be set out to support the case? Is the law based on statute or case law or both?
- *the evidence and records in support* – How will the parties assert and prove their entitlements, and what of such issues as causation and quantum of claims?
- *the contract administration and professionalism of the parties involved* – Have correct procedures been followed and can these be evidenced to build a compelling case?

The strength of each party's case is a key indicator of the likely success of a claim/dispute. The first role for any advisor in this field is to gauge this strength and give as realistic an appraisal of this as possible to their client.

16.2 Definition of terms

Dispute resolution tends to have a language of its own, and it is important to be able to distinguish between the options involved and some of the key concepts at work. Many of the ideas discussed in earlier chapters are involved here, and the rules and procedures of such things as contract law and causation are all intricately involved in pursuing a case in most forms of dispute resolution.

A discussion about the cost is never far from the minds of those involved in dispute resolution. It is a much-repeated mantra that the lawyers are the only ones that win when a case is pursued. There is truth in this, and the legal costs can quickly outstrip other professional costs involved.

16.3 Causes of disputes on construction projects

It is a truism to observe that the scarcer a commodity becomes, the more people fight over it. The recent recession in the construction industry saw an upswing in the number of disputes as people were forced to be more adversarial in a tough economic climate. Disputes in the construction industry tend to follow the boom and bust cycle as profit margins shrink as large amounts of people compete for smaller amounts of work. The NBS survey[2] looked at the most common items in dispute. The most common causes of disputes were loss and expense, defective work and the valuation of variations. All three of these causes are preventable with proper preparation and execution of the works. All of the matters relate to money claims in some proportion. The efforts currently being made to break the cycle of disputes are considered in Part 5 where the root causes and initiatives promoting improvement are discussed.

Table 16.1 Terminology in dispute resolution

Adjudication	Adjudication is a dispute resolution procedure available to the construction industry. The procedure can be used at any time to resolve disputes between the parties. The parties are given a decision by the adjudicator within 28 days. Adjudication was introduced by the Housing Grants, Construction and Regeneration Act 1996 (as amended).
Mediation	Mediation is a dispute resolution procedure involving a neutral third party. The procedure is consensual, confidential and without prejudice. The mediator explores ways to encourage the parties to settle their dispute amicably. When ADR (alternative dispute resolution) is referred to, people usually have mediation in mind.
Early neutral evaluation	Early neutral evaluation is another form of dispute resolution. The procedure aims to provide parties with an early and frank evaluation by an objective observer on the merits of a case with a view to encouraging settlement.
Without prejudice	Without prejudice is a term which applies to all forms of dispute resolution. When negotiating, one party may make an offer or admission that it does not necessarily want the decision-maker to know about before the end of the legal process. By marking the letter 'without prejudice', the writer ensures that the letter is not disclosed without the writer's permission. The letter becomes 'without prejudice' to the parties' legal entitlements.
Hybrid techniques	A hybrid technique is a dispute resolution process which adopts more than one procedure to attempt to resolve a dispute.
Claimant/ referring party	This is the party commencing adjudication. It may be helpful to think in terms of the referring party being the aggrieved party in a claim situation.
Defendant/ responding party	This is the party defending adjudication. It may be helpful to think in terms of the responding party being the party against whom the grievance is held.
Arbitration	This is a dispute resolution process traditionally favoured by the construction industry. The right to arbitrate arises from the contract between the parties. The procedure is held in private and is run by an arbitrator. The arbitrator's award is usually final.
Litigation	This dispute resolution process is available through the courts and is run by a judge. Litigation is the procedure also known as 'suing' and 'taking someone to court'. Civil litigation must be distinguished from criminal litigation which serves an entirely different purpose.[*]
Costs	In the context of arbitration and litigation, 'costs' means the legal costs incurred by the parties. This will usually involve the time-related costs incurred by the solicitor, barrister, claims consultants and expert witnesses together with other expenses (court fees, travel costs) and VAT. The usual rule applied is that the loser of the case contributes to the costs incurred by the winning party as well as bearing their own costs and those of the arbitrator.

[*] See Section 1.2.

16.4 Dispute resolution – voluntary and mandatory options

The key to understanding dispute resolution techniques is to appreciate the distinction between encouraging the parties to come to a settlement and imposing a decision, whether temporary or finally binding, upon them. Some types of dispute resolution are voluntary and non-binding. The parties are in full control here and may quit the process at any time. Negotiation and mediation fall into this category.

The other types of dispute resolution procedure considered here – adjudication, litigation and arbitration – involve the determination of the matter by a third party which then becomes binding on the parties. The essential point is that disputing parties are not likely to agree on a consensual approach to resolution. It is quite often necessary for the third party to become involved in deciding the issues.

The order of the dispute resolution choices set out in Figure 16.1 demonstrates their position with regards to one another in terms of the following.

- *Cost of the proceedings* – It is a moot point as to whether arbitration is cheaper than litigation but the other options are certainly more cost-effective. Adjudication is characterised by its relative low cost whilst mediation needs only the services of a mediator and a venue for, typically, a half-day period.
- *Time of proceedings* – Adjudication should last only 28 days once commenced whereas arbitration and litigation can last much longer despite efforts to keep this under control.
- *Quality of the decision* – If it is important to the parties to know the rights and wrongs of their respective claims then the judge/arbitrator is most likely to arrive at the correct result. The vast majority of clients do not focus on this aspect and are content merely to see the dispute resolved.

Negotiation represents a dispute resolution choice and also underpins the others. Negotiation and barter have been around since time immemorial and are feted in many cultures. In some parts of the world, it is seen as an essential part of national culture and identity. In the UK, it is common for the legal position to be considered alongside a negotiated position. Legal advisors are

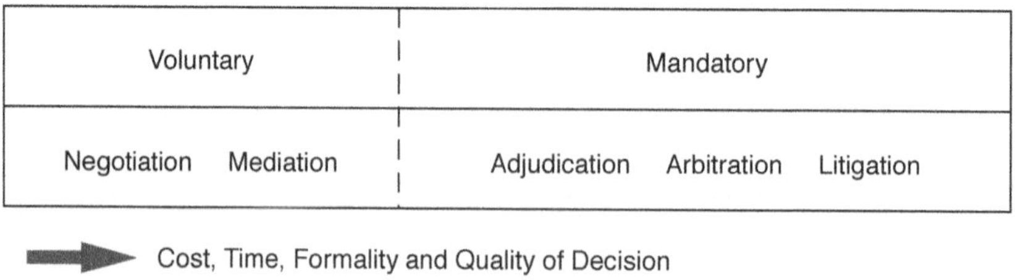

Voluntary	Mandatory		
Negotiation Mediation	Adjudication	Arbitration	Litigation

→ Cost, Time, Formality and Quality of Decision

Figure 16.1 Dispute resolution techniques, ranked by cost, time, formality and quality of decision

well versed in the need to keep open channels for negotiation notwithstanding the existence of formal legal actions. This is usually achieved by the use of 'without prejudice' negotiations.

One of the frustrations for westerners working in the Middle East during the recent boom in construction is the propensity for clients to put negotiations ahead of their contractual entitlements. Despite the many hours and huge cost put in to arranging contract documentation and formalising legal position based on Western-style concepts of statutory and common law, it is quite often in the coffee shop where deals are done, disputes resolved and final accounts settled. There is a sense in which this is no different to the Western negotiation norm. Perhaps it is simply the different location of where the deals are done that is new to the Western advisors.

Quite often, the parties do not wish to negotiate. There are a number of reasons for this, including:

- the timing is not right for negotiation – negotiation requires the disclosure of the parties' cases to one another, and if this has not occurred then negotiation is unlikely to work;
- the amounts of money involved preclude negotiation, which inevitably involves compromise;
- personalities have become involved in a dispute.

It would clearly be asking a great deal of the parties to expect them to be able to resolve a dispute by themselves in a situation involving a complicated situation that can legitimately give rise to more than one interpretation of the facts. It is for situations like these that other forms of dispute resolution have been created.

16.5 Personnel involved in dispute resolution

The following parties may be encountered in the more formal dispute resolution procedures, most notably arbitration and litigation, though some adjudication cases may also involve the full range of personnel.

Claims consultant

Quite often it is necessary to have a claim prepared by a construction industry specialist known as a claims consultant. It makes good economic sense to use the knowledge and experience of a claims consultant in claim preparation rather than to incur the fees of a solicitor or barrister. Claims consultants are well versed in the skills needed to lay out and establish cause and effect in relation to a claim for overruns of time and cost (as detailed in Part 3).

Solicitor

Solicitors prepare cases ready for the barrister to present at trial/arbitration. The solicitor is usually the single central point of contact and receives and passes on information to the other third parties involved in the process, including claims consultants and expert witness. The solicitor also advises on prospects of success and costs issues.

Barrister

Barristers are sometimes referred to as 'counsel'. Barristers present a party's case to the arbitrator/judge and can also be asked to advise on the prospects of success and the correct preparation and approach to a case. Barristers are rarely instructed on small disputes because of the relatively high cost of hiring them.

Expert witness

A judge/arbitrator considers the evidence presented before arriving at a judgement/decision. If technical issues are involved then it may be necessary for the judge to be given evidence by way of specialist opinion(s) so that the case can be properly assessed. In these circumstances, an expert witness would be appointed.

Expert witnesses are usually eminent professionals available to give their opinion either in support of or against a party's case. Obviously, a party would be ill-advised to hire an expert who was against the case being promoted. Expert witnesses must be chosen carefully and transparently to avoid criticism that a party has 'shopped' for an expert. The expert witness must at all times be aware of their duty to the tribunal who they are to assist rather than to take a view of things which is overly supportive of their paymasters. Judges are not known to hold back in their judgements when they suspect that the expert has become too close to their client's case. This can lead to reputational damage on the part of the named expert in the judgement.

16.6 Effective dispute resolution

The widely held view of litigation and arbitration is that they are prohibitively expensive and should only be used as the very last resort. The view is correct on both counts. A telling question to ask anyone involved in the claims industry is whether they would litigate or arbitrate with their own money. The answer is almost always in the negative. However, it is also true to point out that if a dispute is large enough and involves enough money then litigation becomes a real prospect. The hope then is that by properly managing the process, the costs can be limited.

Another reason for the prevalence of litigation is the position taken by insurers in respect of claims they have indemnified their clients against. The insurer will frequently seek to recuperate sums expended by depending on the subrogation rights in the insurance contract to pursue third parties for the loss suffered. Thus cases appearing in the courts are frequently not directly involving the disputants; rather, it is the insurers battling it out using the names of the insured.

Before embarking on litigation/arbitration, the parties should consider whether the dispute can be resolved by any other means. Negotiation skills should always be used first. Another important skill is to possess the foresight not to pursue a claim that ultimately should never have been brought. Quite often in a dispute, one party will know that it is acting unreasonably but is unable to avoid that course of action. Moreover, conflicting parties do not necessarily owe it to one another to act morally or even fairly. Once a dispute has arisen and no informal progress can be made towards resolving it, the time has probably come to seek advice from a third party (usually a solicitor or claims consultant) on the best way to proceed in terms of issuing legal proceedings.

Before making an appointment to see a third party, it is always advisable to weigh up the pros and cons and conduct a cost/benefit analysis of what is achievable. Costs can be limited by preparing files in good chronological order. This is usually the first task for claim preparation and something that the would-be litigant can assist with.

Once professional help has been retained, it is always worth keeping a close eye on cost, asking for specific quotes and not allowing hourly rates to be applied without preordained limits. Obviously, it can be difficult for the advisor to accurately predict the cost of the various stages of the action. However, this does not mean that the advisor should view their client as an open chequebook in terms of their ability and willingness to fund the action. The costs of an action can be prohibitively high and it can sometimes feel as if the costs aspect of the claim has become as important (or more important) than the claim itself. The cost of litigation can easily run into millions of pounds for large cases and into hundreds of thousands for relatively modest claims.

The cost position can give additional weight to the without prejudice negotiations that may occur simultaneously with the court or arbitration action.

16.7 Using the cost position to make without prejudice offers

'Costs following the event' is a term used to describe the situation where the loser makes a contribution to the winner's costs. This is only encountered in those dispute resolution procedures where *inter partes* costs are recoverable. This discussion is not therefore relevant to negotiation, mediation or adjudication where each party bears their own costs. It is a common misconception that once litigation or arbitration starts, they cannot stop. Over 99 per cent of cases settle before trial. The timing of settlement has obvious cost consequences – settling 'at the doors of the court' on the morning of the trial will be expensive for both parties, who will have gone to the trouble of preparing for trial. An early settlement is therefore always desirable. The drawback is that the real strengths and weaknesses of a party's case are not always obvious until the litigation is well developed.

Settlement is encouraged during litigation by the process of making without prejudice offers, known as Part 36 offers. By marking a letter 'without prejudice', the writer is able to ensure that the other party cannot disclose the letter to the court to show that some liability or responsibility is admitted by the opponent. Arbitration has a similar costs position in place.

'Without prejudice' offers have traditionally been made only by defendants. By making an offer to settle, the defendant is effectively saying: 'Although I have said that I do not owe you anything, I really think I owe you no more than £x which I am willing to pay you now.'

The advantage for the defendant who makes an offer of settlement is that if the claimant refuses to accept, the defendant can then proceed with the litigation/arbitration safe in the knowledge that should the claimant be awarded an amount less than or equal to the amount previously offered then the claimant will have to bear his own and the defendant's costs from 21 days after the date of the offer. Such a result will effectively mean that the defendant is the winning party and the claimant is the losing party on costs.

A procedure is now available for the claimant to achieve the same result by making a settlement offer. This entails the claimant 'naming a price'. A claimant's offer says: 'I am claiming this amount, but will settle for this amount'. If the claimant goes on to be awarded more than his offer at trial, the defendant has to pay additional costs and a higher rate of interest than would otherwise have been the case.

The timing and amount of offers to settle is an area of practice where skill and experience in enticing a response from the other side are involved. It is not uncommon to see a series of offers and counter-offers all narrowing down the amounts in dispute. This can continue to the point where both parties finally compromise. A compromise in this sense is something that both parties are neither happy nor unhappy with.

16.8 Case management

Once a third party advisor has been appointed to deal with the case then the disputant needs to ensure that they support the approach being taken. This involves listening and heeding the advice being given. Advisors may describe chances of success in terms of percentage likelihood of win or loss. Disputants quite often become fixated on how good a case they have and overlook the chances of the undesired outcome. Different forms of dispute resolution have different deadlines for the completion of tasks leading up to the final hearing (if any). The timely fulfilment of these tasks is essential as last-minute rushes towards deadlines inevitably lead to mistakes and inadvertent disclosure of material.

Regard should also be had for the approach of the dispute resolver. The conduct and timetable for the procedure are down to the individual appointed. Care should be taken to respond to requests made and the parties should be alive to the signs displayed in terms of how the third party wishes to proceed. Paying insufficient attention to the approach taken can have unwanted consequences.

16.9 Conclusion

A dispute on a construction project is a common occurrence. Once a dispute arises on a construction project, it is difficult – like the genie let out of the bottle – to contain it. Those involved tend to suffer from entrenchment in their position as blaming the other party can be an attractive way to avoid having to answer difficult questions about one's own performance.

In a climate where disputes are common and sometimes seemingly inevitable, the next stage is to consider how they can be managed effectively. This brings into consideration the dispute resolution techniques and their relative strengths and weaknesses, which are considered in the next chapters.

16.10 Further reading

Hibberd, P. and Newman, P. (1999) *ADR and Adjudication in Construction Disputes*, Oxford: Blackwell Science.

Fenn, P. (2012) *Commercial Conflict Management and Dispute Resolution*, Abingdon: Spon Press.

Notes

1 See Section 3.4.
2 NBS National Construction Contracts and Law Survey 2013, available at: www.thenbs. com/pdfs/NBS-NationlC&LReport2013-single.pdf.

17 Mediation and other forms of ADR

17.1 Introduction

Traditionally, arbitration has provided the alternative to the courts as a method of dispute resolution. However, the use of methods known as alternative dispute resolution (ADR) also has a long and distinguished history. If it is true that disputes are largely inevitable on construction disputes then forms of dispute resolution which deal with the problem efficiently and in a cost-effective manner ought to be considered first. Mediation provides a private forum in which the parties can gain a better understanding of each other's positions and work together to explore options for resolution. The parties retain control of the decision on whether or not to settle and on what terms. During the mediation, the mediator meets privately with each party to discuss the problem confidentially. This allows each party to be frank with the mediator and to have a realistic look at their case without fear that any weakness will be communicated to other parties.

Modern mediation has its roots in the USA, first coming to prominence in the early twentieth century. By the 1970s, its potential for reducing the number of court cases was well recognised. In the UK, it was first used in the mediation of family disputes, and a system of court-annexed family mediation was established by the early 1990s. Around the same time, the commercial law courts saw the potential contribution that ADR could make in terms of 'the more efficient use of judicial resources'[1] and consequently began to encourage parties to use ADR mechanisms through court practice statements and pre-action protocols. The major impetus for mediation development was a result of the reform of civil litigation following Lord Woolf's reviews of civil justice.[2]

Mediation in construction took longer to establish itself as a viable alternative to litigation and arbitration. Nowadays, in most construction disputes, a genuine attempt to settle by mediation is accepted as a sensible and necessary step and is invariably encouraged by the UK courts. Mediation is also promoted through EU Mediation Directive 2008/52/EC as a cost-effective and quicker alternative to civil litigation for cross-border commercial disputes.

The key to mediation is that the third party does not decide the dispute and unlike litigation, arbitration or adjudication, the resolution is a matter for the parties' own agreement. The process is voluntary and consensual although the mediation process may be incorporated into the dispute resolution procedures in a contract. The process can take place at any time, including during the course of another dispute resolution procedure. The parties do not have to be bound by or agree to anything at the end of the process unless they record their agreement, in which case a binding agreement can be made.

The process has much to commend it. Consider the following distinguishing characteristics.

- Confidentiality – any settlement reached need not be made public.
- Without prejudice – if no settlement is reached then no account is taken of any admission or offers made.
- The parties have full control over the process.
- The mediator does not decide on the strengths and weaknesses of the cases.
- Costs – each party bears its own costs in mediation.

The driving force behind mediation is the parties' desire to avoid the cost and uncertainty involved in other forms of dispute resolution. The parties need to convince each other that any solution achieved through the mediation is more favourable than could be achieved by either side by any other route available to them.

17.2 Background to mediation

Mediation has existed as a means of resolving commercial disputes since the start of the 1990s. Mediation is largely promoted through two institutions:

- Centre for Effective Dispute Resolution (CEDR);
- ADR Group.

The CEDR is an independent non-profit-making organisation supported by multinational business and leading professional bodies and public sector organisations. It was launched in 1990 with the support of the Confederation of British Industry. Since its introduction, mediation has been acknowledged by the construction industry as being a very good idea. However, take-up of mediations has remained slow and the general consensus is that a good deal of lip service is being paid to the process.

A CEDR survey of the Institute of Directors[3] shows that three-quarters of members would prefer to use mediation and similar techniques rather than go to court. Presumably the remaining quarter had encountered situations where mediation is less likely to be appropriate. These might occur where a party:

- is adamant that they are not at all liable for the damages claimed against them; or
- is in financial trouble with debts piling up and creditors pushing for payment; or
- knows the line manager categorically rejects settlement.

For mediation to work, both parties must be convinced that it is a better option than other forms of dispute resolution. In other words, both parties must want to settle. Settlement involves compromise. One way to think of compromise is a result that hurts both parties a little, both parties giving a little more than they are entirely comfortable with. This, though, is part of the problem. It is precisely because parties are initially uncomfortable with mediation that the take-up has been slow.

Mediation is attractive where there is an ongoing commercial or personal relationship to maintain. Most industry sectors are small, and falling out with a competitor or supplier can cause damage beyond breaking off relations with the other party. The industry is always ready to gossip about warring factions. Mediation is ideal in this situation. It allows both parties to save face and avoid damage to their reputations.

17.3 Building contracts and mediation

Many construction disputes include complex issues, numerous separate claims and multiple parties. Parties can, and frequently do, agree in their contract that a dispute will be referred to mediation before arbitration or litigation proceedings are started. Standard form contracts containing mediation clauses include JCT Standard Building Contract Clause 9.1. One issue arising from the inclusion of a mediation clause is the extent to which it amounts to a condition precedent to litigation or arbitration.

The question of the enforceability of the mediation clause depends on its construction and whether the parties' rights have been defined with sufficient certainty. Thus in one case,[4] a provision to mediate without any reference to process or providers was held not to be enforceable. However, in another case,[5] a contractual requirement that the parties would *first seek to resolve the dispute by friendly discussion* was held as an enforceable clause. The position is more straightforward with regards to mediation and adjudication. The right to adjudicate in the UK is statutory under section 108(2) (a) of the Housing Grants, Construction and Regeneration Act 1996 and is stated to be available 'at any time'. This means that even if the parties agree to a mediation clause in their construction contract, they cannot make complying with that clause a condition precedent to referring a dispute to adjudication.

17.4 The mediation process

The mediation process starts with the parties entering into a mediation agreement with the mediator. The mediation agreement will usually include express provisions about confidentiality, the process itself and the remuneration of the mediator. The latter point is usually dealt with on the basis that the parties will each be responsible for 50 per cent of the fees involved. The mediation agreement should also warrant that the persons attending the mediation have the authority to bind the organisation they represent (if any).

Once appointed, the mediator will set out the procedure to be adopted. This will usually include for each party to prepare and submit a position statement and a core bundle of relevant documents to be provided to the mediator and the other party before the mediation. The mediator is usually booked for a half-day or a whole day. A whole day is recommended in large and complex disputes. Some mediations take longer – up to five days in the most extreme cases. Mediation has a curious habit of growing into the time allowed for it, and the most dramatic events happen in the last 20 minutes or when the parties are putting on their coats ready to leave.

The usual format for mediation is for the parties to hire a mediator and three conference rooms at a neutral venue. The parties usually attend the mediation with their lawyer present. The mediator will have asked in advance for the parties to set out their case in a position statement. The position statement will have been circulated to the mediator and the other party. For each party, the mediation will usually be attended by a senior member of staff from the party itself and a legal representative. In some cases, a member of the insurers may attend. It is important that each party attending has the authority to come to an agreement.

The position statement is unlike a standard court pleading or witness statement. The party sets out:

- the facts in a frank and open manner;
- the key issues in dispute;
- any ground to be conceded at the mediation.

A good position statement allows the mediator to go right to the heart of the dispute and avoid wasting any time. The parties meet in one room. Everyone introduces themselves and says why they are there and what they would like to see achieved at the mediation. The parties split up and go to separate rooms. The mediator then shuttles between the two separate rooms and attempts to make progress towards settlement. In doing so, the mediator adopts a number of negotiating skills and techniques aimed at:

- establishing common ground;
- obtaining permission to disclose relevant material from one party to the other;

- probing perceived strengths;
- exposing perceived weaknesses;
- giving an opinion (if required) about the likelihood of success.

Using these techniques and others, the mediator can build on the predisposition of the parties to reach an agreement or at the very least to narrow the issues outstanding between the parties. In considering the acceptability of a sum being offered to settle the dispute, the mediator will also encourage a party to take account of the risk of proceeding, losing and paying the other party's costs. The courts have expressed the view that adopting an unreasonable position at mediation may lead a party that succeeds at trial to be deprived of its costs;[6] all these factors need to be taken into account when weighing up the acceptability of the best offer to settle on through mediation.

During the course of separate meetings with the mediator, the parties will frequently wish to disclose matters to the mediator that are confidential and not for repeating to the other party. This is a normal part of the process, and the mediator will only divulge to another party details which he has been given permission to share.

If the mediator considers it appropriate, he may call the parties back together in a group meeting from time to time. The parties are then invited to reconvene their initial meeting in the third room. If an agreement has been reached, this can be drawn up by the lawyers present and signed by the parties if they wish to be bound by it. The parties can then shake hands and leave on good terms. A binding agreement is preferable to any period of reflection that either party may wish to instigate. This scenario allows for 'cold feet' and backtracking on the position reached.

17.5 The role of mediator

Occasionally, where there is a mediation clause within a contract, a mediator may be named in advance. It is more common, however, to select a mediator at the time of the dispute, and this gives parties greater flexibility to determine any appropriate expertise which may be desirable.

There are two main forms of mediation: facilitative and evaluative. In facilitative mediation, the mediator assists the parties to reach agreement through a structured process without providing his own opinion on the merits of any party's case. In evaluative mediation, by contrast, the mediator may express a view on the merits of an element of the dispute. In each case, the objective is to reach a binding agreement which settles either all or some of the matters in dispute. Most mediation in the UK is facilitative, and frequently the terms of the mediation agreement prevent the mediator from providing any opinion on the merits of the issues in dispute unless otherwise agreed beforehand by the parties.

In facilitative mediation, it is not essential for the mediator to be an expert in any technical aspects. The mediator is not there to make a decision on the facts of the case. If the mediator's judgement is involved in the case then this may even hamper settlement. That said, construction mediators ought to be well versed in construction contract law and be familiar with contract administration and any restrictions this may impose on the settlement process.

It is important therefore that the mediator should remain impartial throughout the process. The mediator's neutrality provides him with credibility. The mediator should resist the urge to give advice to the parties. Self-determination is a fundamental principle that the mediator should facilitate.

Evaluative mediation is concerned primarily with reaching a deal. This mediation style goes further than facilitation in terms of expressing opinions to the parties about which arguments or evidence are most compelling and which side is most likely to win. This approach can be suitable where the dominant issue dividing the parties is money and one or both parties would benefit from a 'reality check'. An evaluative mediator should have subject matter experience to lend credibility to any opinions or judgements put forward.

17.6 Encouraging mediation

The cost of the procedure is much less than all other forms of dispute resolution, with the possible exception of adjudication. The ADR Group claims that mediation works in 80 per cent of referrals. The hard part, as they would probably acknowledge, is getting the parties to mediate in the first place.

Arbitration and litigation are the traditional alternatives to mediation. Increasingly the trend is developing where parties are being told to stop or suspend these procedures and to mediate instead. The dichotomy is that mediation is by definition a consensual procedure and parties cannot be forced to mediate. Certainly people can be encouraged, but how far should this go?

In litigation, judges promote the use of mediation because of the considerable time and cost saving for the parties. For instance, the more people mediate, the more money the government saves in employing judges and running courts. Judges have the power under the Court Rules (known as the Civil Procedure Rules) to actively encourage mediation. Rule 1.4(2) of the Civil Procedure Rules provides that '[a]ctive case management includes ... encouraging the parties to use an alternative dispute resolution procedure'.

In the commercial court, the move to encourage mediation started in 1993 when it identified cases regarded as appropriate for mediation. In such cases, judges suggested the use of mediation or made an order directing the parties to attempt mediation.

If, following a mediation order, the parties failed to settle their case, they were asked to inform the court of the steps taken towards mediation and why they failed. Although this practice was non-mandatory, mediation orders used in this manner imposed substantial pressure on parties to mediate.

One interesting point emerging out of the mediation orders made in the commercial court was that the timing of the mediation was crucial. Mediation orders made in the early part of the proceedings were less successful in achieving settlement than those orders made later on.

Mediation orders made early on in the case are less likely to result in settlement because:

- the parties may not have had an opportunity to fully consider the strengths and weaknesses of their opponent's case. If they have not then the parties may not yet be convinced that mediation is the best option available;
- the parties tend to be at their most aggressive in the early part of the court case and look forward at this stage to their 'day in court';
- the parties will not yet have incurred most of the legal costs involved in the court case. The financial incentive for mediation may not therefore have materialised in the early stages of the court action.

The real pressure the courts can exert in this area is through costs. Unlike adjudication, a judge has the power in litigation to order that the losing party pays the winning party's costs. The logic for this goes along the lines of the winning party saying to the losing party: 'I was right all along and if you had paid me on day one then I would not have had to sue you nor would I have incurred all these costs'.

The court system supports this and allows the winner to recover costs and interest from the losing party. The losing party also has to bear its own legal costs. However, the judge has the power under Rule 44.5 to penalise unreasonable conduct through an adverse costs order. This has been interpreted to mean that failure without good reason to use mediation may result in a reduction in the winning party's costs.

This principle was taken further in a recent case where the winning party had its costs disallowed completely because it had unreasonably refused to mediate. This appears to be straying into the territory of making mediation compulsory. The landmark decision was made in *Dunnett* v *Railtrack*.[7] Ms Dunnett kept horses in a field next to a railway line. When Railtrack contractors sought to replace a gate, she asked if the new gate could be padlocked; however, she was (wrongly) told by a Railtrack representative that this was illegal. Subsequently some of her horses wandered on to the line and were killed.

Ms Dunnett brought a claim against Railtrack, which was dismissed. She appealed and lost again on appeal. Mediation was not explored before the initial trial. However, in the initial judgement, the judge suggested that mediation be attempted before the appeal was heard. Railtrack rejected this suggestion. Railtrack was confirmed as being the winning party by the Appeal Court, but Ms Dunnett (the losing party) was not ordered to pay Railtrack's (the winning party's) costs. Railtrack found itself in the odd position of having won the case outright but being unable to recover its costs from the losing party. The decision was based on Railtrack's refusal to contemplate mediation after it had won at the first trial.

This case was not the first time the Court of Appeal had used costs to encourage mediation. Lord Woolf (a former leading judge in the country) has said: *Today sufficient should be known about ADR to make the failure to adopt it, in particular when public money is involved, indefensible.*[8]

The *Dunnett* v *Railtrack* decision has some dubious implications.

- Parties may be forced to mediate in unsuitable cases.
- Parties may be prevented from exercising their right to a hearing under Article 6 of the Human Rights Convention.
- The winning party may be expected to 'roll over' even when it is proved right.

Mediation is a good idea in most cases, and the efforts of the judiciary to encourage its use are well founded and well intentioned. Further, parties are not being forced to settle, merely 'forced' to attempt mediation. Unfortunately, situations will always occur where it is unsuitable. These situations usually involve insolvency, intransigence or other restrictions on a party's ability to act freely.

When considering whether a successful party should be deprived of its costs, the court will consider the circumstance of a refusal to mediate and is likely to take account of the nature of the dispute, the merits of the case, and the prospect of success. *Hurst* v *Leeming*[9] is an example of a case where the relationship had broken down totally between a lawyer and his partners. The judge commented that mediation was not suitable 'by reason of the character and attitude of the claimant'.

Further guidance as to when cases might be unsuitable was given by the court in the case of *Halsey* v *Milton Keynes General NHS Trust*.[10] These included cases where injunctive relief was necessary or a point of law had to be resolved. The *Halsey* case featured the argument of whether forcing parties to mediate would constitute a breach of Article 6 of the European Convention of Human Rights, commonly summarised as the right to access to the court. The court thought that compulsory mediation would impose an unacceptable obstruction to this right.

A further round of reforms to civil justice was brought about by the Jackson Reforms taking effect from April 2013. The courts take a firmer attitude to parties who fail to take reasonable and proportionate steps to resolve their dispute. This includes the rule in *The Jackson ADR Handbook*[11] that silence in the face of an invitation to mediate is unreasonable irrespective of the justification. This amounts to even more pressure to mediate and raises the question of whether compulsory mediation has been achieved notwithstanding the *Halsey* decision about the right to have access to the courtroom.

Mediation is also encouraged in litigation by such procedures as the Pre-Action Protocol for Construction and Engineering Disputes.[12] The protocol requires each party to conduct a substantial amount of work up front in order to set out the detail of the claim and the response to the claim. The parties are then required to attend a pre-action meeting. The meeting is similar to a mediation in terms of the brief being to consider whether the dispute might be resolved without recourse to litigation. There is no mediator as such at this meeting. The legal advisors present should attempt to make progress at the meeting or to at least narrow the issues in dispute between the parties.

17.7 Mediation summary

The decision of whether or not to attempt mediation requires careful consideration in light of the costs penalties that a refusing party may face at a final trial. This has resulted in a situation where the parties position themselves with regards to the issue of mediation so as not to appear unreasonable. The time and effort required in this positioning would be much better spent actually participating fully in the mediation as the resolution of the dispute for everyone's benefit is likely to ensue. Table 17.1 summarises the case for mediation and sets out some of the perceived negative points that persist.

17.8 Early neutral evaluation

This form of alternative dispute resolution is not often used in this country and is more popular in the USA. The procedure is akin to mediation in that any result is non-binding on the parties unless adopted by them.

The procedure involves a 'dummy run' on the issues in dispute between the parties. The evaluator studies the parties' cases and gives them a provisional view on the likely outcome if the case were to proceed to a full and final hearing. The parties are then expected to absorb the evaluation and consider its ramifications on their case. The procedure is without prejudice in the same way as a mediation.

Most evaluations focus the parties' minds on the risks of proceeding to trial. An evaluation may also serve to disabuse a party of any misconceptions they may have about their own case.

Table 17.1 Advantages and disadvantages of mediation

Advantages of mediation	Disadvantages of mediation
Opportunity for immediate settlement of the dispute	Achieving settlement requires compromise
A quick and relatively cheap way of settling the dispute	A signed agreement is needed to dispense with the matter
The terms of the agreement remain in the parties' control	A party may feel coerced into settlement for a lesser amount
Gaining an early understanding of the opponent's case	Costs are wasted if no agreement is reached
The mediation is confidential unless otherwise agreed	Mediation can be perceived as a sign of weakness
The mediation will focus on the key points in dispute	

Early neutral evaluation was developed in the late 1980s in response to a request by the courts in the USA to look at ways of lowering the cost of litigation and reducing the burden on the average litigant in progressing a matter through the legal system.

The technology and construction court and commercial court in England and Wales have their own neutral evaluation scheme. The procedure works as follows.

1   The judge is appointed as evaluator.
2   The evaluator studies all material provided in advance of the evaluation.
3   The evaluator performs independent research into relevant case law.
4   The evaluator clarifies the parties' positions through questioning.
5   The evaluator prepares a carefully worded but direct evaluation.
6   The evaluation is delivered to the parties (preferably orally).

The evaluation is non-binding on the parties and is usually in written form. The evaluation focuses specifically on the key issues identified, the wider facts and the relevant law. Neither party is usually ordered to pay the other party's costs in an evaluation process. The benefit for the parties is that they can see at this early stage how the case is likely to pan out. As with mediation, if the risk of an adverse outcome at trial is too great then a settlement on the best terms available becomes very attractive. The downside of an evaluation is that the process might be lengthy if there are multiple and/or complex issues involved. Each party bears its own costs in relation to the evaluation, which might prove expensive.

Alternatives to evaluation include obtaining a leading barrister's opinion and the trial of a preliminary issue. Dispensing with a preliminary issue in formal legal proceedings so as to create a binding decision may allow the remaining parts of a dispute to be settled.

17.9 Dispute review boards

A dispute review board is a project-specific process which facilitates co-operation by using active dispute management to help the parties reach a compromise. The proceedings are confidential and the board is created by agreement between the parties. The board is given the power to hear and advise the parties on issues and disputes as they arise. FIDIC and the other standard forms establish a tripartite agreement between the board member and the parties to the contract. There are typically three board members appointed at the outset of the project.

The World Bank and other funders insist on the use of the board and its rules to be adopted. The key issue is whether the determination issued by the board is a non-binding recommendation or an interim-binding decision. The project size has to be large to justify the cost of the review board. International projects such as Hong Kong Airport and the hydroelectric project at Ertan in China have used the boards successfully. The boards have been seen closer to home on projects such as the Docklands Light Railway and the Channel Tunnel.

One of the key advantages of the dispute board is the familiarity of board members with the project. This is established in the course of occasional visits to the project to ascertain whether there are any issues arising. Another advantage is that the dispute board is not restricted by procedural formalities. Referral to the board is straightforward and may be informal. The board's determination should be quicker than the more conventional methods of dispute resolution.

17.10 Conclusion

Mediation and ADR are essentially about compromise and enabling those involved to take a realistic view regarding their prospects should they proceed to a more formal dispute resolution procedure. The subsequent chapters in Part 4 cross the divide between the consensual approach to dispute resolution and the mandatory. This reflects the all too common occurrence that mediation is rejected as one or both parties do not wish to compromise and wish instead to assert their entitlements, often in the strongest terms possible. The final word on mediation is, therefore, to use an old saying: 'you can lead a horse to water, but you cannot make him drink'.

17.11 Further reading

Uff, J. (2013) *Construction Law*, Eleventh Edition, London: Sweet & Maxwell, Chapter 2.
Hughes, W., Champion, R. and Murdoch, J. (2015) *Construction Contracts: Law and Management*, Fifth Edition, Abingdon: Routledge, Chapter 23.
Richbell, D. (2008) *Mediation of Construction Disputes*, Oxford: Blackwell Publishing.

Notes

1 HM Courts & Tribunals Service (2014) *The Admiralty & Commercial Courts Guide*, Ninth Edition, Paragraph G1.2. Available at: www.justice.gov.uk.
2 Lord Woolf (1995) *Access to Justice: The Final Report to the Lord Chancellor on the Civil Justice System in England and Wales*, London: HMSO.
3 See www.cedr.com.
4 *Sulamerica CIA Nacional De Seguros v Enesa Engenharia* [2012] EWCA Civ 638.
5 *Emirates Trading Agency v Prime Mineral Exports Private Ltd* [2014] EWHC 2104.
6 *Earl of Malmesbury v Strutt and Parker* [2008] EWHC.
7 [2002] EWCA Civ 30.
8 *Cowl and Others v Plymouth City Council* [2001] EWCA Civ 1935X.
9 [2002] EWHC 1051 (Ch).
10 [2004] EWCA Civ 576.
11 Blake, S., Browne, J. and Sime, S. (2013) *The Jackson ADR Handbook*, Oxford: OUP.
12 Available from www.justice.gov.uk.

18 Litigation

18.1 Introduction

There are disagreements in all relationships from time to time. Sometimes such disputes can be sorted out by agreement or mediation. But if they cannot then the parties have to resort to some outside agency. In earlier times, and in more primitive societies, that agency tended to be the ruler – a feudal lord, a tribal chief or possibly the king. In all modern societies, the outside agency provided for dispute resolution takes the form of a court system. Litigation is the court-based procedure allowing a party to seek legal redress for infringement of its contractual rights. Access to justice is one of the basic underlying principles of any civilised society.

The first chapter in this book discussed the rule of law and the wish list of what a legal system should deliver. Providing a tribunal where grievances could be aired and justice meted out featured high on the wish list. This is the role of litigation, which can be described as the process of dispute resolution before a court. In the construction industry, it is not the only, nor even the most common, process. However, litigation remains the fallback method of dispute resolution. The existence of litigation underpins the efficacy of the other models of dispute resolution.

Construction litigation in England and Wales usually takes place in the technology and construction court (TCC), a specialist division of the high court. There have traditionally been around six permanent TCC judges who sit in London at any one time. In addition, there are some 20 circuit judges based in other major cities who sit as TCC judges when the need arises in their area.

All citizens have the right to conduct their own cases in court. The complexity of construction disputes normally precludes this, and almost all litigants in the TCC are represented. The administrative aspects of litigation, such as the issuing of the claim form, must be undertaken by a solicitor, who almost invariably instructs a barrister to act as advocate in the TCC. There are around 200 barristers who have specialist experience of construction work and are members of the Technology and Construction Bar Association.

An alternative venue for civil disputes is the county court. This is where a construction professional might issue proceedings for the non-payment of fees. Litigation has a well-earned reputation for being expensive and slow. This is not new and Charles Dickens' 1853 novel *Bleak House* featured the fictional case of '*Jarndyce* v *Jarndyce*' concerning a contested inheritance. The actual case on which the novel was based was in many ways stranger than the fiction. The case of *Jennens* v *Jennens* commenced in 1798 and was abandoned in 1915 (117 years later) when the legal fees had exhausted the Jennens estate in dispute.

The government has long been concerned with access to justice and taken steps to remove this reputation by issuing procedural rules. The importance of the cost implications of litigation is now uppermost in the minds of litigants and their attention is never allowed to move far from the costs both incurred and projected through the remainder of the court action.

18.2 Efforts to improve access to justice

18.2.1 The Civil Procedure Rules

Litigation in the courts today is a thoroughly modern procedure. The Civil Procedure Rules (CPR) came into force in April 1999 and these implemented ideas proposed by Lord Woolf. These rules apply to both the high court and the county court. The Rules improved on the situation that already existed by making litigation:

- faster;
- cheaper;
- more user-friendly;
- fairer.

Proportionality is a key concept in post-Woolf litigation. This concept was first used in European law. As the word suggests, this concept is based on ensuring that the time or resources used for any purpose are proportionate to the end result sought. An example of proportionality is a technique used by some judges that involves adding up all the small items in a larger claim, dividing the total by two, and making that figure part of the judgement in respect of those smaller items. This approach is certainly proportionate given that it would probably cost more to discuss the issues than either party would stand to gain or lose by arguing about them.

A significant feature of the new regime is the encouragement of settlement. As has been noted in the previous chapter, there are likely to be penalties in the form of higher costs to be paid to the other side if parties unreasonably refuse to take part in mediation or decline to accept an offer of settlement or if they fail to disclose sufficient information at an early stage. Proper use of the Pre-Action Protocol for Construction and Engineering Disputes requires not only the supply in correspondence of details of what parties will be saying but also a without prejudice meeting to narrow the issues between the parties, if not to resolve all matters.

The overriding objective contained in the Civil Procedure Rules is that the courts must deal with cases justly in a way which is proportionate to the amount of money involved, the importance of the case, the complexity of the issues and the financial position of each party. The Rules also draw attention to the resources of the court and how these need to be deployed carefully so as not to prejudice the interests of other cases also before the court.

One consequence of the Civil Procedure Rules is the case management conference where the judge becomes involved at an early stage in deciding on how the case will progress to trial. One tactic here is to give the parties a trial date straight away so that their minds become focused on this rather than being distracted by the steps necessary to arrive at the trial. Cost estimates and time estimates are also required in these interlocutory hearings.

18.2.2 The Jackson reforms

The CPR were overhauled and updated by the 2013 recommendations of Lord Justice Jackson, which were brought into effect through a combination of legislation, regulations and rule changes. The following key changes affect construction litigation.

- *Costs management* – The parties must declare a budget for their costs that is approved by the judge at the case management conference. The court will not depart from a party's agreed budget without good reason.
- *Disclosure* – The duty to disclose all relevant documents is replaced with a 'menu'-type selection.
- *Part 36 offers* – The claimant's offers are strengthened by the addition of 10 per cent of damages.
- *Experts* – Parties must identify the issues on which expert evidence will be sought and give consideration to concurrent expert evidence.
- *Witness statements* – The length of statements and the issues to be addressed can be predetermined.

The reports considered above all achieve a streamlining of the litigation process in terms of time taken to get to trial. Restricting the costs that each party can claim from the other side ensures that the parties are acutely aware of the expense involved.

18.3 Litigation procedure

Usually there are only two parties involved in litigation. The parties are known as the claimant and the defendant. The claimant is the person or company bringing the case, and the defendant is the person or company defending it. More complicated situations arise when:

- there is more than one defendant;
- the defendant wants to bring a cross–claim (known as a counterclaim or Part 20 Claim) against the claimant; or
- the defendant wants to bring a counterclaim against a co-defendant or issue a contribution notice against a third party.

There are seven easily identifiable stages in the litigation process. These are:

1 claimant's statement of case;
2 defendant's statement of case;
3 the claimant's reply to the defendant's statement of case;
4 disclosure of relevant documents;
5 exchange of witness statements;
6 exchange of expert evidence;
7 final hearing/trial.

18.3.1 Claimant's statement of case

Litigation is begun by the claimant setting out its case in a statement of case, comprising a claim form and particulars of claim. The particulars of claim should set out:

- an introduction of the parties and the contract involved;
- the facts in dispute between the parties;
- the alleged breaches of the contract provisions;
- the alleged losses resulting from the breach(es) of contract; and
- the legal remedies sought.

Whenever possible, the statement of case should specify precisely how much is sought in the way of damages and should include a calculation of interest claimed on the damages sought.

18.3.2 Defendant's statement of case

Following the filing and 'service' of the claimant's statement of case, the defendant will have a limited time (usually 28 days) to acknowledge receipt and then to prepare, file and serve its own statement of case, known as a defence. If the defendant wishes to make a claim of its own against the claimant then it can include a counterclaim in its statement of case. This is quite usual in building contracts litigation.

18.3.3 The claimant's reply to the defendant's statement of case

The next stage after the service of the defendant's statement of case is for the claimant to reply to the defence, covering any new material raised by the defendant. The claimant should also take the opportunity to enter a defence to any counterclaim raised by the defendant.

At this stage, the 'pleadings' (the term used for both parties' statements of case) are said to have closed. The issues in the dispute will now have crystallised and the parameters of the possible outcomes will be set for the judge to decide. The judge will then decide what other steps need to be taken to prepare the case for trial and will set out a timetable for achieving them.

In legal terms, the deadlines contained in the timetable are called 'directions'. When deciding how the case should proceed, the judge is said to 'allocate' the case to a 'track' depending on the money at stake and the complexity of the issues.

18.3.4 Disclosure of relevant documents

The next stage involves the disclosure of evidence. The parties will be asked to prepare and disclose all the documents that they consider relevant to their own and their opponent's pleaded case. Each party can call for the documents identified by their opponent to be produced and copied to them.

18.3.5 Exchange of witness statements

Once documents have been exchanged, the next stage is the preparation and exchange of witness statements. The people who were physically involved in the dispute should give statements to support their statements of case. The defendant may wish to call their own project manager and quantity surveyor to demonstrate, for instance, that their site records are accurate and that the claimant was definitely not there on the days identified.

18.3.6 Exchange of expert evidence

Around the time witness statements are filed, the parties should also be in a position to exchange expert evidence, if required. Following the Civil Procedure Rules, a judge must consent to expert witnesses being appointed. This is because of the considerable cost of making the appointment. In deciding on whether to have expert witnesses, the judge will consider:

- whether assistance is required with technical points in the case;
- whether a single expert should be appointed instead of the parties appointing one/more than one each;
- whether the experts' evidence should be in writing or if they should be required to attend trial.

18.3.7 Final hearing/trial

Once all the evidence has been prepared and exchanged, the case is ready for trial. The court fixes a date on which the parties, their witnesses and legal representatives are required to attend a hearing. The usual procedure at a trial is:

- parties' opening statements;
- submission of evidence in chief;
- cross-examination of witnesses;
- re-examination of witnesses;
- summing up by the parties;
- judge gives judgement.

Cross-examination is the procedure where the defendant's barrister questions the claimant's witnesses on their witness statements and vice versa. Being cross-examined can be a gruelling experience because the barrister asking the questions will have prepared a series of questions aimed at discrediting the evidence given or exploiting any inconsistencies they see in the evidence. The witness cannot refuse to give answers, or if they do then the judge may draw adverse inferences from what has not been disclosed.

The cross-examination stage is usually very important in trials. Often the judge has to make a choice between two credible sets of evidence presented. Any insight gained from reactions to cross-examination can make all the difference to the decision. The skill of the barrister is obviously also very important to the outcome of cross-examination.

A judgement usually awards damages from one party to another. The usual test the judge will apply is whether, on the balance of probabilities, either party has made a compelling case based on sustainable legal argument and reliable evidence. A party who feels aggrieved by a judgement against it can appeal to a higher court. The vast majority of losing parties do not appeal, preferring instead to pay the judgement debt and put the experience behind them.

Following the introduction of the Civil Procedural Rules, it is now common to have a final hearing within one year of the issue of proceedings. However, this approximate timetable can be upset by a number of factors, such as:

- judge and court availability;
- a party deliberately delaying the process;
- the amount of evidence needed; or
- the complexity of the issues involved.

The litigation procedure is very expensive. This is particularly true of building contract litigation where the pleadings are long and hard to digest. Hearings can be lengthy and multiple factual and expert witnesses may be called. It is not unheard of for the costs of the litigation to outweigh the damages being sought. Clearly, this would be a disproportionate outcome.

Due to the expense involved, costs are a key concept in litigation. In general, costs are said to 'follow the event'. This means that the losing party pays the winning party's costs. This is the basic position but several factors can change the position.

- A losing party rarely pays all the winning party's costs; a useful rule of thumb is to think in terms of the winning party recovering around two-thirds of its expenditure.
- Any of the winning party's costs deemed to be exceptionally high or 'disproportionate' will not be paid by the losing party.
- The winning and losing positions on costs can be reversed by offers of settlement.

18.4 Particular features of construction litigation

18.4.1 *Scott Schedule*

Scott Schedules are common in construction and engineering disputes. They are widely used in construction and engineering arbitration, adjudication and TCC cases, particularly where a claim involves:

- a contractor's final account;
- extras;
- numerous alleged defects;
- items of disrepair;
- individual items of loss and damage; or
- delay and disruption.

A Scott Schedule may be used to support a claim, a defence or a Part 20 claim (i.e. a claim made by the defendant against the claimant or another party, such as a subcontractor or other construction professional). The use of the Schedules is necessitated by the frequent lengthy pleadings and many different detailed claims and counterclaims. For example, a claim featuring a project with 5,000 variation orders will potentially need to establish entitlement and factual/quantum support for each element. This generates a good deal of documentation. It is not uncommon for courtrooms to run out of space for document management when the walls given over for lever arch file storage run out. Advances made in terms of electronic documents and databases relieve this pressure to a certain point although there remains a morass of things to look at.

This is where the Scott Schedule has proved useful in drawing together the documents and providing a central point for consideration of the items in dispute. The judge can call for the Schedule to be prepared and there follows input from both the claimant and the defendant. Typically, the claimant will set out their argument first, then pass the Schedule to the defendant to set out their response. This requires a degree of collaboration between the parties. The Scott Schedule allows the parties to identify the items in dispute at a glance. The parties then may focus their attention on the large items, if agreed. Another common approach is to simply halve the value of the smaller items and agree the figure. The recognition here is that the time taken to argue about the small items results in the expenditure of more costs than the item is worth. Table 18.1 is an example of a layout of a Scott Schedule. Some versions have an additional column where the judge can record comments.

18.4.2 Multiple defendants

Historically, the number of parties involved tended to be high in construction disputes. The litigation becomes complicated as a consequence. The Civil Procedure Rules permit any number of claimants or defendants and any number of claims to be covered by one claim form.[1] The test is whether all of those claims can be conveniently disposed of in the same proceedings. If there is more than one claimant then there should be no conflict of interest between their positions.

More common in construction are multiple defendants as in the case of *Lindenberg* v *Canning and Others*.[2] This case involved the employer wanting recourse against either or both the surveyor and builder. The position with regards to one defendant and another becomes interesting with the interplay of without prejudice offers between defendants. If the defendants can pre-agree their share of liability then they may be able to present a joint arrangement to the claimant.

Another scenario is where the party being sued considers that another party is wholly or partially to blame for the situation. The defendant may wish to 'join' the third party in the proceedings and may apply under the Civil Procedure Rules for an order to this effect. The proposed joining party may be naturally reluctant and an interlocutory hearing will be necessary to decide the issue.

18.4.3 Stay to arbitration

The contract between the parties may contain a clause referring disputes to an arbitrator. In this case, if one party commences proceedings in a court of law then the other can usually obtain a stay of the proceedings. The stay will be

Table 18.1 Typical layout of a Scott Schedule

ITEM	Claimant's commentary	Value	Evidence	Defendant's commentary	Value	Evidence

ordered upon the finding that there is a binding agreement to arbitrate. The party erroneously starting the proceedings in court will most likely be ordered to pay the costs incurred in the litigation to date.

18.5 Conclusion

Abraham Lincoln's oft-quoted phrase about litigation still resonates today as very good advice to give to warring parties:

> Discourage litigation. Persuade your neighbors to compromise whenever you can. Point out to them how the nominal winner is often a real loser – in fees, expenses, and waste of time. As a peacemaker the lawyer has a superior opportunity of being a good man. There will still be business enough.[3]

Litigation remains expensive and long despite the continuing efforts of the government and the legal community to make savings in the cost and length of time taken. Most lawyers, when asked, say that they would not litigate if they had to pay for the process with their own money. This is a telling condemnation of the process. The next chapter considers the process that was, for a long time, the alternative to litigation – arbitration.

18.6 Further reading

Hughes, W., Champion, R. and Murdoch, J. (2015) *Construction Contracts: Law and Management*, Fifth Edition, Abingdon: Routledge, Chapter 24.

Newman, P. (1996) *Construction Litigation Tactics*, Birmingham: CLT Professional Publishing.

Notes

1 CPR 19.1 and CPR 7.3.
2 [1992] 62 BLR 147.
3 Lincoln, A. (1999:1850) Notes for a Law Lecture: 1850? in S. Sheppard (Ed), *The History of Legal Education in the United States: Commentaries and Primary Sources, Volume 1*, Pasedena, CA and Hackensack, NJ: Salem Press Inc, pp. 489–90; p. 489.

19 Arbitration

19.1 Introduction

Arbitration is an alternative to litigation as a means of resolving disputes. Arbitration is the name given to the process whereby parties agree to refer an existing or future dispute to the determination of one or more independent person(s) acting in a judicial manner. The decision of the arbitrator is expressed in an award which will be binding on the parties and enforceable in law. English law recognises and supports the arbitral process by providing a statutory framework for arbitrations set out in the Arbitration Act 1996.

Arbitration is a consensual process in that unless the parties have agreed to refer their dispute to arbitration then there can be no arbitration. Any arbitrator should comply with any procedure the parties have agreed for the arbitration. The parties are given procedural control and the arbitration offers a flexible method of resolving disputes that fits the circumstances of the case. Further, the parties can choose the decision-maker for his or her particular skill or expertise relevant to the case.

Arbitration is common in construction disputes for the following two reasons.

- In the standard forms of agreement used in the construction industry, provision is made for any dispute to be determined by arbitration and not by the courts.
- The construction industry has developed its own model rules for arbitration – the Construction Industry Model Arbitration Rules (CIMAR) published by the Joint Contracts Tribunal.[1]

Arbitration is similar to litigation in many ways and has been described as 'litigation in suits rather than wigs'. The similarities are that both are intended to be final and the parties usually prepare statements of case similar to litigation. Interlocutory stages including the disclosure of documents and witness statements are commonly seen. The final hearing is approached in the same way as a trial and the arbitrator applies the same evidential tests as a judge. The same costs principles apply to arbitration and litigation.

As a result of the similar approach taken in arbitration, an equivalent amount of time and costs can be spent as in litigation. The advantages offered by arbitration are therefore found in its other strengths.

19.2 Arbitration advantages

The advantages offered by arbitration over litigation include those covered in the remainder of this section.

19.2.1 International application

Litigation is limited by national laws. Arbitration can be used for international disputes. The implications here are considered in more detail below. The main advantage lies in the enforcement of awards. In cases involving foreign parties, it is usually easier to enforce the arbitration award in the foreign courts than would be the case with a judgement of the courts. Many countries are parties to the New York Convention on The Recognition and Enforcement of Foreign Arbitral Awards (1958) (The New York Convention) and have agreed to recognise and enforce foreign arbitration awards on a reciprocal basis. This is preferable to trying to enforce an English judgement in a jurisdiction that does not recognise it.

19.2.2 Specialist skill set

Arbitrators do not need to be lawyers. Many arbitrators are construction industry professionals who can use their professional experience to resolve disputes and dispense with the need for expert witnesses. The parties may feel that they would prefer technical disputes to be decided by an arbitrator with the relevant technical expertise. This is a point of particular relevance to the construction process, which has a variety of different specialist areas such as those relating to technical, architectural, engineering or quantity surveying/ valuation issues. Notwithstanding their knowledge, the arbitrator should only decide the dispute in accordance with the evidence presented to him by the parties.

Most arbitrators appointed in construction disputes are generally experienced legal practitioners or have an engineering/surveying background (some hold both legal and technical industry qualifications). Professional bodies such as the Chartered Institute of Arbitrators, the Royal Institute of British Architects or the Royal Institution of Chartered Surveyors maintain a list of arbitrators. It is likely that the arbitrator will have undertaken further practical training as an arbitrator.

19.2.3 *Private and confidential nature*

Arbitrations are held in private and are confidential. It is often important that the issue of proceedings does not appear as a matter of public record or give the impression that 'dirty washing' is being aired.

19.2.4 *Flexibility*

In litigation, the rules are determined by the court in accordance with the Civil Procedure Rules. In arbitration, the parties have flexibility over the process, not only in terms of choosing the arbitrator but also in selecting the venue for the hearing and the timetable for the dispute. The arbitrator also has wider powers than those given to a judge, such as the power to award compound interest.

Examples of the issues that the parties can decide are:

- whether the costs recoverable in the arbitration should be limited;
- whether the parties wish to dispense with the need for formal statements of case;
- whether the parties wish the arbitration to proceed on paper only or with a hearing.

19.2.5 *Finality of the decision*

The parties are sometimes keen that whatever the decision on a particular dispute may be, it should be final and binding in the sense that it is not subject to appeal. This is not possible in relation to court actions.

19.3 The Arbitration Act 1996

Arbitration in England and Wales is governed by the Arbitration Act 1996 ('the Act'). The Act introduced welcome reform of the law relating to arbitration, which before the Act was piecemeal and uncertain. At this time, it was common for awards by arbitrators to be challenged and to become the subject of scrutiny by the courts despite the fact that arbitration is intended to produce a final decision. The Act has severely restricted the grounds on which arbitral decision can be challenged in the courts. The latter is limited to serious procedural irregularity affecting the arbitral process. Judges have largely fallen into line and challenges to arbitrators' decisions are usually given short shrift.

There are clear echoes in the Arbitration Act of the changes introduced by Lord Woolf into litigation in the form of the Civil Procedure Rules. Lord Woolf has acknowledged that the Arbitration Act was a forerunner of his changes to litigation and that the Act is 'Woolf compliant'.

The Act sets out the law in a simple and logical manner that makes the law accessible to all. The key provision of the Act is to obtain the fair resolution of disputes by an impartial tribunal without unnecessary delay or expense. The Act adopted many principles found in arbitration laws internationally, inspired by the United Nations Commission on International Trade Law (UNCITRAL) Model Law on International Commercial Arbitration 1985 (published under the auspices of UNCITRAL and since amended in 2006). These principles include party autonomy and, crucially, that arbitration takes precedence over litigation.

19.4 Arbitration characteristics

Arbitration has the following characteristics, which can be contrasted to other methods of resolving disputes.

- There must be a valid agreement to arbitrate. In legal terms, an arbitration agreement must either form part of a valid, binding contract between the parties or it must amount to such a contract itself. For an arbitration clause in a building contract to be effective, it has to be in writing.
- The decision made by the process will be a final binding determination of the parties' legal right enforceable in law. This can be contrasted with an agreement to mediate whereby any view expressed by the mediator will not be binding on the parties.
- The arbitrator is obliged to act impartially. He should be independent of the parties even though potentially nominated by one of the parties.
- The arbitrator must carry out his functions in a judicial manner and in accordance with the rules of 'natural justice'. Natural justice refers to the notion that each party should have the opportunity to plead its case and to be heard by the arbitrator. This judicial feature is different to other procedures such as adjudication where the person appointed can make their own enquiries. An arbitrator should not do this. The requirement to act judicially is not the same as the exercise of the duties of the architect/ contract administrator in relation to certifying. In the latter case the A/CA must act fairly; this is not the same as acting judicially.

A disadvantage of the arbitration proceedings is the lack of an effective means of dealing with dispute involving more than two parties. This is an area of particular importance to the construction industry. Construction dispute may arise between the employer, architect, contractor and subcontractor, all relating to the same subject matter. Unless special provision is made in the arbitration agreements of all relevant contracts and subcontracts by which the various parties agree that the entire separate dispute can be determined by the same tribunal then they will not be heard together. The right of joinder does not apply.

There is risk that a party (usually the main contractor) will find itself caught in the middle of arbitrations and may incur a liability which it cannot pass on or recover from a third party. In this situation, a multi-party arbitration can be desirable, and various standard form contracts (including the JCT) seek to provide for multi-party arbitrations in certain circumstances.

Another disadvantage of arbitration is that it may be less effective than litigation in dealing with the reluctant defendant. Defendants may raise a number of weak defences or counterclaims simply as a means of delaying the day when they have to pay their creditors. The courts provide procedures for dealing with this situation, including applications to 'strike out' part of a case or to ask for early determination through summary judgement in addition to other sanctions aimed at preventing parties dragging their feet during litigation. The arbitrator has fewer powers to act summarily but may penalise negative and time-wasting tactics in the costs allowed to the parties at the conclusion of the matter. It should be noted that the arbitrator's fees become part of the costs of the action. A judge is effectively a civil servant and the parties are not required to pay for the judge or the court's upkeep.

Notwithstanding the drawbacks, arbitration can offer an efficient and economical way of resolving a dispute, and its international importance cannot be overstated.

19.5 Arbitration procedure

A party can then seek to refer a dispute under the contract to arbitration by issuing a notice of arbitration. The notice of arbitration requires careful drafting as the arbitrator will only have the ability (referred to as jurisdiction) to deal with issues arising out of the notice.

Confirming the jurisdiction of the tribunal is an important starting point to address pre-commencement. It is an obvious challenge for the defending party to make that the arbitration agreement was never incorporated into any contract the parties may have entered into; for example, where the final building contract was never signed and the work progressed on a letter of intent. Expensive and time-consuming satellite litigation can ensue as to whether the right to arbitrate has been incorporated. The Act allows the arbitrator the power to rule on its own substantive jurisdiction. However, this may not satisfy a party who wants to be absolutely sure of this point before proceeding.

In arbitrations involving foreign parties, the agreement to arbitrate should refer to the applicable law, the language of the arbitration and the place of arbitration. The law of the 'seat' of the arbitration governs the procedural aspects of any subsequent proceedings.

Once commenced, the arbitration will be conducted in accordance with the procedure of rules agreed between the parties or, failing agreement, as determined by the tribunal. In either case, the tribunal must act fairly and impartially between the parties; each party must have a reasonable opportunity to put its case and to deal with that of its opponent, adopting procedures which are appropriate to the circumstances of the case.

One of the main tasks of the tribunal is to establish the facts of the case. The way in which facts are proved will depend on the procedure adopted by the tribunal. The Act allows a proactive arbitrator to take a much more 'inquisitorial' approach to the collection of evidence and the issues in dispute. However, arbitrations generally remain 'adversarial' in nature with both parties advancing their case and the decision-maker picking the most convincing case rather than taking the lead from the parties by establishing the cases and the law out of his own initiative.

Construction disputes frequently raise important issues that turn on opinion evidence and not just evidence of fact. Most arbitration rules provide for the possibility of expert evidence being given on matters of opinion as it would be in litigation. Experts will generally advance a position that benefits the party that instructs them; however, they are meant to be independent and do in fact owe a duty to the tribunal rather than to that party. The collection and service of expert evidence is usually an expensive part of the arbitral process with the tribunal having to assess and weigh the evidence of experts to decide the dispute. Where each side uses an expert, there is usually a direction that they meet with a view to narrowing and defining the issues in dispute.

The hearing is the final stage in the arbitration. Work in the period leading up to the hearing is intensive. Nowadays, tribunals increasingly require the parties to put in written submissions of law and take steps to reduce the length of the oral hearing. The tribunal reads up on the case in advance and uses the time at the hearing to hear cross-examination of witnesses and experts and, frequently, to question the parties on the basis of the cases advanced so far.

After the arbitrator has heard the evidence, the representatives of the parties will make their closing submissions. These may be made in writing and they may lead to an adjournment before further questions that the arbitrator has regarding the written submissions.

The decision is then made in the form of an award. The award is usually in writing and contains reasons. The award should state the seat and the date when it is made and be signed by all of the arbitrators if there is more than one. The arbitrator should state all his findings of fact and briefly state his reasoning on the issues of law. All awards should be certain and final (unless clearly intended to be an interim award). If the situation calls for an award of money then the award should be in a suitable format to allow it to be enforced as if it were a judgement of the high court.

The costs of arbitration include the arbitrator's fees and expenses and those of any institution involved and the legal or other costs of the parties, which includes the professional fees of expert witnesses. The tribunal has a wide discretion but the general rule is that 'costs follow the event' and the loser pays unless it appears to the tribunal that this would not be appropriate.

Following the publication of an award, there are strict time limits for bringing any challenge – usually within 28 days of the award. The grounds for challenge are:

- that the award is not complete and has failed to address an issue in dispute;
- that the award contains a clerical mistake, an error or an ambiguity;

- that the award is made by a tribunal lacking substantive jurisdiction; or
- that there has been a serious irregularity (e.g. fraud or the granted powers being exceeded).

In each case, the applicant must show that the issue has caused substantial injustice. The first step is to ensure that all other forms of recourse within the arbitral process have been exhausted. Applying to the court requires leave to appeal. Once granted, the court will apply the governing law to the contract.

19.6 The arbitrator's jurisdiction

The extent of the arbitrator's jurisdiction depends on the various powers given by the Act and any institutional rules that may apply. The arbitrator typically has the same powers as a judge to award remedies. This usually results in an award of damages and costs from one party to the other. An arbitrator's award is enforceable through the courts as if it were a judgement.

The arbitrator can take steps to handle the dispute in the best way he considers possible. These powers for the arbitrator are set out at section 33 of the Arbitration Act, which requires an arbitrator to 'adopt procedures suitable to the circumstances of the particular case, avoiding any unnecessary delay or expense, so as to provide a fair means for the resolution of the matters falling to be determined'. Operation of this power by the arbitrator might result in the arbitrator deciding to make a provisional decision by letting the parties know at an early stage what the merits of the case appear to be. The arbitrator may also make an interim award; for instance, ordering that some damages are paid by one party to the other whilst the arbitration proceeds with the exact amount of damages being calculated later.

The underlying theme behind these principles is to encourage settlement. By making the parties discuss the case and how they wish to approach it, the parties may well find that the issues in dispute between them have narrowed and that settlement is possible. However, the arbitration procedure could be abused by unscrupulous parties who, for their own reasons, are not really interested in resolving the dispute. This has led to calls for arbitration to be made even faster and cheaper if it is to hold on to its popularity within the construction industry in the face of competition from adjudication and alternative dispute resolution.

19.7 International arbitration

19.7.1 Introduction

International commercial arbitration follows the same judicial process as domestic arbitration. The context is different in that some elements transcend state boundaries, thereby making it 'international' and subject to other legal processes. In England, the law does not now distinguish between domestic and international arbitration except when considering the enforcement of foreign awards made in states which are signatories to the New York Convention.

There are three legal regimes relevant in international arbitration.

1 The substantive law of the contract. Which system of law is specified in the contract as applying? In the absence of any statement on this, the system will be determined by the arbitral tribunal.
2 The place of the arbitration is referred to as the 'seat' of the arbitration. The law of the seat usually governs issues such as the interpretation, recognition and enforcement of the arbitration agreement. Many states have adopted, in whole or in part, the UNCITRAL Model Law (see above), which addresses many of these issues.
3 The law governing the recognition and enforcement of awards is the third issue to consider. This will usually be the law of the place where a party is seeking to enforce an award.

Each of these different legal regimes ought to be considered before a cross-border project is undertaken by a non-national contractor. Many British firms operate in the Middle East, and careful advice would have been taken by these firms concerning the various scenarios and legal systems that would be involved in the event of a dispute.

19.7.2 Background

International arbitration clauses first appeared in the FIDIC Conditions of Contract in 1957. This move was driven by the concern that a dispute in an international construction project might lead to parallel litigation in courts of different states, each declaring jurisdiction to determine the dispute, with the risk of expensive and inconsistent judgements. Other standard form providers followed this lead and the use of arbitration clauses in international construction projects is now widespread and includes professional services contracts.

International arbitration remains as relevant today. One aspect that has changed is the imposition of mandatory intermediate steps aimed at resolving the dispute before it escalates to this final step. The updated FIDIC forms include provision for a dispute review board (which investigates and recommends provisional decisions to resolve disputes) and a mediation–type stage.

19.7.3 Features of international arbitration

The same advantages for domestic arbitration apply equally to the international arena. The privacy of the proceedings and the technical expertise of the arbitrators remain attractive, particularly in locations where such expertise might be lacking. The flexibility the procedure affords the parties is prized highly. Parties involved in an international project are often from distinct legal backgrounds and are unfamiliar with the different legal procedures known to the other party. Where there are parties coming from a common law and a civil law background, they may struggle with each other's concepts around the manner and method of proving or defending a case.

The most compelling reason for selecting international arbitration is the lack of any real global equivalent for the recognition and enforcement of domestic court judgements. The New York Convention has been important in facilitating the creation of international construction projects and their funding.

Consideration should always be given to the implications of the choice of seat for it determines the supportive and supervisory role of the national courts and any mandatory laws that shall apply to any international arbitration. In Saudi Arabia, for example, the arbitral law provides that the tribunal must be composed of men only, who must also be of Muslim faith. Failure to comply with this mandatory requirement of the seat could amount to a procedural ground on which the enforcement of an award might be refused.

Institutional arbitration involves the parties signing up for the arbitration to be administered by one of the established arbitration institutions and for the arbitration rules of the institution to apply. The advantage here is that the institutional arbitration includes the application of established international arbitration rules along with the availability of trained staff to appoint the tribunal, to ensure time limits are observed, to review the award and to assure the general smooth running of the arbitration. The International Chamber of Commerce (ICC) is the biggest institutional arbitration organisation and typically recommends that a clause be incorporated in the building contract along the lines of: 'All disputes arising out of or in connection with the present contract shall be finally settled under the ICC Rules of Arbitration by one or more arbitrators appointed in accordance with the said Rules'.

The ICC headquarters are in Paris, but it appoints arbitrators of many different nationalities, sitting in any place and using any language. The ICC also handles the costs associated with the running of an international arbitration, which in ICC arbitration can be extremely large.

19.7.4 International arbitration procedure

If the dispute is to be referred to three arbitrators, the usual practice is that each party may appoint one arbitrator, and the third arbitrator is then decided by those two party-appointed arbitrators.

All arbitrators are obliged to be independent of the parties and to act fairly and impartially in the conduct of the proceedings.

The terms of reference summarise each party's claims and the relief sought, with sufficient detail in support. The terms of reference should also include a list of the issues in dispute so as to enable the arbitral tribunal to ensure that all matters are considered in the arbitration and decided when it comes to drafting the award.

After the appointment of the tribunal, the parties attend a procedural meeting with the tribunal. At this meeting, the tribunal will usually seek to establish a timetable for the arbitration, providing for such things as disclosure of documents and the preparation and exchange of witness statements and expert reports.

Subsequent procedural hearings may deal with issues that arise in the immediate run-up to the main arbitration hearing or during the conduct of the arbitration hearing itself.

The procedure for the arbitration hearing will usually include written submissions exhibiting the evidence on which each party intends to rely. These submissions are usually presented consecutively: the claimant first, followed by the respondent, after which permission may be given for further 'reply' submissions. Some arbitral rules provide that consideration should be given to whether an oral hearing is necessary at all, but in most cases it will be important for arbitrators to hear directly from key witnesses and any experts to be able to judge the issue.

The ICC Court of Arbitration reviews all international arbitration awards before they are published to the parties. This review is not related to the merits of the decision but is to ensure that the tribunal has addressed all of the issues before it. This is thought to give greater international acceptability to the award than might be the case for an award issued by an *ad hoc* arbitral tribunal. This may be particularly useful in jurisdictions such as the People's Republic of China where enforcement of international awards has historically been not without difficulty.

The use of expert witnesses in international arbitration is not as common as in proceedings in the English courts but in complex construction arbitrations, the opinion of experts is often required because of the technical nature of the matters in dispute. A novel way to introduce expert evidence is the process known as 'hot tubbing'. This is now being seen in litigation as well as in arbitration and involves the tribunal holding a conference with the expert witnesses at the same time. The judge observes and contributes to the discussion around the key issues and is able to form a judicial opinion based on what he sees and hears. A saving can be made in terms of the time and money required in preparation of the expert's written report and the live examination required at the final hearing. This approach is popular where one or more issues have great importance for the tribunal reaching their final determination. Hot tubbing enables the tribunal to hear immediately where the witnesses are in agreement and where their accounts differ.

19.7.5 Recognition and enforcement

The tribunal will publish its award at the conclusion of an international arbitration. The award will generally be immediately final and binding. Typically the tribunal will require payment of all its fees and expenses before publication. The law of the seat may provide for certain limited rights of challenge or appeal. It is common for parties to an international dispute to contract out of the rights of appeal to the fullest extent possible under the law of the seat, and many institutional rules including those of the ICC provide for this. If, prior to the conclusion of the arbitration, the parties reach a settlement, most international arbitration rules provide for the tribunal (if requested by both parties) to record the settlement in the form of an award, which need not contain reasons.

The award, if not carried out voluntarily, may be enforced by legal proceedings through the courts. There are a number of regional and international treaties and conventions that relate to the enforcement of awards, but the most important of these is the New York Convention, which is recognised in over 140 countries.

It requires the local courts of the contracting states to give effect to an agreement to arbitrate when seized of an action in a matter covered by an arbitration agreement by staying any court proceedings which are brought in breach of that agreement; and also to recognise and enforce awards made in the territory of a state other than the state in which recognition and enforcement is sought and to awards not considered as domestic in the state in which enforcement is sought. It has one principal formal requirement stipulated in Article II: that the arbitration agreement be in writing.

The grounds for resisting enforcement are relatively limited and largely concerned with procedural irregularities, which must be proved by the applicant. The exact procedure to be followed and the way in which the New York Convention is interpreted are matters for the national law and national courts of the country in which recognition and enforcement is sought.

Most states support the arbitral process and construe the bases for refusing enforcement fairly narrowly. In these states (among which the UK can be included), a refusal to recognise and enforce an award is very rare. Other signatory states have perhaps embraced the spirit of the convention less wholeheartedly and delays are not uncommon.

A decision by a state judge not to enforce an award is potentially detrimental to their international trade and ability to attract overseas investors. The knowledge that obtaining enforcement is not straightforward may deter investment.

19.8 Conclusion

Until relatively recently, arbitration and litigation were readily distinguishable on several different grounds. Litigation was seen as an inflexible, judge-led procedure where the use of expert witnesses on technical matters was commonplace. Arbitration offered a more flexible procedure using the arbitrator's own expertise. In recent times, the differences have become less pronounced. The Arbitration Act 1996 and the Civil Procedure Rules demonstrated Parliament's intention to cut down on the delay and expense of arbitration and litigation respectively. Some success has been achieved – litigation and arbitration are now both primarily focused on limiting the time taken and expenditure incurred, and both sets of procedures are now relatively flexible.

The difficulty as far as arbitration and litigation are concerned is that two other forms of dispute resolution – adjudication and mediation – have become extremely relevant to the construction industry and are establishing themselves as more popular methods of dispute resolution.

The current trend therefore is for litigation and arbitration to be used after or as an appeal from adjudication or mediation. Litigation and arbitration still have a role if final determination of any given issue is required. If litigation and arbitration want to compete with the relative newcomers then they will need to continue to explore ways in which they can become quicker and cheaper.

The AEC industry ensures that arbitration is used instead of litigation by including arbitration clauses in building contracts. Where the contract specifies arbitration then the other party is prevented from referring the matter to court. The inverse is equally true – where one party sues another in the courts without reference to the arbitration clause in the contract then the other party can stop the court proceedings and insist that arbitration is used, provided that it has not taken a step in the court proceedings.

Arbitration clauses in modern contracts are very widely drawn and give the parties a choice of which dispute resolution procedure they wish to use. A key feature of arbitration is that the parties can choose not only the arbitration procedure but the arbitrator as well. This is an example of a wider difference between arbitration and litigation where the parties can decide themselves how their dispute should be handled. Retaining this control is of vital importance to many stakeholders, and this is especially true in the international context.

19.9 Further reading

Uff, J. (2013) *Construction Law*, Eleventh Edition, London: Sweet & Maxwell, Chapter 3.
Stevenson, D. A. (2001) *Arbitration Practice in Construction Contracts*, Fifth Edition, Oxford: Blackwell Science.

Note

1 Joint Contracts Tribunal (2011) *Construction Industry Model Arbitration Rules 2011*, London: Sweet & Maxwell.

20 Adjudication

20.1 Introduction

Adjudication is a procedure for obtaining a quick and impartial decision on a construction dispute. The decision is not intended to be final but it is enforceable and binds the parties unless and until the dispute is finally resolved by litigation, arbitration or agreement. In most cases, the adjudication is the final word on the matter in contention. Adjudication is not classed in this book as a form of alternative dispute resolution. The parties will, unlike in mediation and other consensual approaches, have a decision imposed on them which they need to conform with.

Construction contracts, as defined by the Housing Grants, Construction and Regeneration Act 1996 (HGCRA), must give the parties to the contract the right to refer dispute to adjudication. There are some limited exceptions to this requirement. This right cannot be excluded by agreement and if a construction contract does not contain the required provision then the provision for adjudication contained in the Scheme for Construction Contracts applies by default.

Adjudication is directly related to Sir Michael Latham's report *Constructing the Team*, published in 1994. The Latham report contained 30 key recommendations for how the construction industry could be modernised and improved. The report identified that the time and cost of disputes within the industry was an enormous drain on the talents, productivity and profits of the industry.

The problems, as Latham perceived them, were that too much time was being spent arguing and that unscrupulous parties were abusing the existing system to hold on to money which they should, by rights, be passing on to another party. The Latham report recognised that cash flow was the lifeblood of the construction industry. A blockage in the system can lead to companies becoming insolvent and people being left emotionally and financially scarred by the experience.

Sir Michael therefore recommended that a new form of dispute resolution should be introduced. The new procedure would give a decision on a dispute in a fraction of the time and at a fraction of the cost that litigation and arbitration had been delivering. The new procedure would be binding, but if anyone disagreed with the result then they could challenge it later. The procedure became adjudication.

20.2 Key features of adjudication

- *It is speedy* – A decision can be given in just 28 days. The construction contract must provide a timetable designed to secure the appointment of an adjudicator within seven days of a party giving notice of his intention to refer a dispute to adjudication. It must require the adjudicator to reach a decision within 28 days of referral. This period can be extended either by up to 14 days with the consent of the party who referred the dispute or by agreement of the parties.
- *It is relatively cheap* – The limited time should keep the legal costs down. The fees of the adjudicator are an additional expense which one or both of the parties will have to meet. There are no orders in adjudication for *inter partes* costs. This means that any legal costs expended in presenting the case must be deducted from any recovery made. In other words, the losing party does not pay the winning party's costs. The losing party is usually ordered to pay the adjudicator's costs.
- *An adjudicator, or a nominating body, can be named in the contract* – The saying 'better the devil you know than the one you don't' can be applied to this situation. Nominating bodies such as the RICS (Royal Institute of Chartered Surveyors) or CIArb (Chartered Institute of Arbitrators) are popular choices for nomination.
- *The adjudicator takes the initiative in ascertaining the facts and the law* – The adjudicator is only going to have a limited time to reach his decision once the cases have been submitted. The adjudicator is not bound by litigation-style formalities in terms of awaiting further submissions of documents and evidence from the parties. The adjudicator may dispense with the need for a hearing and reach a decision based solely on the submitted material.
- *An adjudicator's decision is binding until overturned by arbitration or litigation* – The decision is enforceable notwithstanding the fact that it is provisional. It is also likely to be enforceable notwithstanding the fact that it contains errors of fact or law and perhaps procedural irregularities.

20.3 The role of the adjudicator

The adjudicator must address the issues of the dispute straight away and cannot be too concerned about the formalities prevalent in arbitration and litigation. The idea is for a trained and skilled industry specialist to 'roll their sleeves up' and get down to the business of resolving the dispute.

Speed and the possibility of enforcement despite error have resulted in adjudication being referred to as a 'quick and dirty' procedure. As the Court of Appeal has said, the need to have the 'right' answer has been subordinated to the need to have the answer quickly.[1] Adjudication is nevertheless a popular procedure in the construction industry and parties often accept the adjudicator's decision and do not seek to challenge it by subsequent litigation or arbitration.

Adjudicators are trained and accredited by adjudicator nominating bodies. Most of the professional institutions such as the RICS, CIArb, RIBA (Royal Institute of British Architects) and CIOB (Chartered Institute of Building) have adjudicator panels. Legally trained adjudicators are also available for nomination by the Law Society and the Chartered Institute of Arbitrators.

As to the factors that influence the choice of an adjudicator, the following should be considered.

- Agreeing on an adjudicator is always preferable to an appointment of an unknown – ask around for opinions on names that come forward.
- Different adjudicators suit different disputes – try and match the two where possible.
- Avoid any possible allegations of bias in terms of the adjudicator who is put forward.

The adjudicator is under a duty to act impartially and to avoid unnecessary expense. They may take the initiative in ascertaining the facts and the law necessary to determine the dispute. This can involve:

- requesting documents;
- meeting and questioning the parties and their representatives;
- making site visits;
- carrying out tests and experiments;
- setting deadlines inside the adjudication framework.

20.4 Adjudication and the law

20.4.1 Introduction

There are four key components to the introduction and development of adjudication.

1 The Housing Grants, Construction and Regeneration Act 1996 ('the Act').
2 The Scheme for Construction Contracts (England and Wales) Regulations 1998 (as amended by the Scheme for Construction Contracts (England and Wales) Regulations 1998 (Amendment) (England) Regulations 2011 and the Scheme for Construction Contracts (England and Wales) Regulations 1998 (Amendment) (Wales) Regulations 2011).
3 The Local Democracy, Economic Development and Construction Act 2009 ('The 2009 Act').
4 Decided cases.

The first three are interdependent. The Act as amended by the 2009 Act introduces adjudication and sets out when it applies. The Scheme sets out the component parts of adjudication and amounts to a best practice specification for adjudication. The requirements of the Scheme are mandatory in that any adjudication procedure which does not comply with it is struck out and the Scheme provisions adopted. Similarly, if there is no adjudication procedure in the contract between the parties then the Scheme applies.

A body of case law has now developed which interprets the Act and the Scheme in the light of difficult questions that have arisen from the practice and procedure of adjudication.

Adjudication was introduced by the Housing Grants, Construction and Regeneration Act 1996. As its name suggests, this Act is only partially concerned with construction; only 13 of its 151 clauses deal with these issues.

Rarely have such a small number of clauses had such an impact on an industry. The sections cover:

- revolutionising the payment provisions used by the industry;
- banning 'pay when paid' clauses;
- giving parties the right to suspend their performance for non-payment;
- creating adjudication.

20.4.2 Who can adjudicate?

Any party to a 'construction contract' (section 104 of the Act) can adjudicate. A construction contract is defined as an agreement with another party for any of the following:

- the carrying out of construction operations;
- arranging for the carrying out of construction operations by others, whether subcontracted or otherwise;
- providing labour for construction operations.

Construction operations are defined at section 105 and certain exceptions are identified in sections 106 and 107 where adjudication is not available unless the parties agree to its use. Construction operations include professional appointments. The exceptions include oil/gas extraction and production facilities. Other excluded construction contracts include those with residential occupiers; this principally relates to a dwelling which one of the parties to the contract occupies or intends to occupy as his residence. Some standard form construction contracts for use with residential occupiers do permit adjudication, for example, the JCT Home Occupiers contract.

Part II section 12 of the Scheme states that adjudication does not apply where the construction contract lasts less than 45 days. This has been interpreted to mean 45 person days – if more than one person is used then the reckoning of time is affected accordingly.

Contract terms dealing with adjudication may be set out in the contract itself or may be incorporated into it by reference. Many organisations concerned with the construction industry have standard terms which regulate adjudications and can be incorporated by reference. If there are contract terms which fully satisfy the requirements of the Act then they will govern the adjudication. If the terms are in any way inconsistent then the Scheme applies. The risk of inconsistency has resulted in some standard forms (including the JCT) using the Scheme provisions and not a bespoke set of rules.

20.4.3 When can you adjudicate?

Section 108 of the Act states that a party has a right to refer a dispute to adjudication at any time. However, the ability to adjudicate is restricted. There must be a dispute before the referral to adjudication can take place. The dispute must have 'crystallised'.

A dispute only crystallises once a claim has been notified and rejected. The contract between the parties may well set out a procedure for the notification of disputes. Where this is the case, the procedure must be followed. Otherwise, a letter setting out the details of the claim may be enough. If no response is received then that is likely to be enough to set up a dispute.

Arguments about the prematurity of the adjudication have been used by disgruntled losing parties. This is usually an attempt to avoid having to pay over the amount decided on by the adjudicator. This is an area where case law has stepped in to clarify the position and interpret the Act and the Scheme. The courts have identified some areas where adjudication was premature but have tried not to restrict the general freedom to adjudicate at any time. Adjudications commenced before the contractual mechanisms were exhausted are bound to fail as are 'disputes' commenced before a payment is actually due. Judge Coulson said in one case involving previously undisclosed claims that it was *simply not possible for a dispute ... to have crystallised.*[2]

Enforcement proceedings will not necessarily fail if only part of the adjudicator's decision is tainted by procedural error. Severability is an approach open for the judges to take in order to seek to keep the cash flow moving in relation to that part of the decision which was properly decided. In the case of *Working Environments Ltd* v *Greencoat Construction Ltd,*[3] Judge Akenhead held that not all the issues referred to the adjudicator had crystallised into a dispute at the time of the subcontractor's notice of adjudication. Therefore, the adjudicator did not have jurisdiction to consider the additional items referred to him and the court severed them from his decision, enforcing the remainder.

A major fear around at the time of the introduction of adjudication was that 'ambushes' would occur. The concern was that the time given to respond to an adjudication would be insufficient for the responding party to prepare their response, leading to a situation akin to an ambush where the side being attacked is caught unawares and without the wherewithal to defend themselves. Main contractors feared that subcontractors might collaborate and co-ordinate actions against them, meaning they had to fight battles on several fronts at once. An ambush situation can be exacerbated by adding in the element of timing; for example, commencement of proceedings on the day before the start of the annual Christmas vacation period. There have been few reported incidences of these tactics being used. The adjudicators are sensitive to the time pressures of adjudications and usually suggest that extensions be given where the situation may warrant it. The adjudicator cannot force an extension of the period, but it is never a good idea to refuse a reasonable suggestion outright.

Where insufficient regard has been paid to the issue of timing, the enforcement action may fail. In the case of *Beck Interiors Ltd* v *UK Flooring Contractors Ltd*,[4] Judge Akenhead held that not all the issues referred to the adjudicator had crystallised into a dispute at the time of the referring party's notice of adjudication. This was because the claim was first intimated in a letter sent after close of business on the last working day before the Easter weekend, and the notice of adjudication was then issued on the Tuesday after that weekend. Silence over the Easter bank holiday weekend could not amount to rejection of the claim. Therefore, the court severed those elements from the adjudicator's decision and gave summary judgement for the remainder of the issues.

The right to adjudicate can be exercised at any time. This includes during the currency of other legal proceedings. There is nothing to stop a party in arbitration proceedings from calling a (temporary) halt to the arbitration and taking its chances in adjudication instead. Halting the arbitration proceedings may leave the party requesting this hiatus in a vulnerable position in terms of having to meet the costs of the process incurred up until that date.

20.5 Adjudication procedure

20.5.1 Adjudication timetable

To be Scheme-compliant, an adjudication procedure must contain the following minimum requirements.

- A notice of intention to refer to adjudication must be given to every other party to the contract.
- The referring party shall request the appointment of the person named in the contract to act as adjudicator or request that a nominating body selects a person to act as adjudicator.
- The referring party shall, no later than seven days from the notice, refer the dispute in writing to the adjudicator.
- The adjudicator shall reach a decision no later than 28 days (or 42 days if the referring party consents) from the date of the referral.

20.5.2 Notice of adjudication

Once a dispute has crystallised, the first step (as identified above) is to prepare a notice of adjudication. According to section 1 (3) of the Scheme, this should set out:

- the nature and brief description of the dispute and the parties involved;
- details of where and when the dispute arose;
- the nature of the redress sought;
- the names and addresses of the parties to the contract.

The most important requirement is to specify the nature of the redress sought. This seems straightforward, but parties have been known not to actually ask for an award of money as part of the referral. Precision here is, therefore, vitally important. The notice establishes the limits of the adjudicator's jurisdiction and must be carefully drafted.

The possibility of an ambush has already been noted. A claimant may spend considerable time preparing a case for adjudication. When he is fully prepared, he serves the notice of adjudication. The tight timetable imposed by HGCRA then operates and limits the time available for the defendant to prepare an answer to the claim.

Once the notice has been given to the responding party and a period not exceeding seven days has passed, the whole of the referring party's case is given to the adjudicator and responding party. This document is known as the referral.

20.5.3 The referral notice

The timetable requires the referral of the dispute to the adjudicator within seven days of the notice of adjudication. The referral notice required by the Scheme must be accompanied by copies of, or relevant extracts from, the construction contract and such other documents as the referring party intends to rely on. The referral notice cannot go beyond the parameters of the dispute identified in the notice of adjudication. The referral notice is the only submission that the claimant may definitely make unless leave is given to file a reply to the response. The referral should therefore include all the documents needed. It may well be that the referring party will not be allowed to put in later documents to supplement the referral notice.

Upon receipt of the referral notice, the adjudicator will give directions for the service of a written response. The response is likely to be required within a very short period because of the tight time constraints. The adjudicator may give other directions at this stage for the conduct of the adjudication including whether there should be subsequent pleadings (reply to the response and rejoinder) or an oral hearing and, if so, its date and length and whether there is a requirement for a viewing of the site or tests to be carried out.

20.5.4 The response

The response is the respondent's answer to the claim raised. It should answer the claim as fully as is reasonably possible. All documents on which the defendant relies, in addition to those served with the referral notice, should be served with the response.

The defendant may also seek to raise jurisdictional challenges before the submission of the response or may include the challenge in the body of the response. Jurisdictional challenges include the contentions that the contract is excluded from HGCRA or that the adjudicator has not been validly appointed or that no dispute has crystallised.

It is not advisable for a respondent with a jurisdictional challenge to ignore the adjudication and wait to raise the challenge in subsequent enforcement proceedings. The challenge may fail and the respondent will have lost the ability to put forward a defence on the merits of the claim. The respondent can seek a declaration from the court that the adjudicator has no jurisdiction. This may work but unless and until the declaration is granted, the adjudication is likely to proceed, and a defence on the merits can only be raised if the respondent participates in the adjudication.

The best course of action is for the respondent to raise the challenge in addition to any defence on the merits in the response. The respondent should make it clear that he will maintain the challenge even if the adjudicator decides to continue to act. This reserves the respondent's right to resist later enforcement proceedings on the ground that the adjudicator lacked jurisdiction.

Adjudication is not confidential like arbitration, and the adjudication documents can be considered in any court proceedings concerning the adjudication.

20.5.5 The decision and enforcement

The Scheme requires a written decision and entitles the parties to ask for reasons to be provided. It is rare that the parties do not ask for reasons to accompany the decision. The decision must be circulated to the parties within 28 days of the referral notice or any extended period. If the adjudicator fails to reach a decision within the period then decision is probably null and void. In the case of *Richie Brothers (PWC) Ltd* v *David Philip (Commercials) Ltd*,[5] the adjudicator withheld his decision until he had been paid. However, it is now the case (following *Cubitt Building Interiors Ltd* v *Fleetglade Limited*)[6] that the adjudicator is not entitled to a lien and can no longer delay delivery of his decision pending payment of his fees. If the losing party does not pay then the successful party will need to enforce the decision. This is an example where the underlying power of the court supports other forms of dispute resolution. The successful party must start proceedings in the technology and construction court (TCC). The court staff (particularly in London) are familiar with the enforcement route and are able to progress things very quickly. A claim form and an application for summary judgement are issued simultaneously. The summary judgement application is, in effect, saying that there is no possible defence to this case as the adjudication was properly run. The claimant is confident therefore of defeating any potential defence and truncates the usual litigation procedure to have the final hearing as soon as possible to dispose of any frivolous defence raised. The procedure, which is designed to produce a judgement within 31 days of the issue of proceedings, is outlined in Section 9 of the Technology and Construction Court Guide.

The grounds for resisting enforcement are now quite narrow. This has not stopped many disgruntled parties from contesting the decision. The vast majority of these litigants have been told by the judge to *pay now and argue later*. The judiciary have stayed firmly on the side of upholding the adjudicator's decisions where they can and have aimed to stamp out any undermining. The net result, therefore, is usually to simply add to the losing party's legal costs together with an obligation to pay interest on the adjudicator's decision.

Enforcements can be resisted on the ground that the adjudicator had no jurisdiction or exceeded his jurisdiction and on the ground that there was a breach of the rules of natural justice. This amounts to an allegation that the respondent did not have an adequate opportunity for their case to be heard. However, enforcement will be allowed where the adjudicator came to the wrong decision on the facts or made factual errors in his decision or made an error of law. Even procedural errors are not necessarily fatal to the enforceability of a decision if no prejudice has resulted to the respondent.

The respondent is not normally permitted to raise a counterclaim, cross-claim or set-off as a defence in enforcement proceedings with the possible exception of setting-off an entitlement to liquidated damages that follows logically from the adjudicator's decision. Neither will a stay of execution be ordered merely because the successful party is in financial difficulties. This is one of the main concerns of the losing party – if I pay now, will the successful party still be around when I review the decision and obtain the right result from the judge? Evidence of insolvency on the part of the successful party probably is enough to prevent enforcement.

It follows from the last point that a party dissatisfied with the adjudicator's decision can commence legal proceedings or arbitration, if appropriate, to obtain a final decision on the dispute. In practice, the adjudicator's decision often becomes the final decision of the dispute. For most business people, a bad or indifferent decision in adjudication on day 28 is better than a good decision in litigation on day 365. Parties may not be happy with the decision, but the cost and general hassle of reopening the issues is usually a sufficient disincentive to put parties off further action.

20.6 Adjudication in practice

Adjudication has been well received by the construction industry. Subcontractors benefit most from adjudication in terms of securing disputed payments. Main contractors are less keen to use adjudication with developers and other end-user clients given the potential to adversely affect business relationships. Some stakeholders in the industry have misgivings about the quality of the decisions and the legal knowledge of the adjudicators. The speed of adjudication can also be seen as an obstacle to conducting parallel negotiations.

The government consulted widely about the need to reform the statutory provisions before the introduction of the 2009 Act. The feedback received was on the whole complimentary. The changes made are relatively minor in nature and included the following.

- *The right to refer oral contracts to adjudication* – Adjudicating in these circumstances presents the tribunal with the conundrum of resolving what the contract was before and then using what time remains of the 28 days to resolve the dispute.
- *The introduction of a statutory 'slip' rule* – The adjudicator is now able to correct a clerical or typographical error arising by accident or omission. This power probably existed by virtue of the cases decided on the issue, but the clarification is helpful.
- *The banning of Tolent clauses* – The practice had grown up amongst some main contractors of seeking to penalise a referral to adjudication by making the referring party (typically the subcontractor wanting payment) pay the adjudicators and the other party's costs regardless of the outcome of the adjudication. This tactic has been outlawed as such a decision would now be ineffective pursuant to section 141 of the 2009 Act. The parties may still agree liability for costs but only after the notice of adjudication has been given.

As noted, the changes here are minor in nature, which is an endorsement of a law working well and a dispute resolution technique in tune with its users. The final section of this chapter considers hybrid techniques where more than one approach from those covered above apply.

20.7 Hybrid techniques

The most recent developments in dispute resolution have involved hybrid techniques. These techniques combine mediation on one hand and arbitration or adjudication on the other. The techniques are referred to as 'med/arb' and 'arb/med'.

The hybrid procedures involve the 'dispute resolver' performing more than one role – or wearing more than one hat. In 'med/arb', the dispute resolver sets out as a mediator and when an impasse is reached, swaps hats and delivers a decision on that aspect of the claim. The dispute resolver can then swap hats again to make further progress towards settlement.

However, this role has a major difficulty. When a mediator shuttles between parties, information may be gleaned that may be confidential and would not have been disclosed in an adjudication/arbitration. The difficulty comes when the role changes to arbitrator/adjudicator. Any decision made in this role could be challenged on the grounds that the dispute resolver wrongly took account of confidential information in reaching a decision. An aggrieved party could legitimately complain that this has unfairly prejudiced the whole process against them.

This is where 'arb/med' comes in. Here the dispute resolver allows the parties to present their case with evidence and legal argument and then makes a decision. The decision is written down and placed in a sealed envelope on the table in front of the parties.

The parties are then invited to mediate, knowing that a decision has already been made and can be opened by either party at any time. The mediator cannot change the decision on the table regardless of any additional information (confidential or otherwise) disclosed during the course of the ensuing mediation.

These new techniques basically condense the various options of dispute resolution techniques already available. This gives the parties the benefit of the best points of the techniques available in a single procedure. Whether or not the new techniques are a success or not depends on how well the construction industry responds to their introduction.

20.8 Conclusion

Dispute resolution is a multibillion-pound sub-industry in the built environment sector. Lawyers are regularly vilified for their parasitical role and aggravation of difficult situations. In their defence, lawyers merely do the bidding of their clients, and it is to the latter that searching questions as to the need for disputes should be addressed. The construction industry has multiple different approaches at its disposal as covered in Part 4. Ultimately, all disputes should be avoided if possible. Part 5 examines some of the more collaborative practices where the occurrence of disputes should be much reduced.

20.9 Further reading

Riches, J. and Dancaster, C. (2004) *Construction Adjudication*, Second Edition, Oxford: Blackwell Publishing.

Eaton, T. G. (1998) *Adjudication in Construction Contracts: The Basics*, Lizard: the author.

Mills, R. (2005) *Construction Adjudication*, London: RICS Books.

Notes

1 *Carillion Construction Limited* v *Devonport* [2006] BLR 15.
2 *Enterprise Managed Services Ltd* v *McFadden Utilities Ltd* [2009] EWHC 3222 (TCC).
3 [2012] EWHC 1039 (TCC).
4 [2012] EWHC 1808 (TCC).
5 [2005] SLT 341.
6 [2006] EWHC 3413 (TCC).

Part 5

New directions in construction law

21 The agenda for change

21.1 Introduction

The search for best practice is writ large across most walks of life and the AEC sector is no different in this regard. To stand still in the way that things are done is to stagnate, and the quest for betterment is a natural state. This is summed up succinctly in the Heraclitus quote: 'nothing endures except change'.

The role of law in seeking to facilitate the delivery of new initiatives is of central importance. Law and lawyers are often derided as being conservative and reactionary in character. However, legal acceptance and verification of new initiatives is a vitally important step in the establishment and bedding down of new practices into the realms of sound practice. The role of the law is to validate the means by which new ideas are delivered and to ensure that they are fit for purpose and that the stakeholders concerned have had the opportunity to check their own position with regards the new departure. Law can provide the 360-degree review of the new idea and ultimately translate the same into concrete and reliable laws.

This is not to say that the passage of new law and new initiatives is a smooth process. The protection of vested interests can manifest itself in challenges to the law, and it can be the law which becomes the fall guy or scapegoat for why a seemingly workable initiative fails to culminate in an accepted development. In this sense, law is extremely practical – if a new idea does not translate into real-world interests and behaviour then it is unlikely to become established. The reader will recognise that certainty and clarity are the values that lawmakers and contract writers prioritise above others. This echoes the sentiment of the writers of both the JCT and NEC contracts referred to in Chapter 9.

Part 5 charts the development of the agenda for change within construction before examining some of the current best practice ideas and the changes in law required to accommodate them. The concept of good faith is a case in point where we can see an example of where a widely accepted positive development in the law comes up against the requirement for clarity and certainty. Part 5 also investigates how the law can keep abreast of some of the technological developments reflected in legislation, both current and hinted at for future consideration.

21.2 Government interventions in general

The history of government interventions in the construction industry is well established and detailed. The government usually commissions a report from a savant or guru and asks them to write about the ills of the industry. The reports have unerringly succeeded in identifying the problems and have made extremely sensible recommendations for improvement of the situation. However, diagnosis is not the same as cure. The early reports were allowed to gather dust on government department shelves as the appetite for statutory intervention was evidently not strong enough to administer the medicine required. It was not until the mid to late 1990s that the reports actually resulted in action; this was both surgical and dramatic in its intervention, causing the industry to continue to adjust to those changes some 20 years later. Interventions since have been more measured and used best practice ideas rather than the invasive surgery tried back then.

21.3 Early reports

The first major broad-based report into construction in the UK was published during the Second World War. The urgent need to rebuild the country was doubtless in Prime Minister Churchill's mind. The Simon report[1] looked into the placing and management of contracts across the industry, focusing mainly on procurement routes and labour. The next government report, known as the Emmerson report,[2] identified a lack of cohesion between all parties to a construction contract. It also urged that consideration should be given to the adoption of a common form of contract covering both civil and building engineering work. Further suggestion was made that the standardisation should also apply to subcontracts.

These early reports had, therefore, drawn attention to the fragmentation and lack of consistency in approach in construction contract law. This was an observation repeated again in the next government report, written by Sir H. Banwell in 1964.[3] The report reiterated that the most urgent problem with the construction industry was the need for attitudes and procedures to change and suggested that such changes would be useless unless all of those in the industry act in unison. The changes to practice and procedure included a limited first step towards a common form of contract for building and a common form for civil engineering.

One challenge to the task of promoting a single contract lay in the vested interests of the different professional bodies and interest groups. To an extent, the different contracts were a reaction to the different types of work. One size did not necessarily fit all, with chemical plants (ICHemE), civil engineering (ICE) and construction (JCT) activity being, to the contract writers of the time, manifestly different. This is a challenge that has now been allegedly overcome by contract writers, who are now confident that one size can fit all. The NEC, by taking its options approach, supposes that it can cater for the requirements by creating a contract profile from the options capable of addressing any use.

21.3.1 The Latham report

The next report of considerable note to address the ills of the industry was written by Sir Michael Latham in 1994.[4] Concern was again expressed at the continuing proliferation of standard forms being used in the industry and the problems associated with them. The Latham report went on to suggest that one of the options of dealing with the associated problems could be to 'try to define what a modern construction contract ought to contain' and then either to amend the standard forms to include the requirements or to introduce a new contract. The Latham report listed 13 requirements for a 'most effective form of contract in modern conditions'. The recently produced 1st Edition of the New Engineering Contract was commended as being the closest standard form of contract 'containing virtually all these assumptions of best practice'.[5]

This accolade did much to propel the NEC to the forefront of best practice in the public sector.

21.3.2 The Egan report

Sir John Egan[6] wrote in 1998 that there were five key drivers for change including 'integrated processes and teams'. Substantial changes in the culture and structure of the UK construction industry were required to improve the relationships between companies. Sir John was ostensibly reporting on progress since 1994 and stated that the UK Office of Government Commerce (OGC) recommended integration of the project team as an enabler of change with the proposal from the OGC of the adoption of forms of contract that encourage such team integration. Sir John also indicated that the delivery of the vision for integration required collaboration between the various players in the construction industry, including the legal profession and contract writing bodies, in order to prevent an adversarial approach.

21.3.3 The Wolstenholme report

Entitled *Never Waste a Good Crisis*,[7] this report borrowed a line from President Obama's notion that when things are bad, as they indisputably were for construction during the recession, a good opportunity exists to make long-lasting changes that will be of benefit to those concerned once the recovery is under way.

This report was short on actual content but took a rather reflective look at the journey of the Latham/Egan agenda to date and issued a call to arms to the next generation to take up the responsibility of delivering real change. The report was also notable for having forewords from both Latham and Egan. The following quotes[8] give insight into the size of the task and how both men had managed their expectations notwithstanding the reports themselves: 'What has been achieved? More than I expected but less than I hoped' (Sir Michael Latham); 'I would give the industry 4 out of 10' (Sir John Egan).

The central message emanating from this report was therefore that there should be no turning back on the agenda for change. The report is summed up by the graphic featured in Figure 21.1 (which appeared on page 26 of the report).[9] This striking image sets out that to do anything else than progress with best practice amounts to a U-turn in the traffic.

21.3.4 The Government Construction Strategy

The aim of the government strategy issued in 2011 was explicitly stated as being about saving the government money through better processes. The target for the saving was 15 to 20 per cent of the money spent on construction

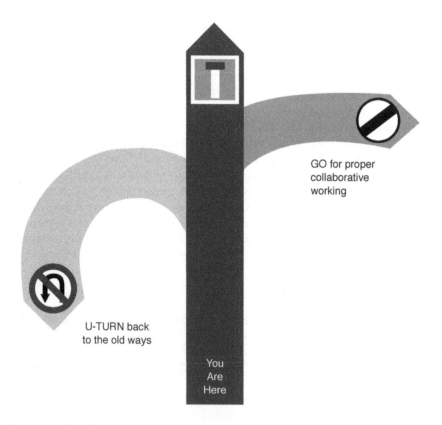

Figure 21.1 The Wolstenholme report

by the end of Parliament. The central notion is that a 'smarter client' is better able to drive out waste and to unlock innovation and growth by using its purchasing power to drive change. The strategy was different in that it recommended methods of delivering change rather than simply repeating the case for change. The desire to trial new models of procurement and supply chain management is of particular interest to construction lawyers. Most notable amongst these initiatives are the requirement to ensure that subcontractors are paid within 30 days, project bank accounts are used and the two-stage open book procurement approach is used. The use of BIM is a central theme in the government strategy.

21.4 Latham's agenda for change

It is helpful to return to the Latham report to examine in detail the requirements set out in terms of the wish list for a modern contract and to comment on the extent to which these items are now provided by construction contracts and their accompanying legal arrangements. Twelve of the thirteen Latham requirements are discussed here. This discussion is based around an academic paper.[10] The final recommendation for advance payment arrangements is not dealt with in detail here given the provision of compliant arrangements in the forms of contract considered.

21.4.1 Requirement 1 – duty of fair dealing with all parties

> A specific duty for all parties to deal fairly with each other, and with their sub-contractors, specialists and suppliers, in an atmosphere of mutual co-operation.

The development and re-emergence of the duty of good faith is discussed in the next chapter. The extent to which the duty is physically provided in the contract is addressed here. The NEC contract obliges the employer, contractor, project manager and supervisor to act 'in a spirit of mutual trust and co-operation'. This obligation is also integrated into the rest of the contract documents forming the NEC suite including the Sub-contract and the Professional Services Contract.

The JCT and FIDIC contracts do not include such a mandatory and specific obligation throughout their suites of contracts. Nevertheless, the parties may elect to add this approach but incorporating either a schedule providing for collaborative working or, as an alternative, the JCT Partnering Charter requiring the obligation to act 'in good faith' in addition to acting fairly and in an open and trusting manner. This charter is suitable for use with almost any standard form of construction contract. It is specifically not a legally binding agreement but simply conducive to creating a collaborative working environment. Similar provisions appear in the framework agreement where this document provides a mechanism for the parties to 'work with each other

… in an open co-operative and collaborative manner and in a spirit of mutual trust and respect'. The importance of this requirement is undone to some extent by the contract draftsmen as, in the event of conflict or discrepancy between the framework agreement and any underlying contracts, the underlying contracts 'will prevail over the conflicting/discrepant provisions of this Framework Agreement and the Parties will be excused compliance with the conflicting/discrepant provisions of the Framework Agreement'.

The strongest obligation can be found in the JCT Constructing Excellence Contract which includes the 'overriding principle' of 'intention to work together with each other … in a co-operative manner in good faith and in the spirit of mutual trust and respect'. There are few reported instances of use of this contract according the RICS Contracts in Use survey.

The PPC contract is predictably compliant with this Latham principle, and in its earlier editions, it actually references the need to comply with the Egan principles and to use best endeavours to achieve this.[11] In summary, this requirement of Latham had been taken on board by contract writers in some shape and form.

21.4.2 Requirement 2 – teamwork and win–win solutions

Firm duties of teamwork, with shared financial motivation to pursue those objectives. These should involve a general presumption to achieve 'win–win' solutions to problems which may arise during the course of the project.

Teamwork and shared financial motivation are cornerstones of the partnering agenda as later developed by the Egan report. The manifestation of these themes is detectable in such ideas as the introduction of a risk register in the NEC contract. The register comprises a list of the risks identified and set out in the contract data before tender and those added to the register by notification during the contract as early warning of such matters assists the parties to share in problem-solving. The risk register is reviewed at risk reduction meetings where, amongst others, the parties who attend will co-operate in 'seeking solutions that will bring advantage to all those who will be affected'.

The NEC also provides shared financial motivation in its target cost contract Option C, where the contractor can look forward to a share of any saving made to the target sum if the contract data has been drafted in such a way as to provide for this.

The JCT and FIDIC forms have shied away from this overtly management style of incentivisation. However, the use of framework agreements does provide for key performance indicators and the measurement of team-like behaviour. Teamwork is writ large in the make-up of the PPC2000 and is central to its character, given the creation of the partnering team.

The lawyer's role of validating new ideas by asking practical questions comes to the fore here. The notion that contracts whose primary role is to protect the interests of the parties involved should present 'win–win solutions' is certainly open to challenge and does not bear too much scrutiny.

21.4.3 Requirement 3 – integrated package of documents

> A wholly interrelated package of documents which clearly defines the roles and duties of all involved, and which is suitable for all types of project and for any procurement route ... standard tender documents and bonds would also be desirable.

The NEC suite of contracts comprises dozens of documents relating to main contracts, subcontractors, consultants and the adjudicator. Each document is in a similar style, wording and format to the others. Sample forms of tender and agreement are provided whilst the most notable absence from the NEC suite are sample forms of bonds or guarantees with the parties remaining free to negotiate their own terms.

The JCT suite comprises a larger range of documents than the NEC, featuring main contracts, subcontracts and appointing an adjudicator. The JCT is unique in being the only one of the three sets of documents to include standard forms of collateral warranties. Whilst none of the JCT forms are drafted specifically to engage a consultant, provision is made within the Constructing Excellence Contract to do so.

Both JCT and NEC include standard forms of contract interrelated with the main contracts as between the employer and the contractor with only NEC incorporating a standard appointment document for a consultant in the form of the Professional Services Contract.

The PPC2000 and FIDIC contracts also take a 'suite' approach. The time when there was the risk of a poor interface between contracts therefore appears to have passed. This Latham requirement has been satisfied by the contract writers.

21.4.4 Requirement 4 – simple language and guidance notes

> Easily comprehensible language and with guidance notes attached.

One of the original drafting aims of the NEC contract was that it should be in ordinary language, thereby being a model of 'clarity and simplicity'. This would have the benefit of making it easier to understand by people who are not used to formal contracts and by people whose first language is not English. The Engineering and Construction Contract Guidance Notes indicate that its use of ordinary language would also make it easier to translate into other languages. Guidance notes and flowcharts have been produced for all the documents in the NEC3 package apart from the Sub-contract and the Short Sub-contract.

Both JCT and FIDIC have retained their traditional styles of wording which, arguably by familiarity within the industry, are readily understood by users of the contracts. Guidance notes have been released by the two contract writing bodies to help explain how they are envisaged to work in practice for practically all the contract documents in their extensive suites. Similarly, the PPC2000 is issued with guidance notes.

The use of guidance notes and their interaction with the contract itself raises an interesting legal question on interpretation. Should the guidance notes be read in conjunction with the contract or simply as background to the forms? Nothing should interfere with the sanctity of the contract itself, which should be interpreted in accordance with the rules of interpretation as discussed in Section 3.6 in the event of ambiguity.

21.4.5 *Requirement 5 – role separation*

> Separation of the roles of contract administrator, project or lead manager and adjudicator. The Project or lead Manager should be clearly defined as client's representative.

The roles of the project manager and adjudicator are clearly separated within the NEC. The project manager is appointed by the employer and becomes the principal point of contact with the contractor under the contract, being able to give instructions and acceptances, issue certificates, assess amounts due for work done to date including assessment of compensation events, amongst other things. The extent to which the project manager has a requirement to act impartially was addressed in the case of *Costain Ltd* v *Bechtel Ltd*.[12] The adjudicator is clearly intended to be independent as he has jurisdiction to resolve disputes under the contract, which may involve an action or inaction of the project manager.

The role of employer's agent is also largely retained within the JCT contracts as having 'full authority to receive and issue applications … and otherwise to act for the Employer'. The role is redefined within the Constructing Excellence Contract as the 'purchaser's representative' with full authority to 'act on the Purchaser's behalf in relation to the Project and who shall be the point of first contact for the Supplier'. The FIDIC and PPC also retain the notion of the overseer. In FIDIC, the role is that of the engineer and in PPC, it is the client's representative.

It is difficult to see how Latham envisaged the reconciliation of the conflicting role for the overseer beyond the imperfect arrangement with which the parties operate. As previously stated, the view of the contractor is likely to acknowledge that the certifier is likely to err on the side of the client but to recognise that the protection offered by their professionalism is better than nothing.

21.4.6 Requirement 6 – risk allocation

> A choice of allocation of risks, to be decided as appropriate to each project but then allocated to the party best able to manage, estimate and carry the risk.

This restatement of Abrahamson's Principles[13] reinforces their status as the fundamental guiding notion of building contracts. All building contracts provide for the standard allocation of risks. The point made in this Latham requirement is that the risk profile should be capable of being altered depending on the project. This can be achieved by amendment or through the selection of optional clauses.

Within the NEC, the employer's base risks are clearly set out with all other risks being carried by the contractor by exception. Additional risks to be borne by the employer can be set out in the contract data with other risks to be borne by the contractor being allocated using amended or additional clauses under Option Z.

The risk register allows for a project-specific consideration of anticipated issues and how the risks will be managed (i.e. who will take what action to manage or minimise them) and identification of the time and cost consequences of doing so. The JCT, FIDIC and PPC forms of contract generally leave the risk and solution of problems occurring on-site up to the contractor with a certain amount of control or direction from the agent of the employer or the contract administrator.

The JCT's introduction of a risk register within the Constructing Excellence Contract can be seen as a nod to compliance with the Latham requirement. The extent to which the discussion of project-specific issues occurs varies from project to project. Providing a mechanism for the recording and mitigation of these risks makes good practical sense whatever the contract.

21.4.7 Requirement 7 – variations

> Taking all reasonable steps to avoid changes to pre-planned works information. But, where variations do occur, they should be priced in advance, with provision for independent adjudication if agreement cannot be reached.

The NEC contract envisages the pre-planned works information being as complete as possible. Nevertheless, the contract also envisages changes being made to the works information by instruction from or a change in an earlier decision by the project manager with further provision for quotations being submitted before the varied work starts. The NEC treats any changes to the pre-planned works information as compensation events rather than claims for extensions of time and/or money.

Within the JCT, FIDIC and PPC contracts, the parties are urged to agree the cost of variations without necessarily setting out a procedure for agreeing them before work starts. The cost and time implications of variations which are not pre-agreed can, therefore, default to being resolved by the final account. The merits of resolving issues at the time they arise rather than storing problems for later are self-evident. However, in those forms that do not provide for the close management of all matters arising (JCT, FIDIC, PPC), the latter strategy is sometimes unavoidable.

21.4.8 Requirement 8 – mechanisms for assessing interim payments

Express provision for assessing interim payments by methods other than monthly valuation, i.e. milestones, activity schedules or payment schedules. Such arrangements must also be reflected in the related sub-contract documentation. The eventual aim should be to phase out the traditional system of monthly measurement or re-measurement but meanwhile provision should still be made for it.

Latham's hostility to the monthly measure is difficult to fathom. The requirement to provide alternatives is understandable, but it is not to be taken as read that the alternatives are a better arrangement than the monthly measure.

The NEC mechanism for assessing interim payments is based on the assessment of the price for work done to date. Depending upon which of the main options are used, the assessment can be against a bill of quantities or against an activity schedule. These assessments are carried out at regular intervals of no more than five weeks. The NEC contract appears to be yet to embrace payment by milestones or payment schedules.

JCT documents rely upon traditional monthly valuation methods whether they are by measurement or by reference to an activity schedule. Some provision is made within the Design and Build contract for payment by stages. The FIDIC and PPC also rely on monthly valuations as the primary means of reimbursement of the contract.

Construction contracts are, by their nature, measure and value arrangements. Advances in technology such as virtual design and construction will lead to improvements in the predictability of earned value and the amount of payments to be made. Latham's perceived shortcomings in the present arrangements may be resolved by means of this route.

21.4.9 Requirement 9 – payments

Clearly setting out the period within which interim payments must be made to all participants in the process, failing which they will have an automatic right to compensation, involving payment of interest at a sufficiently heavy rate to deter slow payment.

Since publication of the Latham report, statutory intervention has taken place in relation to instalments, stage and periodic payments in the Housing Grants, Construction and Regeneration Act 1996. Payment provisions in all the standard forms have been drafted to comply with the legislative provisions of the 1996 Act. In default, the provisions set out in the Scheme apply.

If payments are late, each of the standard forms incorporates a contractual right to interest on behalf of the payee. The parties are free to agree the contractual interest rate although they are usually fixed by the employer at a rate less than the default statutory rate introduced by the Commercial Debts (Interest) Act 1998.

The importance of releasing cash flow to all parts of the supply chain is a theme to which the *Government Construction Strategy* returned. The government mandated payment to tier 3 contractors within 30 days in a welcome recognition of the massive contribution made to the industry by those not in direct contractual arrangements with the client. This public sector commitment to paying subcontractors and suppliers has not yet manifested itself in standard contract drafting

21.4.10 Requirement 10 – trust funds

Providing for secure trust fund routes of payment.

Trust law is a discrete area of law principally focused on the holding of property by the trustees on behalf of the beneficiaries. The role of the trustees is essentially fiduciary by which it is meant that they do not have any self-interest themselves in the property or money they hold – they are essentially guardians of the asset for the benefit of others. This notion is quite far removed from the cut and thrust of commerce and its application to the construction industry is not altogether obvious. Where the two notions do cross over is in security for payment arrangements. The money earned by the contractor, in the form of the interim payment due and owing following the certification process, is analogous to trust money. The employer's interest in the money and the contractor's interest in the money due to his subcontractors should be as guardian rather than owner. The placing of the money earned into a trust fund-type arrangement gives those expecting payment some comfort in that their money is safe and will, in the normal course of events, be forthcoming to them.

Sir Michael was advocating this type of arrangement. There is some precedent for this where the treatment of retention monies and case law[14] had already established that the contractor could compel, by way of injunction, the employer to place the money in a separate account. The case is easier to make out in relation to retention funds as the contractor has already earned the money and the deduction is made to abide the unlikely event that the contractor will not return to remedy the snags. The natural extension of this, covering the whole of the interim application, was introduced by the project bank account, which has clear origins in the recommendation of Sir Michael.

One issue around creating trusts is the formalities required and the mismatch between a commercial undertaking and a fiduciary role. An acceptable compromise between these competing elements took time to catch on as it now has in the public sector with the project bank account. The NEC sought to accommodate the requirement for a trust in the second version of the NEC. This option disappeared following the drafting of NEC3 in 2005. With the enactment and implementation of the Office of Government Commerce's model Fair Payment Charter in 2007, the NEC drafting committee responded by producing an Option Z clause to allow users to implement the fair payment practices into NEC contracts with the creation of a project bank account, beneficiaries of the account being designated by execution of a trust deed.

21.4.11 Requirement 11 – speedy dispute resolution

> While taking all possible steps to avoid conflict on-site, providing for speedy dispute resolution if any conflict arises, by a pre-determined impartial adjudicator/referee/expert.

Sir Michael implicitly acknowledged in this requirement that there would be some disputes that mediation could not resolve. This has been covered in section 16.4 where the coexistence of voluntary and mandatory forms of dispute resolution has been discussed. Once the need for mandatory forms of dispute resolution is acknowledged then top of the wish list is that they be speedy and cost-effective. This, along with the payment mechanisms and minimum security for payment arrangements, found form in the Construction Act of 1996. Adjudication's key feature is its speed, thereby complying with the Latham requirement. A decision on the dispute from the adjudicator is normally expected within four weeks, but this may be extended to give the adjudicator further time to receive information and/or to come to his decision on the dispute.

The details of adjudication have been discussed. The contractual arrangements rendering compliance with the Act can be found in all the forms of contract. Adjudication procedures are introduced in the NEC contract by invoking Main Option Clause W2 within the UK. The JCT contracts introduce a compliant arrangement for the statutory Scheme for Construction Contracts. PPC2000 and FIDIC provide for adjudication. The introduction in the latter contract raises the issue of enforceability in international cases. It has been observed that the key strength of adjudication is in the force that the English courts give the decisions and the clear message: pay now, argue later. This enforcement is particular to the UK courts, and a party expecting the same enforcement in another jurisdiction may find some surprise outcomes.

Dispute prevention, rather than cure, is writ large in attempts to try and avoid conflict on-site by the introduction of risk registers/risk allocation schedules and regular meetings to discuss risks. To reiterate the point, dispute resolution should only be used where a negotiated settlement is not, for whatever reason, proving possible or realistic. Issues can quickly become intractable and appear insurmountable in a pressured environment where autonomy is often compromised.

21.4.12 Requirement 12 – incentives

Providing for incentives for exceptional performance.

This requirement appears prescient in terms of the management approach seized upon by the NEC writers. The management approach has a good deal to commend it but can be contrasted to the JCT outlook whereby the contractor is paid the price he asks for in return for the job done. The word 'exceptional' stands out somewhat in the above quote and begets the question as to whether the performance rendered by the contractor is ever intended to be less than exceptional. This is a semantic point, however, and it has to be borne in mind that the report was not a blueprint for contract wording but, rather, a very welcome incursion into the hitherto reactionary industry.

Nevertheless, the NEC embraced the requirement in a number of ways, including the target cost and by the use of key performance indicators (KPIs) contained in a pre-agreed incentive schedule setting out payments to be made if a particular KPI is achieved or exceeded. Another NEC innovation is that bonuses can also be won for early completion of the works.

The JCT do not generally provide incentives, nor do they set down provisions for measuring performance. Only the framework agreement and the Constructing Excellence Contract provide mechanisms for performance indicators assessed against the framework objectives and performance monitoring against key performance indicators.

The PPC2000 provides for incentives by linking achievement and financial reward. The contractor is encouraged to contribute proposals for cost savings and added value deriving from the value engineering of designs and the reduction of risks.

FIDIC is more circumspect and does not include for incentives. Rather, it is grounded, like the JCT, in capturing the essence of the bargain rather than adding a management element.

Bonds can be required with an appropriate entry within the contract particulars. Forms of bond are helpfully provided for both general advanced payments and for off-site prefabricated materials (the listed items) for most of the main contracts.

21.5 Partnering

It comes as no surprise to learn from the previous section that the contracts most compliant with the Latham agenda are the NEC and PPC2000, which were written very much with the Latham report in mind. The JCT and FIDIC contracts have their origins in a different approach, namely the overarching principle to provide benchmark provisions on which parties can contract in relative certainty of their safeguards should the other party not perform.

The PPC2000 contract goes further than the NEC in that it specifically seeks to address the Egan legacy in addition to the Latham approach. Partnering is a discernible influence on and component of the Latham report. The concept of partnering was identified as a 'structured management approach to facilitate team working across contractual boundaries'.[15] This can be contrasted with the Egan report which is ostensibly chiefly concerned with the *promotion* of partnering. Partnering is lauded internationally as the silver bullet for many of the construction industry's woes. In spite of this, attempts to instill the new collaborative platform have not been as successful as expected.

Project failure has been common over the past decades, and amongst the notable common causes of failure is the lack of effective project team integration across the supply. A lack of shared understanding of key partnering concepts, missing initial effort to establish shared ground rules, communication difficulties in inter-organisational relationships and unclear (perceived) roles have also been identified as contributing to the lack of success.

The rather gloomy picture established has not prevented the spread of the concepts involved or their establishment into the fabric of the industry. The latest NBS survey records that 41 per cent of the industry adopted collaborative techniques in some of their projects in 2012. Furthermore, 32 per cent of respondents used a formal partnering contract, with 20 per cent using an informal one.[16] These figures are difficult to reconcile with the experiences of most people in the industry for whom little has changed in terms of the way business is done in the industry. The lack of any clear definition as to what partnering actually entails partially explains this anomaly.

Defining partnering has proved to be an elusive concept to grasp, but it is usually indicative of a process involving collaboration and sometimes the use of an open book arrangement where there is some degree of transparency in the contractor's dealings with the employer.

The philosophy of partnering is dispute prevention, conflict resolution and equitable risk allocation rather than the legalistic and confrontational approach taken by many in the industry. Partnering also involves the alignment of values and working practices by all supply chain members to meet client objectives. This is achieved by a shift of emphasis from hard issues, such as price, to softer issues, such as attitudes, culture, commitment and capability. The process usually starts with kick-off workshops to establish mutual objectives.

The benefits of partnering have been expressed in terms of the different participants. The employer stands to gain from the effective utilisation of personnel resources and increased flexibility and responsiveness. The design team and supply side are able to refine and develop new skills in a controlled and low-risk way where repeat work will ensure being given a voice in design decisions. The overall outcome is highlighted as being improved project delivery for all concerned.

Foremost in the partnering landscape are issues already touched upon including:

- collaborative practice;
- key performance indicators;
- framework agreements;
- open book accounting;
- supply chain integration and early contractor involvement.

21.5.1 Collaborative practice

In keeping with the notion that something can be explained by discussing the negatives, the following can be identified as reasons why the industry is reluctant to collaborate.

- The industry involves too great risks.
- The industry is characterised by focus on short-term profit making.
- The competition in the industry is too fierce.
- It is unlikely that this group of people will work together again.
- The multiplicity of stakeholders in the industry leads to issues.
- Silo mentality and vested interests prevent collaboration.
- Construction is very complicated.
- There is always someone to blame.

Of these reasons, it is possible to expose the majority as symptoms and not causes of the *ennui* in the industry. The central resistance to partnering is to observe that individuals are in business to make money for themselves or their investors/shareholders. Putting themselves first and looking for ways to protect and grow their investment is the central driving force of business. Many people therefore find it unconscionable to contemplate any other approach. Lip service is frequently paid to the concepts of partnering to camouflage a wider mistrust of supposed 'win–win' solutions.

Assuming that other industries have successfully overcome the above problem, it is difficult to see what prevents similar progress in construction. Of the issues listed above, the only one which translates into a truly different set of circumstances from those of any other industry is the multiplicity of stakeholders. This factor can prevent collaboration unless a strategic approach is taken to overcome the issues presented. Manufacturing and process-driven industries need only satisfy their client and government stipulations. In the built environment, many different interested parties are involved. Taking a residential block of housing, for example, the stakeholders involved are not just the financing client but the initial users of the building and the subsequent owners. People feel a strong link to their built environment and the effect it can have on their daily existence. Construction and maintenance of a building tend therefore to generate strong feelings that can be difficult to reconcile with the capital appreciation targets of the builder/financier.

21.5.2 Key performance indicators

Key performance indicators (KPIs) are a type of performance measurement. KPIs evaluate the success of an organisation or of a particular activity in which it engages. Often success is simply the repeated, periodic achievement of some levels of operational goal (e.g. zero defects, 10/10 customer satisfaction, etc.), and sometimes success is defined in terms of making progress toward strategic goals. Accordingly, choosing the right KPIs relies upon a good understanding of what is important to the organisation. What is important often depends on the employer's business. These assessments often lead to the identification of potential improvements, so performance indicators are routinely associated with performance improvement initiatives. A very common way to choose KPIs is to apply a management framework and to apply targets for improvement.

KPIs on a construction project can include the criteria shown in Figure 21.2. Improvement to safety records and defects recorded are two commonly

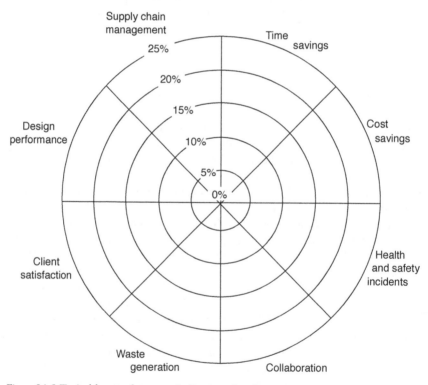

Figure 21.2 Typical key performance indicator radar chart

used indicators. A more sustainable approach to construction can also be achieved by measuring power used and waste generated. The target for the next project can be set at between 5 per cent and 20 per cent depending on the saving sought. KPIs are a good idea and many clients have embraced the approach. However, the proper implementation of the approach requires resourcing and the partiality of the measurement arrangements also needs to be considered. The notion that continual improvement is possible is open to challenge. A point probably exists where the contractor is encouraged to look for ways around the indicators and his attention is thus diverted from providing the construction services.

21.5.3 Framework agreements

Framework agreements are essentially the means by which repeat work is awarded to a preselected number of contractors (see Section 7.8.5). The employer is able to select a small number of contractors to whom its projects will be awarded during a fixed period of time. The framework contractors are then awarded the work and 'draw down' contracts are entered into. The obligations under the draw down contract usually stand alone from the overarching framework agreement. It is possible for the framework agreement to provide for the measurement of KPIs across the performance of one or all of the framework contractors. This essentially leads to competition amongst the framework contractors who can also be encouraged to share best practice between themselves. A framework contractor usually has the expectation rather than the right to be awarded work. The work can be awarded to the framework contractors in rotation or based on previous performance.

21.5.4 Open book accounting

Open book accounting is a partnering style arrangement whereby the contractor provides the employer with an appraisal of the financial information relating to the project that is more in depth than usual. Essentially, the contractor shows the employer what his costs are, thereby effectively opening his books to scrutiny by the employer. This arrangement is similar to a cost reimbursable form of contract. The contractor is reassured that the information will not be used to affect his profit margins as this element is usually ring fenced to encourage openness on the part of the contractor. The contractor therefore is awarded a profit percentage (usually around 10 per cent) for his management and co-ordination of the subcontractors and suppliers, the information about which he is content to provide. The contractor typically must provide details of discounts and savings he has negotiated with his subcontractors and suppliers.

The 'smart client' is able to examine the open book accounts provided by the contractor, and together they can look to make savings in future with the contractor safe in the knowledge that his profit is protected. There are diverging views of these arrangements, including some scepticism about the desirability and usefulness of the arrangement. One view is that the arrangement is ultimately self-defeating as the employer wants the benefits of savings that become apparent over time. Ultimately, therefore, even though profit is ring fenced, there will be a smaller sum on which the profit element percentage is applied. Some arrangements avoid this problem by fixing the profit as a sum rather than a percentage of the contract sum as detailed in the open book approach.

21.5.5 Supply chain integration and early contractor involvement

The importance of subcontracting to the construction industry has already been discussed in Section 11.6. Supply chain management has been a vitally important component of construction management for a long time. The Egan partnering approach reinforced the fundamental importance of treating the supply chain well and of involving them in the design process as much as possible. The extremely valuable contribution that subcontractors bring to projects in terms of know-how and design input is routinely captured on modern projects. Subcontracts involving design obligations are increasingly popular and the integration of important subcontract providers into the briefing stage of the build is commonplace.

One notion in the Egan approach is to move away from talk of 'supply chains' with their image of hierarchical arrangements where the end of the chain is not party to the decisions made higher up. The restriction on the cash flow of the lower components is also discernible in a chain approach. The preferred approach is to view the supply community as a network where the interdependence of the disparate elements is recognised.

21.5.6 Legal issues around partnering

The PPC2000 is a multi-party contract which the main participants sign, thereby undertaking contractual duties vis-à-vis each other. This has resulted in nervousness amongst some legal commentators as to the position of the contractor, who may be at risk of receiving claims in contract from more than one party in respect of the same circumstances. This situation is known as 'double jeopardy'. A situation may arise where the contractor causes delay on a project and then has liability not only to the employer but any other party who can establish that they too have suffered loss as a result of the contractor's default. The meaning and weight attributed to 'good faith' clauses also give rise to issues of interpretation. This is explored in the next chapter. Lawyers are loath to admit that there is insufficient clarity about the standing of these clauses and are concerned about their inability to advise their clients on the exact meaning of the terms in the contract.

21.5.7 Range of partnering arrangements

Several contractual arrangements are available in terms of delivering a partnering approach to contractual arrangements. As well as the multi-party PPC2000, there are free-standing non-binding charters for single projects. The JCT Partnering Charter is an example of this. Another approach is to use 'bolt on' clauses for standard agreements. This is the result if the NEC Option X12 is used to augment the partnering credentials of the NEC form. Another approach is to use an umbrella-type agreement consistent with the framework approach. The operation of this contract is by sitting on top of but not interfering with the operation of the contract.

21.5.8 Conclusion

Twenty years after Egan's call for collaboration to be the norm, it continues to be treated with suspicion in some quarters of the industry. This knee-jerk reaction to reject new initiatives is commonplace in the construction industry. This raises the doubt as to whether commitment to partnering at all levels is possible or even desirable. Another misgiving around the partnering agenda is whether it works as well in a recession as it does in a positive market. Certainly, the anecdotal evidence from the last recession was that many employers were very quick to turn their back on the progressive approach and exploit their dominant market position in seeking ever-lower competitive prices from the providers.

The NBS survey sees the formal partnering arrangements having very little market share. This contrasts with the familiarity many in the industry have with the concepts, which can now be seen as being well 'bedded in' to the industry. Those who embrace this approach rarely wish to contemplate a return to the alternative ways of working, including competitively tendered work with one-off relationships.

The latest developments in technology together with the latest government pronouncements may see a resurgence of interest in partnering arrangements. This is certainly envisaged in the Saxon report which makes the statement: 'What partnering needed to succeed was BIM'.[17]

21.6 Government Construction Strategy

21.6.1 Introduction

The *Government Construction Strategy*[18] reviewed the landscape from the point of view of the public sector and made some observations before prescribing some remedies for the sector. The report starts with familiar ground in terms of lamenting the status quo and the slow nature of establishing any changes. Most of the shortcomings identified have already been alluded to in the previous reports. The latest diagnoses pointed at the following ills:

- construction underperforming in terms of delivering value;
- lack of investment in efficiency and growth opportunities;
- waste and inefficiency;
- low levels of standardisation and fragmentation of public sector client base;
- lack of integration;
- perception that partnering is a barrier to true competition and does not involve supply chain.

The successes of the partnering approach were acknowledged whilst also remarking on the less desirable aspects, including the creation of a situation in which there are fewer suppliers and fewer opportunities for small and medium-sized enterprises (SMEs). The government's desire to include these providers in its initiatives moving forward is a key theme in the strategy. Further steps are to be taken towards integration and collaboration but not at the expense of locking out the rest of the market and only giving limited opportunities to non-participants.

The tension exists, therefore, between the benefits felt by the fewer suppliers in long-term relationships and the desire to make the market accessible to SMEs. The arrival of SMEs on the scene as a major stakeholder is something that is discernible over the last decade or so. This would encourage the subcontractors and their supporters in that they appear at last to have found a platform on which their interests can be taken seriously and addressed.

Ideas that benefit the SME agenda are set out below.

21.6.2 The pipeline

The government expressed a wish to improve visibility and certainty of forward construction by means of a complete, centrally accessible data source. This involves a rolling two-year forward programme, enabling the entire industry to plan for work. All government departments will maintain the pipeline and identify value-for-money criteria relevant to the project to be converted to standards and specifications for suppliers. The pipeline has been used successfully with over 600 projects worth £40 billion being notified by 2015. The issue about whether SMEs are genuinely able to compete effectively for this work remains uncertain. Steps taken to simplify tender information and framework approval should assist in this regard.

21.6.3 The project bank account

The project bank account can be incorporated into any of the other contract families whereby the existing payment arrangements are effectively bypassed and put through the project account.

The main advantages of the project bank account are felt by the supply side, most notably by the subcontractors and suppliers. The interim (or stage) payment, once certified by the overseer, is transferred into the project bank account. The bank account only pays out the sum to the main contractor and the subcontractor when the former and latter agree upon the amounts properly payable to them. Essentially, the main contractor does not access his payment until he has also agreed how much the designated subcontractors (not all of whom will be project bank account holders) will also receive. It is by this strategy that the government seeks to fulfil its commitment to pay down to tier 3 contractors within 30 days. The deal for the main contractor is essentially that if he does not agree payments both upstream and downstream then he is prevented from accessing his own money. It is early days in terms of whether this arrangement will catch on. It certainly has a good deal to commend it in terms of taking out hidden costs and pressures from the industry, bearing in mind Lord Denning's prophetic words about cash flow being the life blood of industry.

21.6.4 *Two-stage open book procurement*

One of the new methods of procurement referred to in the *Government Construction Strategy* was set out in more detail in a recent paper.[19] Two-stage open book procurement was used successfully at the Ministry of Justice BIM-enabled Cookham Wood project. This project was fully BIM enabled and delivered using the PPC2000 form of contract. The procurement route is essentially a continuation of some of the partnering themes discussed above. One difference to existing arrangements is that the contractor presents open book information during the tender process and the firming up of the price and ring-fenced profit element takes place before the formal full contract is entered into. The supply side organises itself into a consortium brought together for the project, which includes the design team and the specialist subcontractor suppliers. The consortium work together collaboratively to ensure the client is impressed enough to award them the contract. The 'can do' attitude instilled at the tender stage is then embedded in the project delivery.

A consortium approach is currently restricted to larger projects and is unlikely to translate to small and medium-sized projects in the short term. The opportunity for SMEs to be selected by the other consortium members is also not clear. It will be interesting to observe whether the government settles on this procurement route after concluding its remaining trial projects.

21.6.5 *Conclusion*

The government's stated aim of improving the position of SMEs and stripping out some of the hidden costs of the industry were key drivers in its construction strategy. The real improvement in cash flow made by the project bank account and the commitment to pay subcontractors within 30 days are commendable. These ideas only affect the public sector and are unlikely to be entered into voluntarily in the private sector. The result should be more opportunities for SMEs provided they can come to terms with a new landscape. The Strategy has redefined what partnering means in terms of producing a strategic and transparent approach from an integrated and smarter client. This gap between best and actual practice remains pronounced. In a recent survey, 90 per cent of respondents from the Specialist Engineering Group maintained that early contract involvement was something they were not involved in.[20] Without this changing dramatically, it is hard to see how collaboration can really take on the embedded silo approaches to construction.

The other key theme of the *Government Construction Strategy* not yet touched upon is Building Information Modelling, which is already having a huge impact in both public and private sectors.

21.7 Building Information Modelling (BIM)

The recent *Government Construction Strategy* promotes the use of BIM alongside the other initiatives featured in the previous section. The Strategy acknowledged that at the industry's leading edge, there are companies capable of working in a fully collaborative 3D environment on a shared platform with reduced transaction costs and less opportunity for error. The move towards BIM is given additional impetus by the Cabinet Office requiring the submission of a 'fully collaborative 3D BIM (with all project and asset information, documentation and data being electronic) as a minimum by 2016'[21] on public sector projects.

The roll call of government-backed reports aimed at improving construction can be continued on into the Saxon report in 2013. The potential for BIM to provide the means by which the construction industry can achieve its partnering goal is writ large throughout the report. Commentators see collaboration as being at the heart of the BIM process, which is explained below.

21.7.1 BIM explained

BIM involves a widening suite of working methods which become possible or necessary when built environment industries move onto a digital basis or use artificial intelligence. This technology-led approach is already well established in other industries, usually under the heading 'product lifecycle management'.

BIM has been described as 'a digital representation of physical and functional characteristics of a facility creating a shared knowledge resource for information about it forming a reliable basis for decisions during its life cycle, from earliest conception to demolition'.[22] It therefore requires contractors, subcontractors, lead designers, architects, project managers and designers to work together and share information. The very term BIM is in itself not universally accepted. The term first appeared in the USA (also where the partnering approach originated) in 2003. The term may be replaced with something more comprehensive – digital design and construction is not simply about buildings. Civil engineering and infrastructure projects can also benefit massively from this technology.

The state of readiness of any stakeholder within the industry is referred to as BIM 'maturity'. Surveys suggest that there are many exaggerated claims around how much the processes are being used and the level of sophistication achieved. Figures establish that maturity and adoption vary between industry sectors and countries. More contractors are using BIM than architects (adoption by 47 per cent and 24 per cent respectively); 37 per cent of engineers in Western Europe use BIM versus 42 per cent in North America; and 48 per cent of architects who use BIM consider themselves advanced or expert.[23] Notwithstanding these statistics, a recent study concluded that there is a great lack of well-educated and trained BIM professionals.

The government published a BIM Protocol designed to accompany standard forms of contract. The industry has (in theory) the tools it needs to deliver the government's ambitions. The UK's construction industry may, therefore, be on the brink of reaching its partnering nirvana on the back of BIM adoption.

The synergy between BIM and partnering is hard to ignore. Both approaches claim to produce project cost reductions, reduced build time, reduced claims and lower project change costs.

21.7.2 BIM issues

It would not be accurate to portray the construction industry as fully in line with the *Government Construction Strategy*. The lack of compatible systems, standards and protocols has inhibited adoption alongside a general resistance to change. The existing arrangements in the industry are all too often characterised by projects funded by third-party financiers concerned only with protecting their investment. The requirements for lending are expressed in terms of collateral warranties and step-in rights rather than any interest in 3D modelling and asset management data.

The challenge with technology is always to ensure the market exists or can be created for new products. Winning the argument that BIM is essential for construction clients is its real challenge. The employers operating in the construction industry private sector need to feel that they need BIM to properly design and make the most of their building. There are signs that this is happening. The construction law challenges presented by BIM are set out below.

21.7.3 Specific BIM-related legal issues

Three issues are dealt with here:

- integrated project insurance;
- blurring of responsibility;
- intellectual property rights.

Integrated project insurance

One of the main areas of waste in the construction industry is the maintenance of separate policies of insurance by the project stakeholders. Very often the same risks are being covered by a number of different policies, resulting in waste. The argument in favour of replacing the many with a single policy is the simple idea behind integrated project insurance.

An important aspect of existing insurance arrangements is the right of subrogation whereby the insurer can avail itself of any rights against third parties that the insured may have to recover any payout made. This right is seen as being valuable to the insurers. One drawback of integrated project insurance is that there is no right of subrogation as it does not apply to joint policies. Furthermore, the cover the insurer is being asked to give is wider in a BIM-enabled environment where parties cross the divide between delineated responsibilities. Essentially, insurers are worried that they will potentially be responsible for an unknown set of risks with no right to recourse.

Insurers also make an assumption that with the waiving of subrogation might also come a waiver of skill and care. The blame-free environment that might result may see professionals adopt a less professional attitude. Accountability is a fundamental feature of professionalism within the built environment and something that will be very hard to relinquish.

One way which has been found to proceed whilst maintaining some 'blame' in the process has been trialled in the USA where 'wrap' insurance is found. The employer pays the premium on the integrated insurance policy. Each participant pays a contribution by way of a reduction in their contract price. The percentage contribution of the party is not fixed and subject to reallocation in the event of a claim. In essence, any negligent party would be retrospectively allocated more of the premium paid. Further, the participant's reputation would be damaged by any allocation of the premium, leading to all possible steps being taken to avoid being 'at fault'.

Insurers in the UK have been very slow to offer integrated project insurance despite government encouragement. The non-availability of these products is of major concern to the establishment of a fully BIM-compliant industry. The disruption to existing practice is felt most keenly at Level 3 BIM. Level 2 is co-operation with a degree of interaction through middleware. Each designer's input remains distinct, and responsibility in respect of this remains evident. At Level 3, the model moves to being fully open and collaborative with designers inputting directly into the model. Responsibility is less clear and the need for clear protocols more pronounced.

The blurring of responsibility

Chapter 4 outlined the basics of professional negligence and a discussion around duties of care in the construction industry. The development of a new field of practice means that no one is quite sure what amounts to reasonably competent practice. An open collaborative approach between the stakeholders could result in unintended consequences. The resulting question would therefore be whether it is proper to hold a tortfeasor responsible for the outcome of his unintended actions.

At an operational level, the prevention of blurring of responsibility is likely to fall into the province of the information manager who has to manage process and procedures. Protocols can assist further with express duties assigned to the parties in terms of clash detection and level of detail, which states the limits placed on the data and the risks of using the data beyond its specified level. This is not an issue in traditional procurement where no design is being undertaken by the contractor.

The intelligent nature of the model is an issue whereby designers cannot be deemed to be aware of modifications to their design made without their knowledge. In all likelihood, the design team are likely to be under one umbrella (particularly if design and build procurement is used), and interfaces between component parts of the design will remain with the single point of responsibility. The issue may arise when the individual component parts seek to recover from each other in respect of any losses suffered.

Intellectual property rights

Another aspect causing nervousness around BIM from a legal point of view is intellectual property rights. The law here is largely codified in the Copyright, Designs and Patents Act 1988, which protects the right of the author.

The familiar term in appointments and collateral warranties alike is that the client (or third party) is granted a licence to use the drawings. However, the BIM model has more applications and outputs in terms of the post-completion monitoring and maintenance. The model itself has a commercial value, which is not necessarily the case with a 2D drawing. The appointments under a BIM project may wish to make alternative provisions than a mere licence and may include full rights of ownership in the component design being transferred over to the client. This may be resisted by those contributing to the design who may have the legitimate concern that their design will be used without their input for this and similar projects.

21.7.4 Conclusion

Whether BIM will become fully established within the construction industry will only be known over time. The legal issues around its use are unlikely to prove determinative in its adoption. Such legal concerns as exist around BIM can be addressed by the adoption of protocols and the adaptation of existing procurement structures and liability arrangements. BIM is primarily a tool for collaboration that should not be so very different from what exists. New legal solutions will probably be required if the full potential of BIM is realised in terms of being the vehicle by which integrated project delivery is achieved.

21.8 Looking further ahead

Law, as a subject, is reactive rather than proactive in terms of providing solutions to existing issues. This section seeks to buck the trend and considers briefly some of the newest initiatives in the industry, which will give construction lawyers food for thought in the future.

The Heraclitus quote given at the start of this chapter stated that 'nothing endures except change'. Another inalienable truth in this area is that risks must be taken in order to improve. Innovation can be seen as being either disruptive or incremental. Incremental progress is the easiest to explain in terms of simply setting out to do something better than it was done yesterday. Disruptive technology starts off worse than what already exists but eventually outperforms what exists. The typical approach to disruptive technology is flat reaction to the change for five years and then exponential growth. BIM is described as disruptive technology in the Saxon report and could well be set to go through the massive growth period typical of its technology type.

BIM is not the only disruptive technology out in the industry at the present time. One major benefit of BIM is the feedback that the building supplies. Taking this to the next level involves intelligent buildings and effectively installing a nervous system into the building. This provides facility management and performance data but can also help you to design better next time. For example, conventional wisdom may have it that x columns are needed to support a space of y dimensions. Feedback from the as-built materials about the loads they are bearing and their capability may permit a recalculation and the use of fewer columns in the next iteration.

Other examples of innovation and the legal challenges they pose include the following.

- *Augmented and virtual reality* – Augmented reality uses digital information which is overlaid on the real world through design models (BIM). Virtual reality is immersive and allows the user to put on a headset and walk through the building before it is built. This can be extremely useful for the layout of such things as operating theatres in hospitals where the surgeon can pre-agree the layout of the facility. The ownership of designs stored in virtual realities remains an issue for those concerned about the loss of their intellectual property rights.
- *3D printing* – This is a reality in many industries and construction is no exemption. The ability to make a building component on-site for immediate installation is now possible. Neither do 3D printers use solely plastic as the printing medium. Airbus industries apparently print in titanium and carbon fibre. Digital fabrication has been used to make a car and even whole buildings are capable of being printed in 3D. The challenge for the supply community in protecting its products and the warranties that would accompany any 3D-printed product would need to be addressed.
- *Robotic construction* – This is, similarly, no longer in the realm of science

fiction. Driverless trucks operate in mining operations where there is little, if any, human involvement in an autonomously running plant. Once the BIM data is evolved to such an extent that the robots merely need to be programmed to deliver the model, the legal challenges around these issues would mean that negligence-type liability would need to be replaced with the higher fitness for purpose duty.

- *Augmented workers* – Using exoskeletons for enhanced performance, augmented workers are the subject of interest in the construction sector. This would allow the productivity rates of a worker to improve and lead to quicker and cheaper construction. The liability for any injuries caused by the technology would require insurance cover.
- *Smart clothing and hard hats* – The workers' kit can plot stress and heart rate and lead to intervention where rates reach dangerous levels. Location, fatigue levels, humidity and temperature can also be monitored. One positive result that is achieved when the results are shared with the workers is an improvement in fitness. Legal issues arising out of this technology include accessing personal data which is protected by legislation in the form of the Data Protection Act 1998.
- *Flying drones* – These can now be used to measure quantities in a labour-saving breakthrough for the industry. Groundwork measurement of extracted material may currently take weeks to measure using conventional methods whereas a drone can perform this task in a matter of minutes. This challenge to the role of existing professionals is writ large in the techno-logical advances. The professional roles will need to evolve to meet the new landscape of construction or be consigned to history. The recalibration of what is in the province for each profession to supply will necessitate the governing bodies co-operating in a way not hitherto experienced.
- *Organic construction materials* – The most 'out there' idea currently being considered is the ability to grow buildings from organic material. Stem cell bacteria have been used successfully to grow replacement body parts, and a logical application for this would be to grow building material. The crossover here is into the fields of biomimicry, material science and nanotechnology in construction. The environmental credentials for such a breakthrough would be considerable. The legal issues around liability and the performance of any such material would need to be addressed.

In conclusion of this point, it is probably accurate to say that not all of the ideas set out above will translate into usable everyday practice. This misses the point, which is that some of them will work and revolutionise the way things are done. New generations of constructors will either engage with them or get left behind. Change does not happen overnight but change is constant. The construction industry is guilty of faddism and the desire to come up with new shiny concepts before the previous ones have been properly embedded. Where technology produces measurable real benefits and savings on a personal, societal and global level then it should be embraced.

21.9 Conclusion

This chapter has examined the government-sponsored initiatives in construction and taken a glimpse at possible future developments. Predicting future trends is not an exact art, and it will be a considerable length of time before legal solutions are needed to regulate what will be, by then, established practice. The frustration at the inability of government reports to make any significant impact on the private sector of the industry can be detected from the following quote from John Egan:

> I am disappointed that the levels of improvement we asked for have not been achieved but pleased we are at least making progress. Right from the start I said most government reports end up in the waste bin so the fact *Rethinking Construction* had any impact at all is an achievement.[24]

Legal issues remain in relation to those initiatives that are established and are in the course of becoming established. Multi-party contracts present issues around the ability of each participant to bring and defend legal actions where liability needs to be established. The protection of intellectual property rights and business intelligence is a key concern for the supply side of the industry moving forwards. Underpinning fundamentals of further integration such as integrated project insurance need to become more widely available for further progress to be made. Legal problems with new approaches are rarely the determinative issue in the success of an undertaking.

The next chapter approaches the question of encouraging change in the industry through a different aspect of legal means. Good faith clauses have already been mentioned in the context of promoting a more collaborative approach to construction. Ethical improvement is another area where construction law has had a bearing.

21.10 Further reading

Mosey, D. (2009) *Early Contractor Involvement in Building Procurement: Contracts, Partnering and Project Management*, Chichester: Wiley-Blackwell.
Bennett, J. and Baird, A. (2001) *NEC and Partnering: The Guide to Building Winning Teams*, London: Thomas Telford.

Notes

1 Simon, E. (1944) *The Placing and Management of Building Contracts: Report of the Central Council for Works and Buildings*, London: HMSO.
2 Emmerson, H. C. (1962) *Survey of Problems Before the Construction Industries: Report Prepared for the Minister of Works*, London: HMSO.
3 Banwell, H. (1964) *The Placing and Management of Contracts for Building and Civil Engineering Work*, London: HMSO.
4 Latham, M. (1994) *Constructing the Team: Final Report of the Government/Industry/Review*

of Procurement and Contractual Arrangements in the UK Construction Industry, London: Department of the Environment.

5   Lord, W. E. (2008) Embracing a modern contract progression since Latham? In *COBRA 2008: The Construction and Building Research Conference of the Royal Institution of Chartered Surveyors*, Dublin, 4–5 September. Available at: https://dspace.lboro.ac.uk/dspace-jspui/bitstream/2134/5959/1/lordCOBRA%202008.pdf.

6   Egan, J. (1998) *Rethinking Construction*, London: Department of the Environment, Transport and the Regions.

7   Wolstenholme, A. (2009) *Never Waste a Good Crisis: A Review of Progress Since Rethinking Construction and Thoughts for Our Future*, London: Constructing Excellence.

8   The Wolstenholme Report, p. 8.

9   The graphic was reproduced from Constructing Excellence (2009) *Survival Guide – Working Out of an Industry Downturn*, London: Constructing Excellence.

10 Lord, W. E. (2008) Embracing a modern contract.

11 PPC2000 Clause 4.2.

12 [2005] TCLR 6.

13 See Section 8.9.

14 *Rayack Construction Ltd v Lampeter Meat Co Ltd* (1979) 12 BLR 30.

15 Construction Industry Board (1997) *Partnering in the Team: A Report by the Working Group 12 of the Construction Industry Group*, London: Thomas Telford, p.1.

16 NBS National Construction Contracts and Law Survey 2013.

17 Saxon, R. G. (2013) *Growth through BIM*, London: Construction Industry Council, p. 68.

18 Cabinet Office (2011) *Government Construction Strategy*, London: Cabinet Office.

19 Cabinet Office (2014) *New Models of Construction Procurement*, London: HMSO.

20 Klein, R. (2015) An Unfair Burden, *Building* [online] 14 September. Available at: www.building.co.uk/an-unfair-burden/5077550.article.

21 *Government Construction Strategy*, p.14.

22 Snook, K. (n.d.) Drawing is Dead – Long Live Modelling, *Construction Project Information Committee* [online]. Available at: www.cpic.org.uk/publications/drawing-is-dead/.

23 McGraw-Hill Construction (2012) *The Business Value of BIM in Europe*, SmartMarket Report, Bedford, MA: McGraw-Hill.

24 McMeeken, R. (2008) Egan 10 Years On, *Building* [online] 8 May. Available at: www.building.co.uk/egan-10-years-on/3113047.article.

22 Ethical improvement and the duty of good faith[1]

22.1 Introduction

One reason identified in the previous chapter for the slow uptake of partnering and the associated agenda for change was the desire to stick with the existing situation. The challenge for the policymakers is to move the industry from its existing practices towards the new preferred routes to project success. It is common in this situation to discuss push and pull factors. The extent to which law can support these approaches is discussed here in the context of two push factors. These are the moves towards ethical improvement and the re-establishment of the duty of good faith. A fact frequently overlooked by those who advocate maintaining existing approaches to construction industry practice is the evidence of poor ethical behaviour in some facets of the industry.

This chapter starts by addressing the question of whether it is realistic to expect ethical improvement in an industry. On the assumption that it is, the chapter considers the legal interventions into the sometimes nefarious practices that take place and the results of those interventions. The attention is then turned to consider the duty of good faith, which, despite its inclusion in many contracts, has a far from certain legal meaning. The link between these subjects is that mandating good ethical approaches may 'lead the horse to water'. Put another way, if ethical improvement is required then one way to achieve it is to give more guidance on how the stakeholders of the construction industry should behave.

22.2 The ethics of an industry

To expect ethical values to be prominent in any industry or within a particular profession is to form an optimistic view about the nature of things. There is no provision in definitions of either industry or profession for ethical or high moral standards in the work being undertaken. Ethical behaviour is not necessarily part and parcel of professional conduct.[2]

Indeed, across different industries, professional ethics regularly receive criticism. Prominent news stories about lack of professionalism in the banking sector and more recently in athletics are commonplace. Anticipation of high ethical standards in any industry therefore is an expectation formed in the face of the available evidence. The construction industry is much more diverse and much nearer to the collective breadline than those involved than any banking venture. The stakeholders no doubt appreciate the hard work they have enlisted themselves to provide in the fulfilment of the industry ambitions. Have they signed up for ethical standards as well? Is it presumptuous for those distant from the day-to-day dealings to prescribe that ethical stances should be taken against certain engrained practices? The answer is perhaps; but then to attempt to do nothing is tantamount to allowing the bad practices to spread and to go unchallenged. The least that can be hoped for is that if stakeholders are alerted to the choices they have to 'do the right thing' in various situations, they may not simply perpetuate poor practice. The best that can be hoped for is to stop trillions of dollars being wasted because of criminal activity.

Further, if the only point to industry is to make as much money as possible then the outlook is indeed bleak. An unchecked rush towards maximum production takes a heavy toll on the planet's resources and creates some unpleasant living conditions. China is experiencing what was common-place in a post-Industrial Revolution UK. A sustainable approach, whether involving eco-credentials or not, is the start of an ethical approach to business. Sustainability involves the careful use of resources with a view to the needs of future generations.

The presumption therefore is that there needs to be an ethical dimension to the work undertaken by the construction industry. A career spent in an industry making money for other people is likely to be unrewarding without this element of ethical standards. Ethics are relevant at the individual and organisational level. Ethics is defined in the *Oxford English Dictionary* as the moral principles by which a person is guided. In the context of professional behaviour, the same source refers to the duties owed by professionals to the public, to each other, and to themselves in relation to the exercise of their profession. In the construction context, ethical behaviour is measured by the degree of trustworthiness and integrity with which companies and individuals conduct business.

There has been debate as to whether some construction professionals are more ethical than others. The proper analysis would appear to be that the closer a professional is to the harsh realities of business, as indicated by their position in the supply chain, the harder it is to maintain ethical standards. A harassed subcontractor constantly on the edge of insolvency will clearly have more pressing concerns than the performance of his components against the energy plan. Nevertheless, ethics has a role to play in general business practice. In recent times, there appears to be a greater consensus on this issue. It is now commonly recognised that the general concepts of ethics are appli-cable to business on the grounds that business exists not solely to suit certain individuals; it also serves society and, in addition, meets collective and social needs. In other words, the altruistic spirit of a genuine profession cannot be achieved without an ethics component.[2]

As mentioned, the most recent expression of this altruistic/social agenda has been evident in the promotion and regulation of sustainability and environmental aspects of the construction industry's activity. There is more to professionalism in the construction industry than self-interest. The extent to which ethical behaviour is promoted by professional codes is moot. The adoption of such codes and commitments to corporate social responsibility may be seen as 'window dressing' and self-serving as simply public relations efforts.[2]

Unethical behaviour is widespread in the construction industry and has been demonstrated as being damaging to the reputation of all those involved. This chapter seeks to expose the depth of the issues and the legal attempts being made to improve things. The next sections address the reasons for and the scale of the problem.

22.3 Reasons for the situation

The predisposition of the construction industry to unethical behaviour can be attributed in part to the following factors.

- *Construction products are simple and not differentiated* – The same basic materials can be provided by any number of suppliers. This creates a situation where the suppliers may be more aware of the need to offer additional benefits in order to secure orders. Equally, the buyer is aware of their purchasing power and the unethical may seek to exploit this.
- *Lowest price wins* – In markets where there is little differentiation of materials and suppliers, cost becomes the key factor and there has to be a good reason why the lowest price is not taken. This knowledge means that efforts can be made to manipulate the market, focusing on the pricing element.
- *Business is highly cyclical* – The propensity to behave unethically is also encouraged by the cyclical nature of business. The construction industry has had a repetition of the boom and bust approach throughout the twentieth century. This lack of a steady flow of work and likelihood of a reversal of fortunes can lead to focus on short-term pragmatism and maximising of business opportunities, whether legitimately encountered or otherwise.
- *Claimsmanship* – Another factor in the propensity for unethical behaviour is the all too common occurrence of claims. The bid price is frequently a long way removed from the actual cost of construction. The reasons for this can lie in poor planning and late design changes. The claims culture within the industry has been addressed in Chapter 15. The defence to the allegation put forward by the contractors would be that the ultra-competitive tendering procedures and focus on lowest cost to the exclusion of all other factors leave them no choice but to seek to make a margin by bringing claims.
- *The prevalence of subcontracting* – The industry is made up of subcontractors. The presumption is that the bids put forward by the different contractors

rely on completely separate sets of subcontractors. This is frequently not the case. Key subcontractors can be approached by more than one contractor, making it relatively straightforward to identify the winning contractor. Supply chains can then migrate to the winning contractor. There is nothing wrong with this on its own, but it has the potential to lead to collusion.

- *Partnering encourages collaboration* – Another unforeseen potential issue resulting from the partnering agenda is the crossover from collaboration into collusion. If no genuine competition exists between framework contractors then there is the chance for illegitimate cartel activity aimed at carving up the market for the benefit of the contractors rather than the employers.

22.4 The scale of the problem

Several factors in the fabric of construction projects lend themselves to illegal bidding activity. The fragmentation and one-off nature of the projects together with the lack of accountability of public and private bodies are of major concern to governments worldwide. The differential in size between the smallest and largest organisations is similarly noteworthy. The firms involved are overwhelmingly small local concerns of fewer than 20 employees. At the other extreme, the competition between the large firms is oligopolistic. This can manifest itself in cartels whereby the small number of big firms have the opportunity to determine which company wins which project by manipulating and co-ordinating the bidding.

The evidence points to unethical behaviour, and a number of studies have made some depressing findings in relation to poor practice.

- A Chartered Institute of Building study of practice in the UK[3] discovered that of the 700 professionals surveyed, nearly half perceived that corruption in the industry was either fairly or extremely common, with one in three indicating that they had been offered a bribe or incentive on at least one occasion.
- The approach taken by a South African[4] study was to ask respondents about the incidence of unethical behaviour, particularly collusive tendering. The results were shocking but not untypical. The responses indicated that 100 per cent of the construction managers questioned had either witnessed or experienced collusive tendering, with 88 per cent of quantity surveyors having been in the same position. Over half of the architects questioned had also seen such collusion. Overall, the figures amount to 79 per cent of all respondents being involved with unethical behaviour. Neither was the incidence of unethical behaviour reducing. When asked whether the problem had increased in the last ten years, 32 per cent said yes, 64 per cent said it had stayed the same, and only 4 per cent felt it had decreased. In their interpretation of the results, the authors identify the severe depression in the South African construction industry during the period and suggest that

local contractors may have formed groups to spread the work in an attempt to see off financial disaster.[2]

- An American study[5] collected thoughts on the ethical state of the industry from 270 architects, engineers, construction managers, general contractors and subcontractors. When asked if they had experienced, encountered or observed construction industry-related acts or transactions that they would consider unethical in the past year, 84 per cent answered yes; in addition, 34 per cent said they had experienced unethical acts many times and 61 per cent said that the construction industry was 'tainted' by unethical acts.[2]

These studies are only likely to have scratched the surface of the true position. The reports available consider the issues in First World economies where ethical codes are promoted and commitment made to improvements. The position in developing countries is unregulated and potentially provides much more fertile ground for transgression. The work of Transparency International is important in this regard. Their studies indicate that construction is the most corrupt sector in the world, with 2.6 trillion US dollars being wasted on these practices. The predisposition of construction to illegal activity results from its complexity and fragmentation. Opportunities for bribes and illicit behaviour arise out of such interfaces as planning permission, the overstatement of interim accounts and manipulation of payment applications. Other illegal activity involves the hiring of illegal labour and claiming full rates, the theft of materials and tax evasion.

This chapter concentrates its attention on the unethical and illegal practices arising from the tendering process discussed in Chapter 8. This is the area where legal intervention, particularly in the UK, has been focused.

22.5 Unethical behaviours in relation to tendering

Examples of unethical and illegal behaviour in construction practice include the following.

- *Simple price-fixing* – This involves the contractors tendering for work agreeing not to charge less than a certain amount in respect of their services. The prices submitted ought to be arrived at by each contractor individually pricing the work and adding their profit and overhead element to it. Price-fixing starts with a figure that the contractors will not go below and they then stick to this.
- *Bid rotation and bid suppression* – These are two examples of collusion between the tendering contractors. Bid rotation involves the contractors taking turns to win work either in rotation or by reference to another factor such as geographical location. Bid suppression covers a situation where certain contractors agree not to bid where they know or suspect that certain other contractors are also bidding. This knowledge may be made available

through the medium of the subcontractors, who may have been approached by both contractors.

- *Cover pricing* – This form of unethical behaviour previously divided opinion as to the rights and wrongs of the practice. A cover price is a tender bid put in by a contractor which is priced deliberately high so that the contractor does not win the work. The reasons why a contractor would deliberately tender high are not initially all that clear. However, it is quite common for the employer to have a tender list of, say, six to eight contractors and to be subject to requirement that all bids are received. contractors not wishing to bid quite often find themselves in the position where they could be excluded from future tender competitions that they do wish to be involved in – hence the practice of submitting a high bid in order to stay on the list but not to actually win the work. Anecdotally, it is not unheard of to find that the cover pricing contractor actually wins the contract. Presumably the work flow of the contractor is internally managed in such situation as to deliver the project in accordance with the employer's expectations.

- *Reverse auctions* – This unethical practice originates from the employers rather than the contractors. This is where the lowest bidder, rather than the highest bidder, is successful. This may appear to be no different from a traditional lowest price competition until the open nature of the competition is factored in. Rather than each party submitting a single bid, the employer gives the tenderers access to their electronic tenders, which are ranked in accordance with their value. The tenderers have the opportunity to resubmit a lower bid at any time before the deadline passes. Many contractors come under enormous pressure to try and win the work, often at prices below the actual costs they will themselves incur on the project. The winning tenderer is the one who gives the lowest price at the point at which the time expires. This form of tendering was used in the UK in the supermarket sector but has recently fallen out of fashion. The process remains prevalent in the USA with clients viewing reverse auctions as important and valuable in the procurement process. The contractors disagree and see them as unethical. The employer would counter that the bidding rules are clear and upfront for all parties and discount the argument.

Unethical practices range from the blatantly dishonest to the marginally sharp in terms of business practices. Drawing the line as to what is acceptable and what is not is one of the purposes of the legal system. The regulation of this spectrum of activity is considered in the following sections. The first discussion is around the role of ethical codes in alleviating the situation.

22.6 Ethical codes

The cause of ethical failure in an organisation can often be traced to its organisational culture and failure on the part of the leadership to actively promote ethical practices. Whilst personal ethics are a reflection of beliefs, values, personality and background, any propensity a person may have towards ethical conduct is strongly influenced by the value systems reflected by their employing organisation. This often results in a personal sense of what is right and wrong becoming buried amongst an organisation's non-observance of professional ethics.

Surveys have been carried out into the ethical state of health of different construction industries. An Australian study[6] demonstrates the popularity of the use of ethics codes. Of the 31 people surveyed, most subscribed to a professional code of ethics (90 per cent) and many (45 per cent) had an ethical code of conduct in their employing organisations. Despite this high incidence, for half the respondents, the subject of ethics never cropped up in business meetings.

The South African study mentioned above was based on the work done in Australia that surveyed 63 professionals, a large proportion of which were governed by ethical codes. A new suite of professional Acts promulgated in 2000 in South Africa had boosted the profile of ethics. An important consequence of the new legislation was the official recognition of construction management as a professional discipline.

The emerging position is of an industry in a rut with a continuing and recurring acceptance of poor practices. The ethical codes that exist are not achieving their stated aim of protecting the public from unethical behaviour. All professionals in the construction industry subscribe to their institutions' professional ethics. Ethical codes are rarely specific about what amounts to poor ethical practices. Codes are usually written in aspirational language with nebulous concepts such as 'do nothing to impinge on the good name of the profession'. As such, ethics on their own are not going to lead to improvement as they do not go far enough and are not enforced rigorously enough, if at all. The role of law is to set the standards and ensure that messages are communicated as to what the standards are and what happens in the event of non-observance.

22.7 Legal intervention

Bid rigging and bid rotation are obviously illegal. Both involve processes whereby the bidders collaborate (in the bad sense of the word) to predetermine who will win the bid. Similarly, procedures such as 'loser's fees' and bid suppression are also examples of illicit activity.

The UK's construction industry was caught by surprise to find out that the practice of cover pricing – something which was very close to being seen as acceptable and actually mentioned in some construction textbooks as being standard practice – was in fact illegal. Cover pricing involves a tenderer putting in a higher than normal price in order not to win the work being tendered for. The reasons for this usually involve wanting to remain on a tender list (this being of value to the contractor) whilst being too busy to undertake this particular job or not wanting it for some other reason such as location or specialism required.

According to the law of offer and acceptance, the activity of putting in a high bid – provided there was no collusion with any other tenderer – seemed to be on the right side of the line. The employer is not obliged to accept the price if they do not approve of it. This is not the position at law. Chapter I of the Competition Act 1998 prohibits agreements or concerted practices which have the object or effect of preventing, restricting or distorting competition in the UK. The Office of Fair Trading (OFT) can impose a maximum penalty of 10 per cent of worldwide turnover of the relevant undertaking in its last business year for infringement of the prohibition. Cover pricing falls within this definition as the cover price has artificially inflated the price that would otherwise have been paid, based on an average of all the prices given.

This misunderstanding of the legal position was the backdrop to the OFT investigation in 2007 when 103 firms in the UK were fined a total of £129.5 million following a four-year operation. A total of 57 raids were carried out and 37 leniency applications received. The investigation was curtailed as the agency was discovering more offences than it could handle. On appeal, these fines were reduced and it appeared to be accepted that the industry was genuinely ignorant of the illegality of these actions. Follow-up research has established that the message has been put across and cover pricing as a practice has reduced dramatically.

The investigation had uncovered widespread cover pricing on public contracts. More worrying was the incidence of more than one cover price on the same project. The wisdom of prosecuting the construction industry at a time when it was suffering from a recession was questioned. The fines levied were subsequently reduced on appeal with legal costs being awarded against the OFT. Criticisms were levied at the organisation for being overzealous. Such observations sought to gloss over the fact of endemic and systematic illegal activity that required correction. The extent to which correction has been achieved was the subject of a review that established that cover pricing was now frowned upon more severely by the stakeholders in the industry. Domestically, the OFT has not ruled out further crackdowns on the construction industry with the threat of the imposition of custodial prison sentences for reoffenders. Further legislative measures have also been introduced in the form of the Bribery Act 2010 and the encouragement of whistle-blowing.

Enforcement of anti-collusion legislation is relatively common in Europe and the USA. Enforcement agencies look into irregular bidding patterns and can detect unethical and illegal processes. Patterns such as qualified bidders never bidding against each other and subcontractors from one unsuccessful bid being transferred to the winning bidder can indicate something untoward. An investigation in the USA revealed that the contractors operated a strict geographical cartel with bids being co-ordinated so that each contractor had a territory in which work was won. Each project was represented by a coloured pin denoting the winning contractor in a multistate area. The concentration of coloured pins in certain areas was unmistakable.

Enforcement activity is less common in international markets despite the higher incidence of corruption. Improving the situation internationally is a difficult challenge in the absence of political will to effect change. As previously mentioned, it is also presumptuous to impose Western values in other jurisdictions where cultural and legal attitudes are different. This returns the debate to the ethical and moral compass by which one's actions are governed and the interconnection with the legal measures in force.

The unethical practice studies highlight the key issue, which is people's different understanding of what the rules are and what is right/wrong in any given situation. Ethics can be viewed as being more personal than laws and involving an element of choice. The element of choice is removed in professional ethics and the lines between what is acceptable and what is not are clearer. Ethical codes are not always as clear as they might be and are often expressed in general terms relating to societal and professional values. Where ethical codes are not enough, it is the role of lawmakers to set out clearly what amounts to a transgression of ethics and law. It also falls to enforcement agencies to remind participants of the need to exhibit compliant behaviour.

The adoption of the new approach to partnering as set out in the *Government Construction Strategy*[7] should also remove the need to behave opportunistically in relation to project finances. A smarter, better informed client is a massive disincentive for the supply side to 'chance their arm' in this regard. Another way to encourage a more ethical approach in construction is to set out the standard required in the building contracts themselves. This involves a consideration of the recent moves to include good faith clauses as a matter of course in contracts. However, adding the clauses to the contracts is only part of the issue. The courts must be willing to enforce the terms included, and this has not been in any way as straightforward as it might at first seem. It is to this area that this chapter now turns.

22.8 Good faith clauses generally

This chapter, and the book itself, concludes with an examination of the law around good faith. It was noted at the outset of this work that different approaches to writing and examining construction law would be used in this narrative. The approaches have taken in a discussion around first principles of law and construction administration commentaries. The processes of construction law have been described and analysed in the form of claims management and dispute resolution. The remaining part of the work returns to a black letter law investigation of an emerging area of law.

Partnering promotes a co-operative approach to contract management with a view to improving performance and reducing disputes. The relationship between a contractor and a client in a partnering contract contains firm elements of trust and reliance. In so far as partnering is delivered through the medium of contracts, those contracts more often than not contain an obligation that the parties act in good faith to facilitate delivery of those aims.

Partnering contracts pose a problem for contract advisors containing as they do 'hard' and 'soft' obligations. Whilst all conditions of contract are equal, some – to misquote George Orwell – are more equal than others. Clients can be advised and terms drafted stipulating hard obligations such as payment and quality standards. But what of the soft obligations and, in particular, the duty of good faith; what are to be made of them? As one commentator put it: 'We in England find it difficult to adopt a general concept of good faith ... we do not know quite what it means'.[8]

The resulting situation is that 'soft' obligations are often overlooked and not given any particular importance. This sentiment was picked up by a report[9] expressing the consensus of construction lawyers as being that duties of good faith are not likely to be newly recognised in law by reason of their introduction into partnering contracts. This consensus of opinion invites the questions of whether this is what the users of construction contracts want and whether an opportunity is being missed for encouragement towards ethical improvement. Parties having taken the trouble of entering into a partnering contract may feel disappointed to learn that their voluntarily assumed mutual obligations are not enforceable. The concretising of the duty of good faith by judiciary and/or Parliament to deliver what the parties have chosen for themselves might achieve progress.

The most popular forms of contract are those that make no mention of partnering obligations in their standard form. The dominance of the Joint Contracts Tribunal (JCT) lump sum and design and build forms remains intact although under threat. However, the growing trend is to use contracts that move away from formal legal 'black letter' contracts to contracts fulfilling a different role which includes viewing the contract as a management tool and a stimulus for collaboration. The challenge for these newer contract forms is to capture this new role whilst providing sufficient contractual certainty in the event that disputes arise.

The link between contracts, partnering and good faith was initially made by organisations such as Associated General Contractors of America, making statements such as: 'Partnering is recognition that every contract includes an implied covenant of good faith'.[10] These connections are relatively straight-forward in the USA where the legal system recognises the duty of good faith in contracting. The principles of partnering are congruent with the doctrine of trust, open communication, shared objectives and keeping disputes to a minimum. Making the connections in the English context is more challenging given the absence of the general duty of good faith. In its absence, it is the partnering contracts themselves that fill the gap.

Partnering contracts have become significantly more sophisticated in terms of the wording of partnering obligations and the conduct expected. There are variations on the exact imposition of the duty to act in good faith. A distinction can be drawn between those which are intended to regulate the parties' behaviour through the contractual terms and conditions (binding) and those which place a non-contractual partnering framework over the top of another contract (non-binding). The latter have been described as seeking to influence rather than to mandate certain behaviour. The parties to the JCT non-binding partnering charter agree to 'act in good faith; in an open and trusting manner, in a co-operative way; in a way to avoid disputes by adopting a "no blame culture"; and valuing the skills and respecting the responsibilities of each other'.[11]

The binding multi-party PPC2000 requires that the parties 'agree to work together and individually in the spirit of trust, fairness and mutual co-operation'. The NEC X12 Partnering Option calls the parties 'partners' and requires that partnering team members shall 'work together to achieve each other's objectives'.

The latest contract to enter the fray is the JCT Be Collaborative Constructing Excellence form. The contract goes further than the other partnering contracts in introducing an overriding principle that includes a duty of good faith, and it stipulates that this principle takes precedence over all other terms.

This contract completes the transition of good faith-type provisions from being somewhere on the undercard of contractual terms to being the main event. A significant proportion of the standard forms of contract now available to the construction industry expressly impose an increasingly onerous duty on the parties to act in good faith. Regrettably, the contract has not caught on into general usage.

The attraction for contract draftsmen to use the phrase 'the parties owe each other a duty of good faith' is understandable. The concept of good faith has great normative appeal. It is the aspiration of every mature legal system to be able to do justice and do it according to law. The duty of good faith is a means of delivery. The phrase resonates with the reader who has an instinctive grasp of what it is the contract is trying to do. This resonance is due, in part, to the long history and high esteem in which the duty is steeped.

22.9 Background to good faith

Good faith has an ancient philosophical lineage and is referred to in the writings of Aristotle and Aquinas. They were concerned with the problems of buying/selling and faced the dilemma of how to achieve fairness while not stifling enterprise in commerce. This dilemma is still an issue today and its successful resolution is a major challenge for those seeking to (re)establish a duty of good faith.

The ancient concept of good faith in a revived form went around Europe, England and the USA like wildfire at the end of the eighteenth century. Lord Mansfield, in 1766, described the principle of good faith as the governing principle applicable to all contracts and dealings.

The duty of good faith subsequently fell into disuse in England in favour of encroaching statute law and the emphasis on the promotion of trade.

Emphasis shifted instead onto contractual certainty in contracting. Contractual certainty has remained the cornerstone of standard form construction contracts since their inception at the start of the twentieth century. Procurement and contracting in the twenty-first century, however, is different. The role of the contract is changing and the re-emergence of the duty of good faith is an important element in this development. The advantages of recognising the legal enforceability of the duty have been presented as:

- safeguarding the expectations of contracting parties by respecting and promoting the spirit of their agreement instead of insisting upon the observance of the literal wording of the contract;
- regulating self-interested dealings;
- reducing costs and promoting economic efficiency;
- filling unforeseen contractual gaps;
- providing a sound theoretical basis to unite what would otherwise appear to be merely a series of disparate rights and obligations.

The support for introducing the duty of good faith amongst industry commentators has not to date been overwhelming. Academic studies in this area tend towards mild encouragement for the judiciary or Parliament to take action and introduce a general duty. Making the case for the imposition of a general duty of good faith is as challenging as attempting a definition. Despite its beguiling simplicity, it has proved to be an elusive term. The attempts to define good faith at best replace it with equally vague and nebulous terms. The danger, as one commentator put it, is that any definition would 'either spiral into the Charybdis of vacuous generality or collide with the Scylla of restrictive specificity'.[12]

The difficulty of defining 'good faith' is not necessarily a problem for partnering contracts which tend to evoke the spirit rather than the letter of the law. However, progress has been made in defining the term, particularly by the Australian judiciary. The parallels here are striking – a common law jurisdiction grappling with the issue of how best to 'concretise' the duty of good faith. The Australian judge Paul Finn made the following useful contribution towards definition in the common law tradition:

> Good faith occupies the middle ground between the principle of unconscionability and fiduciary obligations. Good faith, while permitting a party to act self-interestedly, nonetheless qualifies this by positively requiring that party, in his decisions and action, to have regard to the legitimate interests therein of the other.[8]

Thus far, the English courts have denied themselves the opportunity to engage in this shaping of the meaning of good faith in the modern construction context despite its historical relevance and its resonance with the public and even in light of other recent stimuli to its introduction. English law made a choice to promote trade through contractual certainty rather than through widely drawn concepts. In Europe, the duty of good faith has flourished to the extent that its existence or otherwise in contract law is one of the major divisions between the civilian and common law systems. The great continental civil codes all contain some explicit provision to the effect that contracts must be performed and interpreted in accordance with the requirements of good faith; for example, art.1134 of the French Civil Code and s.242 of the German Code. In Germany, the experience has been that the articulation of a general principle has enabled the identification and solution of problems that the existing rules seem unable to reach. Through the duty of good faith, the German court has developed its doctrines without incurring the reproach of pure judicial decision lawmaking.

It is unsurprising, given the establishment of the good faith doctrine into continental legal systems, to discover that the duty is enshrined within European law. For example, the Unfair Terms in Consumer Contracts Directive 1993 may strike down consumer contracts if they are contrary to the requirements of good faith. The Commercial Agents (Council Directive) Regulations 1986 also makes reference to good faith.

Moves towards the harmonisation of European contract law by the European Contract Commission stopped short of outright commitment to the duty of good faith but did state that regard is to be given to the observance of good faith in international trade.

Neither is good faith a concept unknown to English law. The obvious example is in insurance contracts, which are subject to a duty of utmost good faith owed by the assured to disclose material facts and refrain from making untrue statements while negotiating the contract. The duty of good faith is also apparent in areas of law where there is a special relationship, such as family arrangements and partnerships.

A pattern is discernible towards the re-emergence of the duty of good faith in English law. Despite this encroachment (or possibly because of it), suspicion and hostility abound. In the words of one commentator, the duty of good faith is a 'vague concept of fairness which makes judicial decisions unpredictable'.[8]

Another argument against the imposition of a general duty of good faith is the preference given to *ad hoc* solutions in response to demonstrated problems of unfairness. In other words, good faith outcomes are already being achieved through other means. Examples of these outcomes have been given as the contractor's duty to progress the works regularly and diligently and the employer's duty not to obstruct and to co-operate. However, *ad hoc* solutions can lead to unsatisfactory results. Contract draftsmen have given the judiciary a unique opportunity to create new law based around the key concept of good faith

The grounds for the seeming hostility (with one notable exception) of the judiciary to the concept of good faith have already been stated – suspicion of broad concepts. The approach is, to paraphrase Lord Bingham in *Interfoto Picture Library* v *Stiletto Visual Programme Ltd*,[13] to avoid any commitment to overriding principle in favour of piecemeal solutions in response to demonstrated problems of unfairness.

The judgement of Lord Ackner, in the case of *Walford* v *Miles*,[14] sums up the prevailing sentiment: *the duty to carry on negotiations in good faith is inherently repugnant to the adversarial position of the parties involved. ... how is the court to police such an 'agreement'?*

Any initiative towards the introduction of a general duty of good faith has not found support in the technology and construction court. Judge Lloyd, in the case of *Francois Abballe (t/a GFA)* v *Alstom UK Ltd*,[15] said that:

> *The proposition that 'good faith' may be used as a fall back device tellingly shows why it is wrong but tempting to consider with the advantage of hindsight whether a term should be implied. ... I do not consider I should be a hero and permit the Claimant to advance this term.*

The door seemed to be more firmly closed on the introduction of a general duty of good faith by Judge Seymour: *the development of the law ... would, it seems to me, be fraught with difficulty. ... I should not be prepared to venture into these treacherous waters.*[16]

The judiciary therefore remain concerned about the sea monsters and whirlpools that may lie in store for them. This echoes a point made earlier[17] about law being more comfortable regulating what society should not do rather than what it should do. It is easier to be specific in the negative. Nevertheless, the moves towards concretising the duty of good faith appear to grow apace.

22.10 Good faith clauses in construction cases

The most renowned case where a specific duty to act collaboratively has been considered by the judiciary is that of *Birse Construction Ltd* v *St David Ltd*.[18] The relevant considerations of the case are that it featured a non-binding partnering charter and the judge specifically highlighted that the parties had entered into a partnering arrangement.

Judge Lloyd recognised that the terms of the partnering charter were important in providing the standards of conduct of the parties. Although such terms may not have been otherwise legally binding, the charter was taken seriously as a declaration of assurance. In short, the parties were not allowed to interpret their relationship in a manner that would have been inconsistent with their stated intention to deal with each other collaboratively.

It is possible to discern support from this judgement for the parties' expressed desire to operate in good faith in their dealings with one another. This support means that the expectations of contract users are likely to be met. Increasing numbers of contract draftsmen have been bold enough to include good faith provisions in their contracts. The contracts have been welcomed by their users. If they find themselves getting into difficulties then the users have a reasonable expectation to be bound by their promises to one another. The challenge for the judiciary is to decide on the appropriate level of support to be given to the more prescriptive and onerous terms of contract now employed in the latest construction contracts.

In the more recent judgement in *Yam Seng Pte* v *International Trade Corporation*,[19] Judge Leggatt implied a number of good faith obligations into the parties' agreement.

In this case, the parties entered into an agreement by which the defendant granted the claimant exclusive rights to distribute Manchester United-branded toiletries in various territories in the Middle East, Asia, Africa and Australasia. The contract that the parties signed up to was rather brief and was negotiated by the parties directly without recourse to lawyers. The parties' relationship soured and the claimant made various allegations about the conduct of the defendant, including late shipment of orders, failure to supply products and failure to adhere to agreed minimum retail prices. Many of the matters complained of by the claimant, perhaps as a result of the brief nature of the written agreement between them, were founded upon an allegation that the defendant had breached an implied obligation to act in accordance with principles of good faith. This submission found favour with Judge Leggatt, who considered the question of whether or not a duty of 'good faith' ought to be implied in this case. He concluded: *the traditional English hostility towards a doctrine of good faith in the performance of contracts, to the extent that it still persists, is misplaced.*

Hot on the heels of *Yam Seng Pte* came the judgement of the Court of Appeal in *Mid Essex Hospital Services NHS Trust* v *Compass Group*.[20] This was another case where the court was required to grapple with the concept of 'good faith'. This time the question was not solely whether a duty of good faith ought to be implied into an agreement, but also what the scope of an express good faith contractual clause ought to be.

The facts of the case were that the respondent was engaged by the appellant to provide catering and cleaning services. The agreement provided for the respondent to meet certain agreed service levels with financial consequences for the respondent in the event that the agreed levels were not met. Following a first instance decision that provided for a broad application of an express contractual good faith provision, on appeal, the appellant contended that the good faith obligation ought to be construed narrowly and ought not be applied to the contractual provisions relating to service level failures. The respondent, relying heavily on *Yam Seng Pte*, contended that the contractual good faith clause should be construed widely and applied to the service-level provisions and/or that a general duty of good faith ought to be implied into the contract in any event.

The Court of Appeal decided that the effect of the contractual good faith provision was merely to require the parties to work together honestly to achieve the effective transmission of information and the full benefit of the respondent's services to the appellant. These were the express stated intentions of the good faith provision as set out in the wording of the clause. Lord Justice

Jackson summarised the position in his judgement by stating: *The obligation to co-operate in good faith is specifically focused upon the two purposes... i) the efficient transmission of information and instructions; [and] ii) enabling the Trust or ... any beneficiary to derive the full benefit of the contract.* The Court of Appeal also decided that there was no need to imply a general obligation of good faith into the contract and therefore declined to do so.

In the event that a contract contains an express good faith obligation, parties should pay attention to the judgement in *Mid Essex Hospital Services NHS Trust* to give some guidance as to how the courts might construe the obligation. Following the judgement of Lord Justice Jackson, the specific parameters of the good faith provision will be key. Parties should also take care when negotiating good faith provisions to ensure that their expectations as to the scope of that obligation are clearly reflected in the drafting.

22.11 The future for good faith clauses

Opinions differ around how a good faith clause should be drafted. If practical utility is the key driver then the clause should provide a few clearly understandable action-guiding principles of conduct. Failure to be specific could lead to the same issue experienced with ethical codes, which is that the resulting wooliness is not helpful. On the other hand, the small print solution of listing every possible potential misconduct on the part of any party is not suitable given the complexity of construction contracts and the move away from voluminous forms. One approach would be to allow the judge/ arbitrator/adjudicator a wide discretion so that they might 'concretise' the duty in line with the principles of conduct as they see fit or in line with experiences in other jurisdictions.

Good faith in negotiations could mean an inquiry into the reasons for breaking off negotiations. Examples of bad faith might include negotiating without serious intention to contract, non-disclosure of known defects, abusing a superior bargaining position, arbitrarily disputing facts and adopting weaselling interpretations of contracts and willingly failing to mitigate your own and other parties' losses and abusing a privilege to terminate contractual arrangements.

The effect of the court taking further steps towards recognising the duty of good faith as a hard obligation has been likened to recognising the general duty of care in negligence or the principle of undue enrichment. As a result, the principle may remain relatively latent or continue to be stated in extremely general terms without doing too much damage to the important virtues of certainty and predictability in the law. The principle could also provide a basis on which existing rules can be criticised and reformed.

The alternative way of introducing a duty of good faith is to set down guidelines in a statute. A statutory obligation to act in good faith was recommended by the Latham report as a measure that would lead to the improvement of the performance of the construction industry. The government of the time chose not to move in this direction. The time may have come to revisit this decision.

22.12 Conclusion

Whether concretising good faith would lead to the ethical improvement sought is difficult to fathom. Individuals intent on conducting unethical business practices are unlikely to be dissuaded by good faith provisions in the contracts governing their arrangements. Nevertheless, a more detailed and clearer understanding of what good faith entails would assist in drawing the line between what it acceptable and what is not. Bad faith, in this sense, would become easier to spot and therefore any further enforcement activity could benefit from any such development.

Good faith can no longer be described as it was in *Walford* v *Miles* as *repugnant to the adversarial position of the parties.* The industry is now characterised by and actively pursuing an agenda not of adversarial relations but of collaboration. As such, the industry would benefit from some clear messages from the judiciary as to the enforceability of their collaborative arrangements. The expression of this underlying principle with its uncluttered simplicity may serve to bring clarity to the dense contractual conditions for which the industry is renowned.

22.13 Further reading

Mirsky, R. and Schaufelberger, J. (2015) *Professional Ethics for the Construction Industry*, Abingdon: Routledge.

Fewings, P. (2009) *Ethics for the Built Environment*, Abingdon: Taylor & Francis.

Forte, A. D. M. (1999) *Good Faith in Contract and Property*, Oxford: Hart Publishing.

Notes

1   Extracts from this chapter first appeared in *Ethics for the Built Environment* (Abingdon: Taylor & Francis) by P. Fewings (2009), pp. 281–92.
2   Mason, J. (2009) Ethics in the Construction Industry: The Prospects for a Single Professional Code, *International Journal of Law in the Built Environment*, 1(3): 194–205.
3   Chartered Institute of Building (2013) *A Report Exploring Corruption in the UK Construction Industry*, Ascot, UK: CIOB.
4   Pearl, R., Bowen, P., Makanjee, N., Akintoye, A. and Evans, K. (2007) Professional Ethics in the South African Construction Industry – A Pilot Study, *Journal of Construction Management and Economics* 25(6): 631–48.
5   Doran, D. (2004) *FMI/CMAA Survey of Construction Industry Ethical Practices*, Raleigh, NC: FMI Corporation.
6   Vee, C. and Skitmore, R. M. (2003) Professional Ethics in the Construction Industry, *Engineering Construction and Architectural Management* 10(2): 117–27.
7   See Chapter 21.
8   Goode, R. (1992) The concept of 'good faith' in English law, Centro di Studi e Richerche di Diritto Comparato e Straniero, Saggi, Conferenze e Seminari 2, Rome.

9   Honey, R. and Mort, J. (2004) Partnering Contracts for UK Building Projects: Practical Considerations, *Construction Law Journal* 20(7): 361–79.
10 Heal, A. (1999) Construction Partnering: Good Faith in Theory and Practice, *Construction Law Journal* 15(3): 167–98.
11 JCT (2011) *PC/N 2011 Partnering Charter: Non-binding 2011*, London: Sweet & Maxwell, p. 3.
12 Summers, R. (1968) Good Faith in General Contract Law and the Sales Provisions of the Uniform Commercial Code, *Virginia Law Review* 54(2): 206.
13 [1989] QB 433.
14 [1992] 2 AC 128.
15 LTL 7.8.00 [TCC].
16 *Hadley Design Associates v Lord Mayor and Citizens of the City of Westminster* [2003] EWHC 1617 [TCC].
17 See Section 4.8.3 concerning the duty to warn.
18 [1999] ABC.L.R. 02/12.
19 [2013] EWHC 111 (QB).
20 [2013] EWCA Civ 200.

23 Conclusions and future directions

Construction has an enormous influence on all our lives. It provides our homes, our schools, our hospitals and our places of work. Construction also provides all the infrastructure which is essential for our day-to-day lives, such as roads, electricity supply and broadband. Construction is a truly diverse field; it encompasses projects ranging from domestic builds to large iconic structures. Construction also makes an important contribution to the economy.

The development and internationalisation of the industry saw projects become bigger and more complex. This prompted a need for regulation to ensure that the construction industry operates efficiently, ethically and safely. The work performed by the industry should also be to the highest standard.

Construction law is a highly specialised industry-driven area of law. Construction law is about applying the principles to the process rather than studying the law in terms of the legal principles involved in isolation. It is a dynamic and practical area of the law that is constantly evolving to accommodate advances in technology and practice.

This book has covered a good deal of ground and not all of it directly related to a traditional approach to construction law. This departure from the standard fare of construction law textbooks represents the guiding principle of seeking to share insights from a wide range of relevant material. The approach has been legal in nature and delivered from a variety of viewpoints, giving the reader a holistic view of the material covered.

The following chapter synopsis recaps the content with a view to consolidating the material covered and re-establishing the connections made between the preceding chapters.

23.1 Chapter synopsis

23.1.1 Part 1

The first part of the book introduced the reader to the background law. An understanding of these basic concepts is essential for an appreciation of construction law, which is in effect a specialist user of these approaches. Legal systems and the need for law were discussed in Chapter 1. Distinctions were drawn between civil law/common law and case law/statute law. The historical development of law was discussed alongside the rationale for the court systems that are in place. The different members of the legal profession were introduced and the sources of law considered.

Chapter 2 concentrated on the basic principles of contract law. Construction law practice is heavily reliant upon contracts, and knowledge of their formation and content is essential. As important in the consideration of contract law are the remedies for breach and the circumstances in which contracts are discharged. These important questions were addressed in this chapter and formed the basis for later chapters specifically addressing claims and the establishment of entitlements.

Chapter 3 focused on aspects of commercial law required to understand the business considerations and landscape in which construction operates. The different types of business organisations were set out together with issues concerning the operation of companies. The question of security was introduced as a key requirement for those funding construction works and taking occupation of a new asset. Insolvency issues were also addressed given the all too frequent occurrence of business failure within construction. The chapter concluded with a consideration of the rules of interpretation employed by judges to resolve arguments over the meaning of contractual arrangements entered into.

Chapter 4 introduced the law of torts and focused on those most pertinent to the construction industry. Negligence is a vitally important concept in understanding the approach taken to professional appointments and security arrangements. The reliance placed on professionalism is underscored by the tort of negligence and the insurance carried by professionals to guard against claims. Property-based torts such as nuisance and trespass also help students to appreciate the obligations incumbent on landowners considering construction activity on their land. These themes were expanded in Chapter 5 which examined aspects of property law. The extent to which anyone actually 'owns' land was considered alongside certain types of property rights over neighbouring land.

23.1.2 Part 2

The purpose of Part 2 of the book is to bring into focus some of the particular aspects of the construction industry which have shaped construction law. The aim of Chapter 6 is to simplify, as far as possible, the core obligations of the central building contract. The most popular forms of contract were introduced in this part together with those statutes specifically targeted at the construction industry. The hugely important Housing Grants, Construction and Regeneration Act 1996 was introduced within this context.

Chapter 7 turned attention towards procurement, which is a mixture of law and process. The processes at work have evolved over many years and determine the employer's approach to the construction project. Key decisions are made at this stage concerning personnel and the deliverables sought by the construction project. The time/cost/quality theory was introduced as a tool to assist employers in deciding the most appropriate procurement approach for them to use. The procurement approaches were discussed and examples given of their operation.

Chapter 8 followed the employer's journey from inception to completion of a project by considering the tender arrangements and risk allocation theory. Tenderers can be classed as a specialist user of the standard contract formation rules examined in Chapter 2. Allocating risk is one of the fundamental reasons why construction contracts are used. The Abrahamson approach to deciding which party ought to bear which risk was described in this chapter.

Chapter 9 examined the range of standard form contracts that dominate the construction industry and discussed their different approaches. The JCT and NEC forms were introduced alongside the PPC2000 and FIDIC versions of standard forms.

23.1.3 Part 3

The purpose of Part 3 is to give the detail of how construction contracts operate and to highlight the contract administration tools required for project success. Chapter 10 took a simplified approach to the key obligations of the building contract and added the detail required to be able to demonstrate a working knowledge of contract administration.

Chapter 11 continued this approach of providing detail and examined some of the other areas of contract administration outside the core obligations. The parties' respective positions towards such issues as defects and subcontracting and insurance were discussed here. Chapter 12 widened the field of examination to include matters ancillary to the construction contract itself. The phenomena of letters of intent and third party rights were introduced.

The second part of Part 3 specifically addressed the preparation and presentation of claims. The legal requirements in terms of proving causation were discussed in Chapter 13 alongside the issues of establishing entitlement to consequential loss and a discussion around global claims. The focus on claims was taken further in Chapter 14 with the discussion around types of delay and disruption. Claim preparation and how to address the burden of proof are important aspects of construction law discussed here. The money entitlement attached to claims for delay was discussed in Chapter 15 under the heading of loss and expense.

23.1.4 Part 4

Part 4 introduced the reader to the subject of dispute resolution techniques commonly used in the construction industry. A key role for construction lawyers is to assist in the resolution of disputes by the means available through statute, common law or contractual provision. Chapter 16 introduced the different approaches and their rationale. Chapter 17 discussed consensual forms of dispute resolution commonly known as ADR (alternative dispute resolution) and its key attributes. Chapter 18 provided a consideration of arbitration, which has a close affinity with construction users going back for a lengthy period of time.

Part 4 concluded with an introduction to adjudication in Chapter 19, presented here as a statutorily mandated form of dispute resolution. Adjudication has enjoyed considerable success in resolving disputes more quickly and more cost-effectively than the other formal procedures. Chapter 20 widened the scope of Part 4 by examining the international application of arbitration.

23.1.5 Part 5

New directions in construction law were the subject of Part 5. Construction's dynamic and evolving nature necessitates new solutions and ways of working to ensure gaps do not emerge between practice and the existing legal provision. Chapter 21 introduced the change agenda and governmental steps taken to address the ills of industry practice. The chapter started with considering the role of the Latham report as the central tenet of change within the industry. The focus was then moved to consider the partnering phenomenon and its legacy was discussed. Building Information Modelling was also subjected to scrutiny in terms of its deliverables and application.

Chapter 22 presented an investigation into the need for ethical improvement and the imposition of duties of good faith within the industry. These two notions were prescribed as being necessary for the industry to overcome its existing adversarial and short-term agendas and for the industry to flourish.

23.2 Studying construction law

Construction law forms a component of undergraduate study in construction-related programmes, alongside other subjects. It is only at postgraduate level that it is studied in its own right. The requirement at undergraduate level is to introduce key areas of knowledge in order to support specialism in other fields. The experience of most undergraduate students is to see construction law as a collection of separate topics without underlying themes or connections necessarily being made.

The purpose of this book has been to tie together the themes and to present construction law and process as an interconnected whole. First principles of law are described and applied to this specialist industry user. The detailed appreciation of such topics as loss and expense and delay analysis should not cause the reader to lose sight of basic contract law and causation principles.

Studying construction law at the postgraduate level allows the student to form a deeper appreciation of the concepts at work and their application. Quite often, discrete areas of construction law are studied in detail and judicial pronouncements on the emergent themes discussed. The complexity of construction and the multiple stakeholders involved have resulted in a huge amount of cases decided. Construction law is often the nursery for the development of new departures in the law as is currently being demonstrated in the area of good faith clauses and new payment arrangements.

Whatever the level of study of construction law, the encounter can lead to a fascination with the topic and a lifelong interest in the subject. The creation of new legal questions and challenges in the evolving field of construction law certainly sustains this interest.

23.3 Observations on the agenda for change

The government has spent considerable time and effort in exhorting the construction industry towards better practice. The industry now has a great number of tools to help focus on collaborative working. The rate of the rise of partnering in the pre-recession years was surpassed by its rate of decline during the years when the economy contracted. Many construction clients chose not to circle the wagons and see off the bad times together within their frameworks, but instead dusted off their competitive tendering procedures and put the clock back to the pre-partnering times. This knee-jerk reaction was always going to be exposed as such and, happily, the forces at work to improve the industry have resurfaced.

The agenda for change has some powerful new tools to support it. The emergence of Building Information Modelling (BIM) is the most well known. The integration, communication and mutual benefit in BIM are the ideal match for the collaborative partnering approach. The two are fully compliant and the links can be further enhanced by using a multi-party contract such as the PPC2000.

Integrated project insurance is another driver in the re-emergence of partnering. The good sense of this approach is self-evident – why insure every single participant on a project when you can insure the project itself? Anecdotally, the resistance to change comes from the insurance industry and from those contractors and subcontractors who feel unable to rely on one policy of insurance. No individual insurance is needed if the parties are not going to sue each other. Developments in the USA have taken good faith clauses a stage further. The Consensus Docs.300 contracts have an option for use with a mutual waiver of liability clause. This goes hand in hand with their notion of 'safe harbour' decisions where a no blame culture is created. These clauses have not yet been tested in the courts.

Returning to the domestic arena, we see that in recent years, the law on good faith clauses has remained uncodified despite some encouraging judicial pronouncements. Judges have found behaviour incompatible with the good faith clause to be discreditable and have given judgements accordingly. Good faith clauses do though, one senses, remain deeply unattractive to the establishment and are viewed as a Pandora's box of problems waiting to be unleashed. However, it is hard to see how the judiciary could legitimately refuse to enforce a mutual waiver of liability clause of the type in use in the USA.

A look ahead to the future direction that partnering may take domestically and abroad reveals an appetite for greater collaboration. Certainly, public clients would see the appeal of such an approach. This would be all the more apparent if backed by the type of savings claimed on the recent trial project at Cookham Wood and elsewhere. It is always harder to persuade clients in the private sector to put the effort and resource into setting up a scheme in this manner.

Furthermore, many in construction may prefer their industry red in tooth and claw. A preference for adversarial-type construction practice is certainly detectable. Perhaps this is sticking with nurse for fear of something worse and being wary of change. It remains to be seen whether the arrival of digital modelling will move parties away from the blame culture inherent in reasonable skill and care (i.e. I will prove it was your fault) and into the more anodyne and dispassionate fitness for purpose (does it work or not?). This move would have to be accompanied by developments in project insurance and in removal of the blame culture from the industry.

Change is inevitable and it is only the rate of change which is in question. The development of further collaborative practice could bring benefits in terms of giving participants a clear indication of their duties and expected behaviour. A return of investment is largely compatible with this, and as in a successful partnership, there should be prizes for all. Establishing these ground rules could allow the cycle of persecutor/rescuer/victim to be broken to the undoubted benefit of the construction industry.

23.4 Prospects for the standard forms of contract

The approach of the different contracts and their share of the market have been discussed in Chapter 9. The competition between the forms is a healthy process; it keeps the contract writers striving for improvement and to better serve the construction industry and reflect their needs. The cross-pollination of ideas is discernible amongst contract writers. The inclusion by the JCT of options which can deliver the more forward-looking contract deliverables is commendable.

A turn in the fortunes of the PPC2000 seems to have been the advent of BIM and the government endorsement of its approach. The *Government Construction Strategy* of 2012 advocates new forms of procurement which have been successfully trialled in conjunction with the PPC2000. The preference for two-stage open book procurement appears to achieve the government's goal of dealing with consortia that are preformed and that bring their own ideas and approaches/solutions to the government's projects. The obvious benefit for the government in such an approach is that less time and money is needed for its own professional team in terms of working up the design. It can enter a dialogue with a consortium and firm up the price in the second stage of the tendering exercise.

The team approach of the consortia is encapsulated in the PPC multi-party contract whereby the stakeholders are bound together in a quasi or real joint venture approach. In the meantime, the NEC continues to increase its market share at the expense of the JCT and FIDIC forms. Its success story has been founded upon winning endorsements from public sector authorities both domestically and abroad.

Any industry is a conglomerate of a small group of practitioners seeking to initiate change and a large, and often silent, majority of those for whom business as usual is their *modus operandi*. This majority neither seek out nor expect to have imposed on them the changes discussed in Part 5 of this book. The suitability of the JCT approach and its desire to service the needs of industry ensure that it will retain its large market share and status within the privately led section of the construction industry.

23.5 The future of construction law

The disruptive nature of the changes currently affecting the AEC industry is being felt in many different quarters and is causing nervousness amongst those wondering if their role will become redundant. Construction lawyers appear to be one of the professions who may be wiped out in the first wave of change. Intelligent buildings and smart clients should not require the services of lawyers to make claims. However, the promotion of new initiatives and their adoption as standard practice are a long way removed from each other. The existing picture requires the security of contractual and tortious protection of commercial interests, and the role of construction law is to continue to provide these safeguards. Further, the challenge for professions is to redefine their role in light of new technological departures. In this regard, construction law will continue to adapt to the changing circumstances of the industry and retain its role as sentinel and keeper of the peace.

Statutes

Cases

Government reports

Banwell, H. (1964) *The Placing and Management of Contracts for Building and Civil Engineering Work*, London: HMSO.

Cabinet Office (2011) *Government Construction Strategy*, London: Cabinet Office.

Cabinet Office (2014) *New Models of Construction Procurement*, London: HMSO.

Department for Business Innovation and Skills (2013) *UK Construction: An Economic Analysis of the Sector*, London: Department for Business Innovation and Skills.

Egan, J. (1998) *Rethinking Construction*, London: Department of the Environment, Transport and the Regions.

Emmerson, H. C. (1962) *Survey of Problems Before the Construction Industries: Report Prepared for the Minister of Works*, London: HMSO.

Latham, M. (1994) *Constructing the Team: Final Report of the Government/Industry/Review of Procurement and Contractual Arrangements in the UK Construction Industry*, London: Department of the Environment.

Saxon, R. G. (2013) *Growth through BIM*, London: Construction Industry Council.

Simon, E. (1944) *The Placing and Management of Building Contracts: Report of the Central Council for Works and Buildings*, London: HMSO.

Wolstenholme, A. (2009) *Never Waste a Good Crisis: A Review of Progress Since Rethinking Construction and Thoughts for Our Future*, London: Constructing Excellence.

Woolf, H. (1995) *Access to Justice: The Final Report to the Lord Chancellor on the Civil Justice System in England and Wales*, London: HMSO.

Index